STRESS RESPONSES IN BIOLOGY AND MEDICINE

Stress of Life in Molecules, Cells, Organisms, and Psychosocial Communities

ANNALS OF THE NEW YORK ACADEMY OF SCIENCES
Volume 1113

STRESS RESPONSES IN BIOLOGY AND MEDICINE

Stress of Life in Molecules, Cells, Organisms, and Psychosocial Communities

Edited by Péter Csermely, Tamás Korcsmáros, and Katalin Sulyok

Published by Blackwell Publishing on behalf of the New York Academy of Sciences
Boston, Massachusetts
2007

Library of Congress Cataloging-in-Publication Data

World Conference on Stress (2007 : Budapest, Hungary)
  Stress responses in biology and medicine : stress of life in molecules, cells, organisms, and psychosocial communities / edited by Péter Csermely, Tamás Korcsmáros, Katalin Sulyok conference organizers, Lóránd Bertók ... [et al.] ; advisory board, Nancy E. Adler ... [et al.].
       p. ; cm. – (Annals of the New York Academy of Sciences, ISSN 0077-8923 ; v. 1113)
   Includes bibliographical references and index.
   ISBN-13: 978-1-57331-675-0 (paper)
   ISBN-10: 1-57331-675-X (paper)
   1. Stress (Physiology)–Congresses. 2. Adaptation (Physiology) – Congresses. 3. Stress (Psychology)–Congresses. I. Csermely, Peter. II. Korcsmaros, Tamas. III. Title. IV. Series. V. Sulyok, Katalin
     [DNLM: 1. Stress-immunology–Congresses. 2. Stress–physiopathology–Congresses. 3. Adaptation, Physiological–Congresses. 4. General Adaptation Syndrome–Congresses. 5. Heat-Shock Proteins–physiology–Congresses. 6. Stress, Psychological–physiopathology–Congresses. W1 AN626YL v.1113 2007 / QT 162.S8 W927s 2007]
   QP82.2.S8W67 2007
   616.9'8–dc22
                                                             2007031925

The *Annals of the New York Academy of Sciences* (ISSN: 0077-8923 [print]; ISSN: 1749-6632 [online]) is published 28 times a year on behalf of the New York Academy of Sciences by Blackwell Publishing with offices at 350 Main St., Malden, MA 02148 USA; 9600 Garsington Road, Oxford, OX4 2ZG UK; and 600 North Bridge Rd, #05-01 Parkview Square, 18878 Singapore.

Information for subscribers: For new orders, renewals, sample copy requests, claims, changes of address and all other subscription correspondence please contact the Journals Department at your nearest Blackwell office (address details listed above). UK office phone: +44 (0)1865 778315, fax +44 (0)1865 471775; US office phone: 1-800-835-6770 (toll free US) or 1-781-388-8599; fax: 1-781-388-8232; Asia office phone: +65 6511 8000, fax; +44 (0)1865 471775, Email: customerservices@blackwellpublishing.com

Subscription rates:
Institutional Premium   The Americas: $4043   Rest of World: £2246
The Premium institutional price also includes online access to full-text articles from 1997 to present, where available. For other pricing options or more information about online access to Blackwell Publishing journals, including access information and terms and conditions, please visit www.blackwellpublishing. com/nyas
*Customers in Canada should add 6% GST or provide evidence of entitlement to exemption.
**Customer in the UK or EU: add the appropriate rate for VAT EC for non-registered customers in countries where this is applicable. If you are registered for VAT please supply your registration number.

Mailing: The *Annals of the New York Academy of Sciences* is mailed Standard Rate. Mailing to rest of world by International Mail Express (IMEX). Canadian mail is sent by Canadian publications mail agreement number 40573520. **Postmaster:** Send all address changes to *Annals of the New York Academy of Sciences*, Blackwell Publishing Inc., Journals Subscription Department, 350 Main St., Malden, MA 02148-5020.

Membership information: Members may order copies of *Annals* volumes directly from the Academy by visiting www.nyas.org/annals, emailing membership@nyas.org, faxing

212-298-3650, or calling 800-843-6927 (US only), or 212-298-8640 (International). For more information on becoming a member of the New York Academy of Sciences, please visit www.nyas.org/membership. Claims and inquiries on member orders should be directed to the Academy at email: membership@nyas.org or Tel: 212-298-8640 (International) or 800-843-6927 (US only).

Copyright and Photocopying:
© 2007 The New York Academy of Sciences. All rights reserved. No part of this publication may be reproduced, stored, or transmitted in any form or by any means without the prior permission in writing from the copyright holder. Authorization to photocopy items for internal and personal use is granted by the copyright holder for libraries and other users registered with their local Reproduction Rights Organization (RRO), e.g. Copyright Clearance Center (CCC), 222 Rosewood Drive, Danvers, MA 01923, USA (www.copyright.com), provided the appropriate fee is paid directly to the RRO. This consent does not extend to other kinds of copying such as copying for general distribution, for advertising or promotional purposes, for creating new collective works, or for resale. Special requests should be addressed to Blackwell Publishing at: journalsrights@oxon.blackwellpublishing.com.

Printed in the USA. Printed on acid-free paper.

Disclaimer: The Publisher, the New York Academy of Sciences and the Editors cannot be held responsible for errors or any consequences arising from the use of information contained in this publication; the views and opinions expressed do not necessarily reflect those of the Publisher, the New York Academy of Sciences, or the Editors.

Annals are available to subscribers online at the New York Academy of Sciences and also at Blackwell Synergy. Visit www.blackwell-synergy.com or www.annalsnyas.org to search the articles and register for table of contents e-mail alerts. Access to full text and PDF downloads of *Annals* articles are available to nonmembers and subscribers on a pay-per-view basis at www.blackwell-synergy.com and www.annalsnyas.org.

The paper used in this publication meets the minimum requirements of the National Standard for Information Sciences Permanence of Paper for Printed Library Materials, ANSI Z39.48_1984.

ISSN: 0077-8923 (print); 1749-6632 (online)
ISBN-10: 1-57331-675-X (paper); ISBN-13: 978-1-57331-675-0 (paper)

A catalogue record for this title is available from the British Library.

ANNALS OF THE NEW YORK ACADEMY OF SCIENCES
Volume 1113
October 2007

STRESS RESPONSES IN BIOLOGY AND MEDICINE

Stress of Life in Molecules, Cells, Organisms, and Psychosocial Communities

Editors
PÉTER CSERMELY, TAMÁS KORCSMÁROS, AND KATALIN SULYOK

This volume is the result of a conference entitled **2nd World Conference on Stress**, held on August 23–26, 2007 in Budapest, Hungary.

CONTENTS

Introduction. *By* PÉTER CSERMELY, TAMÁS KORCSMÁROS, AND KATALIN SULYOK ... xi

Part I. Stress Proteins: Molecular Stress

New Tricks for an Old Dog: The Evolving World of Hsp70. *By* KEVIN A. MORANO ... 1

Heat Shock Factors at a Crossroad between Stress and Development. *By* MALIN ÅKERFELT, DIANE TROUILLET, VALÉRIE MEZGER, AND LEA SISTONEN ... 15

Extracellular Heat Shock Proteins in Cell Signaling and Immunity. *By* STUART K. CALDERWOOD, SALAMATU S. MAMBULA, AND PHILLIP J. GRAY JR. ... 28

Part II. Stress and Cellular Functions

Membrane Regulation of the Stress Response from Prokaryotic Models to Mammalian Cells. *By* LASZLO VIGH, HITOSHI NAKAMOTO, JACQUES LANDRY, ANTONIO GOMEZ-MUNOZ, JOHN L. HARWOOD, AND IBOLYA HORVATH ... 40

Temperature Stress: Reacting and Adapting: Lessons from Poikilotherms. *By* JOHN L. HARWOOD 52

Endoplasmic Reticulum Stress. *By* GÁBOR BÁNHEGYI, PETER BAUMEISTER, ANGELO BENEDETTI, DEZHENG DONG, YONG FU, AMY S. LEE, JIANZE LI, CHANGHUI MAO, EVA MARGITTAI, MIN NI, WULF PASCHEN, SIMONA PICCIRELLA, SILVIA SENESI, ROBERTO SITIA, MIAO WANG, AND WEI YANG 58

Chaperones, and Proteases—Guardians of Protein Integrity in Eukaryotic Organelles. *By* CLAUDIA LEIDHOLD AND WOLFGANG VOOS 72

Oxygen, Hypoxia, and Stress. *By* CORMAC T. TAYLOR AND JACQUES POUYSSEGUR 87

Review on Bacterial Stress Topics. *By* ANNA MARIA GIULIODORI, CLAUDIO O. GUALERZI, SARA SOTO, JORDI VILA, AND MARÍA M. TAVÍO . . . 95

Variation in Stress Responses within a Bacterial Species and the Indirect Costs of Stress Resistance. *By* THOMAS FERENCI AND BENY SPIRA 105

Part III. Plant Stress

Molecular Mechanisms of Light Stress of Photosynthesis. *By* IMRE VASS, KRISZTIÁN CSER, AND OTILIA CHEREGI 114

The Plant Host–Pathogen Interface: Cell Wall and Membrane Dynamics of Pathogen-Induced Responses. *By* BRAD DAY AND TERRY GRAHAM 123

Long-Term Acclimation of Plants to Elevated CO_2 and Its Interaction with Stresses. *By* ZOLTÁN TUBA AND HARTMUT K. LICHTENTHALER 135

Part IV. Stress at the Level of the Organism

Heat Shock Proteins and Protection of the Nervous System. *By* IAN R. BROWN . 147

Heavy Metal Ions in Normal Physiology, Toxic Stress, and Cytoprotection. *By* MICHAEL A. LYNES, Y. JAMES KANG, STEFANO L. SENSI, GEORGE A. PERDRIZET, AND LAWRENCE E. HIGHTOWER 159

Interleukin-1 System in CNS Stress: Seizures, Fever, and Neurotrauma. *By* TAMAS BARTFAI, MANUEL ALAVEZ-SANCHEZ, SIV ANDELL-JONSSON, MARIANNE SCHULTZBERG, ANNAMARIA VEZZANI, ERIK DANIELSSON, AND BRUNO CONTI 173

Part V. Stress in Medicine

Chaperonopathies by Defect, Excess, or Mistake. *By* ALBERTO J. L. MACARIO AND EVERLY CONWAY DE MACARIO 178

Heat Shock Proteins in Cancer. *By* MICHAEL SHERMAN AND GABRIELE MULTHOFF 192

Drugging the Cancer Chaperone HSP90: Combinatorial Therapeutic Exploitation of Oncogene Addiction and Tumor Stress. *By* PAUL WORKMAN, FRANCIS BURROWS, LEN NECKERS, AND NEAL ROSEN . . . 202

Stress, Heat Shock Proteins, and Autoimmunity: How Immune Responses to Heat Shock Proteins Are to Be Used for the Control of Chronic Inflammatory Diseases. *By* WILLEM VAN EDEN, GEORGE WICK, SALVOTORE ALBANI, AND IRUN COHEN 217

New Molecular Mechanisms of Duodenal Ulceration. *By* SANDOR SZABO, XIAOMING DENG, TETYANA KHOMENKO, LONGCHUAN CHEN, GANNA TOLSTANOVA, KLARA OSAPAY, ZSUZSANNA SANDOR, AND XIMING XIONG 238

Metabolic Syndrome: Psychosocial, Neuroendocrine, and Classical Risk Factors in Type 2 Diabetes. *By* N.G. ABRAHAM, E.J. RUNNER, J.W. ERIKSSON, AND R.P. ROBERTSON 256

Stress Sensitization in Schizophrenia. *By* KUNIO YUII, MICHIO SUZUKI, AND MASAYOSHI KURACHI 276

Part VI. Psychosocial Stress

Glucocorticoid Hyper- and Hypofunction: Stress Effects on Cognition and Aggression. *By* JEANSOK J. KIM AND JÓZSEF HALLER 291

Cognitive Activation Theory of Stress, Sensitization, and Common Health Complaints. *By* HOLGER URSIN AND HEGE ERIKSEN 304

The Catecholamine–Cytokine Balance: Interaction between the Brain and the Immune System. *By* J. SZELÉNYI AND E.S. VIZI 311

Chronic Stress and Social Changes: Socioeconomic Determination of Chronic Stress. *By* MÁRIA S. KOPP, ÁRPÁD SKRABSKI, ANDRÁS SZÉKELY, ADRIENNE STAUDER, AND REDFORD WILLIAMS 325

Attitude toward Death: Does It Influence Dental Fear? *By* GÁBOR FÁBIÁN, ORSOLYA MÜLLER, SZILVIA KOVÁCS, MINH TÚ NGUYEN, TIBOR KÁROLY FÁBIÁN, PÉTER CSERMELY, AND PÁL FEJÉRDY 339

Stress, Immune Function, and Women's Reproduction. *By* PABLO A. NEPOMNASCHY, EYAL SHEINER, GEORGE MASTRORAKOS, AND PETRA C. ARCK 350

Index of Contributors 365

The New York Academy of Sciences believes it has a responsibility to provide an open forum for discussion of scientific questions. The positions taken by the participants in the reported conferences are their own and not necessarily those of the Academy. The Academy has no intent to influence legislation by providing such forums.

Introduction

We are pleased to present the proceedings of the 2nd World Conference on Stress that was held in August 2007 in Budapest, Hungary. This multidisciplinary meeting was a sequel to the highly successful conference in 1997, which is summarized in Volume 851 of the *Annals,* entitled *Stress of Life: From Molecules to Man,* and which led to the establishment of Cell Stress Society International. The 2nd Conference was the 3rd congress of the society. We organized the conference with the following aims: (1) to help strengthen the membership and ties within the Cell Stress Society; (2) to help establish unusual links between various fields of stress research; (3) to give special opportunities for young scientists and women scientists; (4) to show that science is fun; and (5) to celebrate the centennial of the birth of Hans Selye, the founder of the stress concept, who was born in Hungary in 1907, and (6) to give our readers a taste of the thousand-year-old history of Hungary and Hungarian science.

It was our great pleasure to see the overwhelmingly positive response of the scientific community to this event. We had more than 400 symposium proposals, over 3,000 preregistrants, and well over a thousand participants. The congress introduced the stress concept as proposed by Hans Selye, and had as its core molecular stress. Also included were other congress modules covering various topics in stress research. Importantly, the congress showed the necessity for, and a means of understanding, a systems-level understanding of molecular data. The event had exceptional impact for young scientists in this multidisciplinary field through fostering the expansion of their contacts (for example, in Pub Tours with the most prestigious speakers) as well as by allowing more than one hundred students between the ages of 16 and 19 years to act as organizers and helpers at the event.

To summarize the scientific results, we asked the chairpersons of the symposia to write a brief, comprehensive overview of their research field for a general audience. The papers are organized thematically (from molecules to humans), and the titles of these thematic parts refer to the nature of the stress covered by the adjoining symposia.

We are most thankful to all of the organizations that supported this conference: Akadémiai Kiadó; (Hungary), Alexis, Assay Designs, Biogen Idec, Biological Group of the Hungarian Academy of Sciences – Miskolc Section, Biomol, Blackwell Publishing, Cell Stress Society International, EMBO, Eötvös Loránd University, European Science Foundation, Experimetria, FEBS Letters, INFORMA, International Society for Neurochemistry, IUBMB, Novartis, Randox, Roche, SALIMETRICS, Science, Sociedad Iberoamericana de Información Científica, Softflow Hungary, Springer, StressMarq Biosciences, Taylor and Francis, and WisePress.

We thank our patrons: László Sólyom, president of the Hungarian Republic; Janez Potočnik, member of the Commission of the European Union; Sylvester E. Vizi, president of the Hungarian Academy of Sciences; and Ferenc Hudecz, president of the conference host, Eötvös Loránd University.

In addition, the editors would like to thank Mr. Szilárd Kui for his help in editorial work.

AWARDEES OF THE CONGRESS

An important highlight of the conference was the giving of three distinguished awards. These went to Mary Dallman, R. John Ellis, and Ellen Nollen. We also wish to highlight the life of Hans Selye.

THE 2007 HANS SELYE MEMORIAL DISTINGUISHED LECTURE AWARD

The Hans Selye Foundation (Montreal, Quebec, Canada) selected Professor Mary Dallman as a recipient of the 2007 Hans Selye Memorial Distinguished

MARY DALLMAN

INTRODUCTION

Lecture Award. Previous awards have gone to W. Vale (1994), G.P. Chrousos (1997), M. Palkovics (1998), J. Rivier (2000), F. Holsboer (2002), and S. Szabo (2004).

Mary Dallman studied chemistry, graduating from Smith College, and then obtaining her Ph.D. in physiology from Stanford University. She was a postdoctoral fellow in Stockholm and at the University of California at San Francisco, where she has been a member of the faculty since 1970. Professor Dallman served as a member of NIH Study Sections of the NRC Committee on Space Biology and Medicine and participated in NIH workshops and working groups on perimenopause, drug abuse, stress and cardiovascular disease, and Alzheimer's disease. She was an editor of two sections of the American Journal of Physiology and Endocrinology and served as member of the editorial boards of the journals *Steroids*, *Journal of Neurosciences*, *Stress*, and *Molecular Psychiatry*. Professor Dallman has been given several awards and honors, including a membership in the council of the Association for Psychological Science (Endocrinology and Metabolism), and the Endocrine Society. She was the president of the Women in Endocrinology section of the International Society of Neuroendocrinology. She obtained a Fogarty Travel Fellowship, the MT Jones Prize from the British Neuroendocrine Society, the Levine Lectureship from the University of Trier, a lectureship from the Brain Research Institute in Amsterdam; and the Lifetime Achievement Award of the International Society of Psychoneuroendocrinology.

THE 2007 CELL-STRESS SOCIETY INTERNATIONAL MEDAL

The recipient of the 2007 Cell Stress Society International Medal was Professor R. John Ellis. Previous awards have gone to T. Yura (2000), S. Lindquist (2003), and A. Ciechanover (2005).

JOHN ELLIS

R. John Ellis earned his doctorate in 1960 from King's College, London for research on transamination reactions with Professor Davies. His postdoctoral studies on sulfate reduction in bacteria were done at Oxford in the Biochemistry Department with Professor Pasternak. In 1964 Dr. Ellis joined the Departments of Botany and Biochemistry at the University of Aberdeen. He moved to the newly founded Department of Biological Sciences at the University of Warwick in 1970 as senior lecturer and head of the Chloroplast Research Group. In 1976 he was awarded a personal chair in the department, and in 1983 was elected to the Royal Society for his work on chloroplast biogenesis. In 1980 he discovered the first example of a newly synthesized polypeptide that binds to another protein before it folds and assembles; this binding keeps this polypeptide, the large subunit of rubisco, from aggregating with itself. This finding led to the formulation in 1987 of the molecular chaperone concept as a new general cellular function, the term *molecular chaperone* having being proposed by Ron Laskey in 1978 to describe the properties of a nuclear protein involved in the assembly of nucleosomes. The cloning and sequencing of the rubisco-binding protein led to the discovery of the chaperonin family of chaperones in 1988. Ellis continues to contribute to the development of ideas about the chaperone function.

THE 2007 ALFRED TISSIERES YOUNG INVESTIGATOR AWARD

The recipient of the 2007 Alfred Tissieres Young Investigator Award of the Cell Stress Society International was Dr. Ellen Nollen. In 2005 the award went to A. Vila-Sanjurjo.

Ellen Nollen currently holds a Rosalind Franklin Fellowship in the Department of Genetics, University Medical Centre Groningen, in the Netherlands,

ELLEN NOLAN

where she is studying the molecular basis of Parkinson's disease, and working on *C. elegans* models for protein-misfolding diseases. The results will aid in understanding of the cellular processes underlying age-related, misfolding diseases and may yield potential drug targets for Parkinson's disease and other related disorders. She has previously worked in the Department of Functional Genomics, Hubrecht Laboratory, Utrecht, under Professor Ronald Plasterk, and in the Department of Biochemistry, Northwestern University, Evanston, Illinois, USA, under Professor Rick Morimoto. She completed her doctoral thesis on Hsp70 chaperone functions in stressed cells in 2000 under Professor Harm Kampinga at the University of Groningen.

HANS SELYE AWARDEES OF THE CONGRESS

Eight outstanding young scientists have been awarded the Hans Selye Award of the Congress: Justin L. P. Benesh, Laury Chaerle, Bindi Doshi, Bella Groisman, Maki Kawai-Yamada, Hynek Pikhart, Ronald Ullers, and Cindy Voisine. Three of them (J. L. P. Benesh, B. Doshi, and B. Groisman) also received the Assay Design Student Poster Award.

HANS SELYE
(1907–1982)

Hans Selye was born in Komarno, Slovakia in 1907 (at that time Komárom, Hungary). Selye attended school at a Benedictine monastery, and, since his family had produced four generations of physicians, entered the German Medical School in Prague at the age of 17 years, where he graduated first in his class. He later earned a doctorate in organic chemistry.

As early as his second year of medical school (1926), he began developing his now-famous theory of the influence of stress on people's ability to cope with and adapt to the pressures of injury and disease. He discovered that patients with a variety of ailments manifest many similar symptoms, which he ultimately attributed to the body's efforts to respond to the stresses of being ill. He called this collection of symptoms—this separate stress disease—stress syndrome, or the general adaptation syndrome (GAS). The syndrome details how stress induces autonomic hormonal responses and how, over time, these hormonal changes can lead to ulcers, high blood pressure, arteriosclerosis, arthritis, kidney disease, and allergic reactions. His seminal work "A Syndrome Produced by Diverse Nocuous Agents" was published in 1936 in *Nature*. Selye's multifaceted work and concepts have been used in medicine and in almost all biological disciplines from endocrinology to animal breeding and social psychology.

A physician and endocrinologist with many honorary degrees for his pioneering contributions to science (including 43 honorary doctorates), Selye also

Hans Selye

served as a professor and director of the Institute of Experimental Medicine and Surgery at the University of Montreal. He was an elected member of several dozen of the world's most recognized medical and scientific associations. More than anyone else, Selye has demonstrated the role of emotional responses in causing or combating much of the wear and tear experienced by human beings throughout their lives.

Selye probably received more awards than any other physician (including the highest order of Canada), but not the Nobel Prize, although he was nominated for it several times. Many of his 40 books and over 1,700 publications, including *Stress without Distress* (1974) and *The Stress of Life* (1956), became bestsellers all over the world. At the time of his death in 1982, his work had been cited in several hundred thousand scientific papers, in countless popular magazine stories, and in most major languages in countries worldwide. He is still by far the world's most frequently cited author on stress topics. So impressive have his findings and theories been that some authorities refer to him as the Einstein of medicine.

Selye spent significant parts of his life in Hungary, in Czechoslovakia, in the United States, and in Canada. He died in 1982 in Montreal, Canada. His influence on the scientific community is unabated, and his work contributed to a better scientific and popular understanding of disease and its causes.

—PÉTER CSERMELY, TAMÁS KORCSMÁROS, AND KATALIN SULYOK
Semmelweis University, Budapest, Hungary

New Tricks for an Old Dog

The Evolving World of Hsp70

KEVIN A. MORANO

Department of Microbiology and Molecular Genetics, University of Texas Medical School, Houston, Texas, USA

ABSTRACT: The Hsp70 chaperone is arguably the most studied member of the heat shock protein family, a legacy traced back to the early days of phage genetics. However, much still remains to be learned about this essential protein-folding machine. Its involvement in a number of human pathologies, ranging from cancer to protein aggregation diseases, underscores the need for a comprehensive understanding of the myriad cellular roles Hsp70 plays and the outstanding open questions. This article will explore several exciting avenues of research into the function and biology of the chaperone. Analysis of the many eukaryotic Hsp70 isoforms has demonstrated distinct functional roles for some Hsp70 members, to the point of transition from a protein "foldase" to a chaperone cofactor. New insights gained from structural studies have unveiled a likely model for interdomain communication and thus regulation of substrate binding and processing. Advances in small molecule modulation of Hsp70 activity are likely to have significant clinical impact. There is also a growing realization that Hsp70 participates in distinct functional networks in partnership with other protein chaperones. The field is thus at an exciting time when the substantial successes of the past have provided a solid framework that will be used to fuel both discovery and application—Hsp70, from molecule to man.

KEYWORDS: heat shock protein; Hsp70; Hsp110; chaperone

INTRODUCTION

The Hsp70 family of molecular chaperones is ubiquitous in nature, and plays essential roles in protein biogenesis, transport, and degradation. Indeed, it has been the subject of intense study for nearly 30 years, through genetic and biochemical approaches in model prokaryotic and eukaryotic systems as well as detailed *in vitro* reconstitution and structural analyses.[1,2] There is

---

Address for correspondence: Kevin A. Morano, Ph.D., Department of Microbiology and Molecular Genetics, University of Texas Medical School, 6431 Fannin St., Houston, TX 77030. Voice: 731-500-5890; fax: 713-500-5499.

Kevin.a.morano@uth.tmc.edu

mounting evidence of the involvement of Hsp70 isoforms in cell survival, most notably after pathological damage relevant to issues of human health. Given this wealth of information, one is tempted to believe that we have achieved a full understanding of this molecule. However, the rate at which new findings and revelations are made concerning Hsp70 and its protein partners is decidedly nonasymptotic. In fact, major advances in the field have occurred in the last few years, including the first full-length crystal structure and the realization that Hsp70 is not a simple, monolithic, folding machine. Rather, the Hsp70 family appears to have diverged to include isoforms with general or specific cellular roles and targets. Moreover, more distantly related Hsp70 molecules appear to have lost the ability to fold proteins independently and instead function as cofactors for the "classic" Hsp70s. This overview will highlight these findings and more, following the theme of the 2007 Stress of Life Congress: the complex and intricate world of Hsp70, from molecule to man.

THE HSP70 CHAPERONE FAMILY

Hsp70 Architecture and Function

All Hsp70 molecules share a conserved architecture consisting of an amino-terminal ∼ 44 kDa adenine nucleotide-binding domain (NBD, or ATPase) and a ∼ 27 kDa carboxy-terminal substrate/peptide-binding domain (SBD, or PBD). This region is further broken down into an ∼18 kDa domain consisting of two β sheets that form a pocket for substrate binding (β-sandwich domain), and a pair of α-helices (α-domain, or "lid") whose conformation dictates substrate-binding affinity.[2] Amino acid sequence between the β and α-domains is variable in length in the more divergent Hsp70s (see below), as is the extreme carboxy-terminus distal to the α-domain.[3,4] The interdomain linker connecting the NBD and SBD is highly conserved and in fact plays an important role in the allosteric switch mechanism controlling Hsp70 chaperone function (discussed below). Hsp70s exhibit two stable substrate-binding conformations, governed by nucleotide occupancy within the NBD. With ATP bound, the SBD adopts a conformation that binds substrate molecules with low affinity. In the presence of ADP, substrates are tightly associated with the SBD. Iterative cycling between these two states, controlled by independent rates of nucleotide binding, hydrolysis, and release, results in substrate folding (see FIG. 1).

A number of protein cofactors have been described that modulate Hsp70 ATPase activity. Chief among these is a diverse group of proteins containing so-called "J-domains," homologous to the bacterial DnaJ chaperone.[5] J-domain proteins interact with Hsp70 through an invariant histidine-proline-aspartic acid motif (HPD) and stimulate its ATPase activity, thus accelerating substrate-folding reactions.[5] Moreover, this interaction provides a means to recruit Hsp70 to function in various cellular processes. For example, the Simian

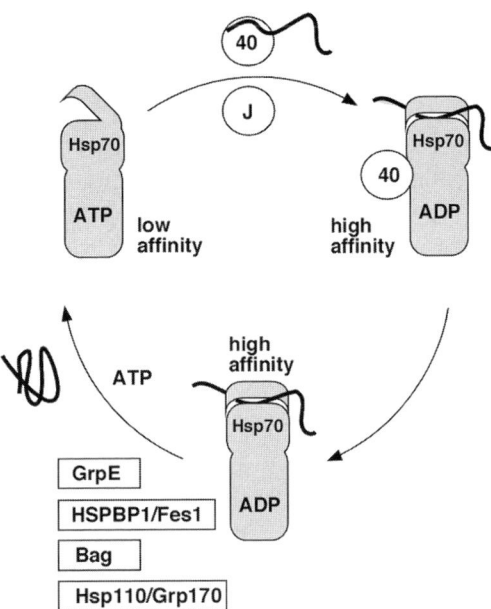

FIGURE 1. The Hsp70 protein-folding cycle. Hsp70 is shown in gray, with the substrate-binding domain on top. The hinged lid is shown open or closed to indicate low- and high-affinity substrate-binding states. Nucleotide status is also indicated. Substrate is depicted as an extended squiggle line in the unfolded" state, and a compact shape in the "folded," or native state. Hsp70 ATPase activating proteins are depicted as circles ("J" indicates any protein with a J-domain), nucleotide exchange factor proteins are shown as rectangles. See text for appropriate citations.

Virus 40 (SV40) virus large T antigen protein contains a J-domain that recruits Hsp70 to drive mitosis and therefore cell transformation.[6] Similarly, the yeast Sec63 component of the endoplasmic reticulum translocation apparatus harbors a functional J-domain that recruits Hsp70 to assist in protein unfolding prior to membrane transport.[7] The Hsp40 class of J-proteins also binds substrates and in at least some folding scenarios is responsible for initial presentation of the unfolded protein to Hsp70. Additional proteins function as nucleotide exchange factors (NEFs), stimulating the release of bound ADP for ATP to initiate a new folding cycle. Interestingly, many Hsp70 NEFs, such as bacterial GrpE and the mammalian Bag and HspBP1 proteins are structurally unrelated and effect nucleotide exchange through both shared and distinct mechanisms.[8,9]

Generalists and Specialists

The Hsp70 family of chaperones appears to have undergone substantial adaptive radiation, and three distinct subfamilies are now recognized: the classic DnaK/Hsp70, the Sse/Hsp110, and the Lhs1/Grp170. To date only the first class has been identified in prokaryotes, while all three are present in the eukaryotic lineage. The groups are distinguished by the variable length of the loop connecting the β-sandwich and α-helical lid domains. This loop consists of approximately 10 residues in the classic Hsp70s, ~100 residues in the Hsp110s, and upward of 135 residues in the Grp170 family. The "extra" sequences present in these isoforms may be of functional consequence, as the Hsp110 and Grp170 proteins are not capable of mediating protein folding on their own.[10,11] Instead, both classes exist in stable heterodimeric complexes with a typical Hsp70 and exert potent nucleotide exchange activity, as will be discussed in detail later in the article.[12-17]

While the apparent transformation of the Hsp110 and Grp170 chaperones from foldases to NEFs clearly represents a major departure from the ancestral Hsp70, recent evidence supports specialization of function even within the classic DnaK/Hsp70 group. This is most obvious in the budding yeast *Saccharomyces cerevisiae*, where the Hsp70 superfamily is amplified to contain 14 homologs, 11 of them classic Hsp70s. These chaperones can be functionally divided into two groups, as proposed by Dr. Elizabeth Craig: the generalists and the specialists. The former group interacts with a broad range of substrates *in vivo* and are the cellular protein-folding "workhorses," as exemplified by the Ssa1-4 proteins in yeast and the cytoplasmic constitutively expressed and inducible Hsp70s in higher eukaryotes.[18,19] The specialists are Hsp70 isoforms whose function is restricted to a single cellular process or even substrate. For example, the *S. cerevisiae* Ssb1 and Ssb2 Hsp70s bind to ribosomes and appear to function exclusively in folding of newly synthesized proteins based on interaction with nascent chains.[20] Perhaps the most dramatic example of Hsp70 specialization is found with the *S. cerevisiae* Ssq1 isoform. Most eukaryotes possess a single "generalist" Hsp70 residing in the mitochondrial matrix that functions in protein import. However, budding yeast and a small group of other fungi possess a second Hsp70, Ssq1, which functions exclusively in the biogenesis of iron–sulfur cluster proteins along with a dedicated J-protein Jac1 and scaffolding protein Isu1.[21-23] Phylogenetic and biochemical analyses indicate that a dedicated Hsp70 for Fe-S biogenesis arose twice in evolution—once in bacteria (HscA) and once in the fungi (Ssq1), underscoring the necessity and utility of such an arrangement.[24]

HSP70 AT THE MOLECULAR LEVEL

Structure and Allosteric Control of Hsp70

Crystal structures of independent NBD and SBD protein fragments from various Hsp70s have allowed characterization of both nucleotide and substrate binding at the atomic level.[25,26] This information has been most useful in understanding the conformational changes in the SBD, specifically the α-helical lid subdomain, that dictate substrate-binding affinity. However, the means by which nucleotide status in the NBD is communicated to the SBD to effect these conformational changes is unknown, owing to lack of solved structures of the intact molecule in different nucleotide states. In the last 2 years work from the Bukau and Sousa laboratories has largely filled this knowledge gap. Difficulties in obtaining crystals of the full-length protein were overcome by Jiang and co-workers through conservative replacement of surface residues and removal of the 10 kDa oligomerization domain carboxy-terminal to the α-helical lid subdomain.[27] Importantly, this modified construct was shown to be functionally indistinguishable from "wild-type" bovine Hsc70. In addition, although the obtained structure does not include nucleotide, information from previous ATP- and ADP-containing NBD structures can be used for interpolation. A number of previously unexpected interdomain interactions were revealed in the full-length structure, including both hydrophobic interactions and a number of salt bridges. Together these contacts serve to bring the SBD into very close proximity to the NBD. Moreover, the well-conserved interdomain linker region, protease-sensitive in the nucleotide-free and ADP states, occupies space between the two domains produced by conformational changes in the NBD upon ATP binding.[27] This may account for observed resistance of Hsp70 molecules to proteolytic cleavage in the ATP-bound state.[28] Domain docking is also ATP-dependent and in turn results in destabilization of residues near the substrate.[29] In addition, the SBD/NBD interface may play a role in stimulation by J-proteins, although definitive support for this hypothesis will require generation of a J-Hsp70 co-crystal structure.

A number of observations support a model whereby Hsp70 operates via an allosteric switch mechanism. It is clear that differential nucleotide binding generates distinct conformational changes within the NBD that are then transmitted to the SBD.[30] Similarly, substrate binding in the SBD stimulates ATP hydrolysis, therefore this communication must be bidirectional.[31,32] Hsp70 molecules do not readily interconvert between the low- and high-affinity substrate-binding states, suggestive of a high energy barrier stabilizing the two conformations.[31,33] A mechanism capable of explaining all of these functional features has remained elusive. Using biophysical analytic techniques on both wild-type and mutant Hsp70 molecules, the Bukau and Mayer laboratories have made significant inroads toward our understanding of the switch. A series of hydrogen bond interactions connect the γ-phosphate of ATP to the surface of the NBD, in close proximity to a proline residue. Replacement of this proline (P143) or a surface-exposed arginine (R151) renders the *E. coli* Hsp70 DnaK nonfunctional and unable to transduce the nucleotide-binding signal to the SBD.[34] Moreover, loss of the proline destabilizes the "switch" mechanism, resulting in mixed populations of Hsp70 molecules exhibiting both substrate-binding affinities.[34] Both the proline and arginine residues are conserved in Hsp70s (although interestingly the arginine is replaced by a lysine residue in *S. cerevisiae* Ssz1, see below), supporting the allosteric switch mechanism as a universal model. Additional mutant analysis suggests that the very well-conserved linker region connecting the two domains is also required for transmitting conformational changes brought about by nucleotide hydrolysis to the SBD.[35] Taken together, we now have a plausible and universally applicable mechanism for Hsp70 chaperone function. A major goal in the near future will be to achieve a similar understanding for how the Hsp70 cofactors, most notably the J-protein family, influence Hsp70 function at the molecular level.

Pharmacological Modification of Hsp70 Function

Given the involvement of Hsp70 chaperones in human diseases including cancer and protein aggregation diseases, such as Parkinson's and Huntington's, and the availability of high-resolution crystal structures, there is much interest in small molecule intervention.[36] In support of this venture, numerous inhibitors of the Hsp90 protein chaperone are currently in development or in clinical trials to assess their efficacies as chemotherapeutic adjuvants.[37] However, very few Hsp70-specific drugs have been reported to date and results from existing compounds have generated only moderate enthusiasm. The compound 15-deoxyspergualin (DSG) binds Hsp70 with micromolar affinity but only modestly (20–40%) stimulates its ATPase activity.[38,39] DSG has a similar affinity for Hsp90, and since the two chaperones share a great deal of functional overlap in many disease pathologies it is difficult to conclusively ascribe its clinical manifestations solely to action on Hsp70. Similarly, 3'-sulfogalactolipids bind the Hsp70 NBD resulting in an approximately 25% decrease in both unstimulated and Hsp40-stimulated ATPase activity.[40,41]

Recent efforts from the Brodsky laboratory have centered on rational drug design and selection using the parent compound NSC-630668-R/1. Importantly, a number of leads have been isolated that specifically inhibit Hsp40-mediated stimulation of Hsp70 ATPase activity.[42] This exciting development may lead to highly specific therapeutic compounds that inhibit cellular Hsp70-dependent processes, as most of these events likely involve a dedicated J-domain co-chaperone *in vivo*. Additionally, such compounds may help to further elucidate the precise mechanism of J-protein action at the biochemical and structural levels.

HSP70 CHAPERONE NETWORKS

Comprehensive genetic, biochemical, and cell biological analyses of Hsp70 biology in yeast and mammalian cells have led to an awareness that these chaperones are organized into functional networks. Systems level analysis suggests that the entire collection of chaperones can be split into stress-inducible and stress-repressible networks, the former involved in protein refolding and the latter dedicated to protein translation, which is also downregulated by stress.[43] While the lines of demarcation between these groups may be blurrier than initially reported, it is clear that some degree of functional specialization has evolved with regard to the Hsp70s. In this section, Hsp70 chaperone networks participating in translation and refolding will be discussed.

Translational Networks

As introduced earlier in this article, the Hsp70 chaperone superfamily has diverged to include homologs that no longer behave as "classic" Hsp70 protein-folding chaperones and have instead acquired new roles as Hsp70 cofactors. One of the earliest recognized examples of this phenomenon is the case of Ssz1. Much like the well-known Ssb Hsp70 homologs in *S. cerevisiae*, Ssz1 is found stably associated with translating ribosomes.[44,45] However, unlike all other Hsp70s characterized to date, Ssz1 mutants lacking their carboxy-terminal substrate-binding domain are fully functional.[46] Moreover, Ssz1 forms a heterodimer with the obligate ribosomal J-protein Zuo1 (zuotin), which together are required for activation of Ssb ATPase activity, and presumably substrate cycling.[47] In support of a functional dependence, mutations in all three genes encoding these proteins share identical phenotypes.[44,46] This theme is partially recapitulated in human cells, where the zuotin homolog Mpp11 is required to recruit the cytoplasmic "multipurpose" Hsc70 to translating ribosomes.[48]

While the role of the Ssb Hsp70s in nascent chain synthesis is well documented, the participation of the "general" Ssa Hsp70 subfamily in budding yeast, and Hsc70 in mammals, is less well understood. The emerging picture now suggests that Hsp70s interact with nascent protein at the ribosome, and that a subset of proteins may also engage nearby but non-ribosome-associated Hsp70 molecules.[49] This category of substrates is enriched in longer proteins with masses of 40 kDa or greater, which makes intuitive sense: small proteins can fold immediately with the help of ribosome-bound Hsp70s, while larger, multidomain proteins may require additional Hsp70 molecules to fold or stabilize regions awaiting interdomain contacts yet to be synthesized.[49] The Sse family of Hsp110 homologs in budding yeast binds and activates both Ssa and Ssb chaperones.[12–16] Loss of Sse1 results in slower transit of large proteins through the Hsp70 translational network, demonstrating that Hsp70 substrate cycling is coupled to protein synthesis.[15] This result further predicts that misfolding, specifically of larger, multidomain proteins, may occur when Hsp70 cofactors are absent or dysregulated. A similar scenario exists in the endoplasmic reticulum, where nascent chains are threaded through the protein translocon and must fold on the trans side of the membrane. The luminal Hsp70 (Kar2/BiP) similar associates with another divergent Hsp70 (Lhs1/Grp170), or a soluble exchange factor (Sls1/Sil1), either of which accelerate substrate cycling.[17] Intriguingly, mutations in human Sil1 appear to be the cause of Marinesco–Sjogren syndrome, an autosomal recessive ataxia, demonstrating the critical role of Hsp70 cofactors in human cell biology.[50,51]

Regulation and Refolding

The discovery that Hsp110 and Grp170 chaperones function *in vivo* as potent Hsp70 nucleotide exchange factors has led to the hypothesis that they may have in fact lost the ability to operate independently as protein-folding chaperones.[4] Several lines of evidence support this belief. First, the *S. cerevisiae* Hsp110 homolog Sse1 does not require ATP catalysis, as demonstrated by the ability of alleles carrying mutations known to inactivate Hsp70 ATPases to functionally complement the *sse1Δ* strain.[52,53] Second, most if not all Sse1 molecules exist in a stable heterodimer with an Ssa or Ssb molecule.[15,16] Third, purified Sse or mammalian Hsp110 is incapable of accelerating folding of model denatured substrates.[10,11] Whether the peptide/substrate-binding domain of these "atypical" Hsp70s is directly involved in protein folding in concert with Hsp70 is not clear. Sse1 lacking its PBD binds Hsp70 but does not enhance protein folding, leading Hartl and colleagues to postulate that the PBDs of both Hsp70 and Hsp110 may interact with substrate.[14] However, Sse1 alleles lacking the PBD or carrying truncations in this domain interact with Hsp70, but do not accelerate nucleotide exchange, nor can they complement the null mutant.[12] Moreover, the Sse1 PBD was shown to interact with the Sse1 NBD *in vivo* when expressed as independent molecules, and this combination can in fact restore near wild-type growth to the *sse1Δ* mutant.[53] Sophisticated biochemical analysis of the interactions occurring in an Hsp70/Hsp110/substrate complex will be required to distinguish between these two models. The observation that the Hsp70/Hsp110 complex is further stimulated by Hsp40/J-proteins suggests that the notion of an Hsp70-folding machine must be broadened to include a maximally activated Hsp40/Hsp70/Hsp110-folding complex. While substantial *in vitro* results support this contention, *in vivo* evidence that eukaryotic cells rely on this assemblage is slowly accumulating. For example, the *S. cerevisiae* mating pheromone α-factor is initially produced in a pre-pro-form, which is posttranslationally inserted into the ER prior to transit and processing in the secretory pathway. Loss of either the Ssa or Sse proteins causes accumulation of the untranslocated form, demonstrating a requirement for Hsp110 function in this cellular event.[53] Similarly, *de novo* folding of heterologously expressed firefly luciferase is hampered by too little or too much Sse1, which when coupled with the observation that overexpression of Sse1 is toxic, suggests not only that this protein is a critical Hsp70 cofactor but also that stoichiometric balance of Hsp110 and Hsp70 is required for optimal folding.[13,14,53] It remains to be determined how Hsp110 is recruited to promote Hsp70 activities in the eukaryotic cytosol. Does Hsp110 participate in all Hsp70 roles or only a subset? Is there differential recruitment of Hsp110/Hsp70 complexes versus solo Hsp70 in a substrate-specific manner?

The Hsp110/Hsp70/Hsp40 network also interacts with other chaperones to engage in higher-order networks. Most notably, Sse1 is required for proper signaling through the Hsp90 chaperone complex.[54] With the recent assignment of Sse1 as a cofactor for Hsp70, it is highly likely that the effects on Hsp90 observed in *sse1*Δ cells are due to Sse1's direct interactions with Hsp70, although this hypothesis remains untested. Some support for this notion is provided by the result that folding and quality control of the von Hippel–Landau (VHL) tumor suppressor is strongly affected by mutations affecting Sse1, Ssa1 (Hsp70), Sti1, and Hsp90 itself.[55] These proteins are all components of the so-called "early" complex in the Hsp90 substrate maturation cycle. In contrast, the "late" stage protein Sba1 (p23) is not required for regulated VHL degradation.[55]

The view of Hsp70 as a core component of multiple, quasi-independent-folding pathways, dictated in part by differential cofactor recruitment, is intriguing. How cells partition the pool of active Hsp70 to support different processes is unknown. The high abundance of the chaperone and many of its partners might suggest that the simplest solution was to increase chaperone availability above the rate-limiting threshold. However, the entire chaperone armamentarium is overexpressed in many aggressive cancers, suggesting that the threshold may be context- or condition-dependent.

HSP70 AND HUMAN DISEASE

Given the central role Hsp70 plays in cellular protein biology, it is not surprising that interactions between the chaperone and human pathophysiological conditions have been uncovered. In this section, recent advances in understanding these connections are presented.

Hsp70 and Cancer

Protein chaperones have long been linked to tumor formation and progression. Indeed, increased expression of Hsp70 and Hsp40 chaperones, as well as others, is indicative of metastasis and poor prognosis in certain cancers.[56] This correlation is now understood to have an important protective effect, since Hsp70 is a potent antiapoptotic protein. Hsp70 inhibits multiple facets of the apoptotic pathway, from the apoptosome to cyctochrome c release to caspase activation.[57] Chaperone overexpression during tumorigenesis thus allows cells to evade apoptosis and continue proliferating. What can be viewed as an important cellular defense strategy, that is, attempting to fix proteotoxic damage before making the final decision to self-terminate, has been co-opted to allow cancer cells to escape surveillance. Cytoprotection offered by Hsp70 overexpression can additionally render cells insensitive to chemotherapeutic compounds, further complicating therapeutic intervention strategies.[58] Knockdown of inducible Hsp70 expression in a range of cancer types triggers apoptosis in diseased cells but not neighboring healthy tissue, suggesting that the chaperone may be a key chemotherapeutic drug target.[59] Although this effort lags behind the development of Hsp90 inhibitors for this purpose, there is clearly both academic and biopharmaceutical interest in this line of investigation.

Hsp70 and Amyloidoises

Over the last few years, work in both model systems and mammalian cells has established a firm link between protein chaperone function and the panoply of protein conformational diseases also known as amyloidoses.[60] These diseases are caused by alternative folding or processing of endogenous proteins, for example, α-synuclein, the causative agent in Parkinson's disease, or the tau protein in Alzheimer's disease. In addition, the range of "triplet expansion" pathologies that result in proteins with long contiguous stretches of glutamine residues (polyQ), similarly result from protein accumulation or aggregation that can be modulated by Hsp70 activity.[61–63] Addition of Hsp70 to *in vitro* aggregation assays has been found to block or at least minimize the formation of insoluble fibrils or plaques, and overexpression of Hsp70 *in vivo* appears to reduce toxicity associated with amyloid generation.[64] Therefore, it is enticing to speculate that again, pharmacological activators of Hsp70 may someday be employed to reduce progression of many of these debilitating diseases. However, work in *S. cerevisiae* suggests that Hsp70 can have both positive and negative effects on propagation of the [PSI+] and [URE3] prions at multiple steps, underscoring the need for a more sophisticated understanding of chaperone interactions.[65]

SUMMARY

A cursory search for the term "Hsp70" in the National Library of Medicine's PubMed database yields over 11,000 papers. Nearly half of these (∼ 4,400) have been published in the last 5 years. In that time we have come to appreciate the central role played by this chaperone in protein biogenesis, stress response, and cellular regulation. We have detailed structural information that may finally explain the protein-folding mechanism and the roles of the numerous (and expanding) family of Hsp70 cofactors. We have begun to explore the clinical ramifications of Hsp70's ability to afford cytoprotection, with the goal of therapeutic intervention to enhance antitumor treatments. We have uncovered human diseases linked to Hsp70 dysfunction. Yet we still know very little about how Hsp70 differentiates between protein targets. These findings, together with the startling realization that the atypical Hsp70s represented by the Hsp110 and Grp170 families appear to have been evolutionarily "repurposed" to become Hsp70 cofactors, suggests that this enigmatic chaperone has still more yet to teach us.

ACKNOWLEDGMENTS

Due to the nature of this review, it was impossible to comprehensively recognize all the work that has led to our current understanding of Hsp70; the author apologizes for any omissions and directs the reader to the many excellent topical reviews cited herein. Work in the author's laboratory was supported by Research Scholar Grant MBC-103134 from the American Cancer Society and NIH grant GM074696.

REFERENCES

1. GEORGOPOULOS, C. 2006. Toothpicks, serendipity and the emergence of the *Escherichia coli* Dnak (hsp70) and Groel (hsp60) chaperone machines. Genetics **174:** 1699–1707.
2. MAYER, M.P. *et al.* 2001. Hsp70 chaperone machines. Adv. Protein Chem. **59:** 1–44.
3. EASTON, D.P., Y. KANEKO & J.R. SUBJECK. 2000. The Hsp110 and Grp170 stress proteins: Newly recognized relatives of the Hsp70s. Cell Stress Chaperones **5:** 276–290.
4. SHANER, L. & K. A. MORANO. 2007. All in the family: atypical Hsp70 chaperones are conserved modulators of Hsp70 activity. Cell Stress Chaperones **12:** 1–8.
5. CHEETHAM, M.E. & A. J. CAPLAN. 1998. Structure, function and evolution of DnaJ: conservation and adaptation of chaperone function. Cell Stress Chaperones **3:** 28–36.
6. FEWELL, S.W., J.M. PIPAS & J.L. BRODSKY. 2002. Mutagenesis of a functional chimeric gene in yeast identifies mutations in the simian virus 40 large T antigen J domain. Proc. Natl. Acad. Sci. USA **99:** 2002–2007.
7. FELDHEIM, D., J. ROTHBLATT & R. SCHEKMAN. 1992. Topology and functional domains of Sec63p, an endoplasmic reticulum membrane protein required for secretory protein translocation. Mol. Cell. Biol. **12:** 3288–3296.
8. SHOMURA, Y. *et al.* 2005. Regulation of Hsp70 function by HspBP1: structural analysis reveals an alternate mechanism for Hsp70 nucleotide exchange. Mol. Cell. **17:** 367–379.
9. SONDERMANN, H. *et al.* 2001. Structure of a Bag/hsc70 complex: convergent functional evolution of Hsp70 nucleotide exchange factors. Science **291:** 1553–1557.
10. OH, H.J. *et al.* 1999. The chaperoning activity of Hsp110. Identification of functional domains by use of targeted deletions. J. Biol. Chem. **274:** 15712–15718.
11. BRODSKY, J.L. *et al.* 1999. The requirement for molecular chaperones during endoplasmic reticulum-associated protein degradation demonstrates that protein export and import are mechanistically distinct. J. Biol. Chem. **274:** 3453–3460.

12. SHANER, L., R. SOUSA & K.A. MORANO. 2006. Characterization of Hsp70 binding and nucleotide exchange by the yeast Hsp110 chaperone Sse1. Biochemistry **45:** 15075–15084.
13. RAVIOL, H. *et al.* 2006. Chaperone network in the yeast cytosol: Hsp110 is revealed as an Hsp70 nucleotide exchange factor. EMBO J. **25:** 2510–2518.
14. DRAGOVIC, Z. *et al.* 2006. Molecular chaperones of the Hsp110 family act as nucleotide exchange factors of Hsp70s. EMBO J. **25:** 2519–2528.
15. YAM, A.Y. *et al.* 2005. Hsp110 cooperates with different cytosolic Hsp70 systems in a pathway for *de novo* folding. J. Biol. Chem. **280:** 41252–41261.
16. SHANER, L. *et al.* 2005. The yeast Hsp110 Sse1 functionally interacts with the Hsp70 chaperones Ssa and Ssb. J. Biol. Chem. **280:** 41262–41269.
17. STEEL, G.J. *et al.* 2004. Coordinated activation of Hsp70 chaperones. Science **303:** 98–101.
18. JAMES, P., C. PFUND & E.A. CRAIG. 1997. Functional specificity among Hsp70 molecular chaperones. Science. **275:** 387–389.
19. CRAIG, E.A. *et al.* 1994. Cytosolic hsp70s of *Saccharomyces cerevisiae*: roles in protein synthesis, protein translocation, proteolysis, and regulation. *In* The Biology of Heat Shock Proteins and Molecular Chaperones, Vol. 26. R.I. Morimoto, A. Tissieres & C. Georgopolous, Eds.: 31–52. Cold Spring Harbor Laboratory Press. New York, NY.
20. PFUND, C. *et al.* 1998. The molecular chaperone Ssb from *Saccharomyces cerevisiae* is a component of the ribosome-nascent chain complex. EMBO J. **17:** 3981–3989.
21. ANDREW, A.J. *et al.* 2006. Characterization of the interaction between the J-protein Jac1p and the scaffold for Fe-S cluster biogenesis, Isu1p. J. Biol. Chem. **281:** 14580–14587.
22. VOISINE, C. *et al.* 2001. Jac1, a mitochondrial J-type chaperone, is involved in the biogenesis of Fe/S clusters in *Saccharomyces cerevisiae*. Proc. Natl. Acad. Sci. USA **98:** 1483–1488.
23. DUTKIEWICZ, R. *et al.* 2003. Ssq1, a mitochondrial hsp70 involved in iron-sulfur (Fe/S) center biogenesis. Similarities to and differences from its bacterial counterpart. J. Biol. Chem. **278:** 29719–29727.
24. SCHILKE, B. *et al.* 2006. Evolution of mitochondrial chaperones utilized in Fe-S cluster biogenesis. Curr. Biol. **16:** 1660–1665.
25. ZHU, X. *et al.* 1996. Structural analysis of substrate binding by the molecular chaperone DnaK. Science **272:** 1606–1614.
26. FLAHERTY, K.M., C. DELUCA-FLAHERTY & D.B. MCKAY. 1990. Three-dimensional structure of the ATPase fragment of a 70k heat-shock cognate protein. Nature **346:** 623–628.
27. JIANG, J. *et al.* 2005. Structural basis of interdomain communication in the Hsc70 chaperone. Mol. Cell. **20:** 513–524.
28. BUCHBERGER, A. *et al.* 1995. Nucleotide-induced conformational changes in the ATPase and substrate binding domains of the DnaK chaperone provide evidence for interdomain communication. J. Biol. Chem. **270:** 16903–16910.
29. SWAIN, J.F. *et al.* 2007. Hsp70 chaperone ligands control domain association via an allosteric mechanism mediated by the interdomain linker. Mol. Cell. **26:** 27–39.
30. SCHMID, D. *et al.* 1994. Kinetics of molecular chaperone action. Science **263:** 971–973.
31. LAUFEN, T. *et al.* 1999. Mechanism of regulation of Hsp70 chaperones by DnaJ cochaperones. Proc. Natl. Acad. Sci. USA **96:** 5452–5457.

32. KARZAI, A.W. & R. MCMACKEN. 1996. A bipartite signaling mechanism involved in DnaJ-mediated activation of the *Escherichia coli* DnaK protein. J. Biol. Chem. **271:** 11236–11246.
33. SOUSA, M.C. & D.B. MCKAY. 1998. The hydroxyl of threonine 13 of the bovine 70-kda heat shock cognate protein is essential for transducing the ATP-induced conformational change. Biochemistry **37:** 15392–15399.
34. VOGEL, M., B. BUKAU & M.P. MAYER. 2006. Allosteric regulation of Hsp70 chaperones by a proline switch. Mol. Cell. **21:** 359–367.
35. VOGEL, M., M.P. MAYER & B. BUKAU. 2006. Allosteric regulation of Hsp70 chaperones involves a conserved interdomain linker. J. Biol. Chem. **281:** 38705–38711.
36. BRODSKY, J.L. & G. CHIOSIS. 2006. Hsp70 molecular chaperones: emerging roles in human disease and identification of small molecule modulators. Curr. Top. Med. Chem. **6:** 1215–1225.
37. WHITESELL, L. & S.L. LINDQUIST. 2005. Hsp90 and the chaperoning of cancer. Nat. Rev. Cancer **5:** 761–772.
38. NADEAU, K. *et al.* 1994. Quantitation of the interaction of the immunosuppressant deoxyspergualin and analogs with Hsc70 and Hsp90. Biochemistry **33:** 2561–2567.
39. BRODSKY, J.L. 1999. Selectivity of the molecular chaperone-specific immunosuppressive agent 15-deoxyspergualin: modulation of Hsc70 ATPase activity without compromising DnaJ chaperone interactions. Biochem. Pharmacol. **57:** 877–880.
40. MAMELAK, D. *et al.* 2001. The aglycone of sulfogalactolipids can alter the sulfate ester substitution position required for Hsc70 recognition. Carbohydr. Res. **335:** 91–100.
41. WHETSTONE, H. & C. LINGWOOD. 2003. 3'-sulfogalactolipid binding specifically inhibits Hsp70 ATPase activity *in vitro*. Biochemistry **42:** 1611–1617.
42. FEWELL, S.W. *et al.* 2004. Small molecule modulators of endogenous and co-chaperone-stimulated Hsp70 ATPase activity. J. Biol. Chem. **279:** 51131–51140.
43. ALBANESE, V. *et al.* 2006. Systems analyses reveal two chaperone networks with distinct functions in eukaryotic cells. Cell. **124:** 75–88.
44. GAUTSCHI, M. *et al.* 2002. A functional chaperone triad on the yeast ribosome. Proc. Natl. Acad. Sci. USA **99:** 4209–4214.
45. GAUTSCHI, M. *et al.* 2001. RAC, a stable ribosome-associated complex in yeast formed by the DnaK-DnaJ homologs Ssz1p and zuotin. Proc. Natl. Acad. Sci. USA **98:** 3762–3767.
46. HUNDLEY, H. *et al.* 2002. The *in vivo* function of the ribosome-associated Hsp70, Ssz1, does not require its putative peptide-binding domain. Proc. Natl. Acad. Sci. USA **99:** 4203–4208.
47. HUANG, P. *et al.* 2005. The Hsp70 Ssz1 modulates the function of the ribosome-associated J-protein Zuo1. Nat. Struct. Mol. Biol. **12:** 497–504.
48. HUNDLEY, H.A. *et al.* 2005. Human Mpp11 J protein: Ribosome-tethered molecular chaperones are ubiquitous. Science **308:** 1032–1034.
49. THULASIRAMAN, V., C.F. YANG & J. FRYDMAN. 1999. *In vivo* newly translated polypeptides are sequestered in a protected folding environment. EMBO J. **18:** 85–95.
50. SENDEREK, J. *et al.* 2005. Mutations in Sil1 cause Marinesco-Sjogren syndrome, a cerebellar ataxia with cataract and myopathy. Nat. Genet. **37:** 1312–1314.
51. ANTTONEN, A.K. *et al.* 2005. The gene disrupted in Marinesco-Sjogren syndrome encodes Sil1, an HspA5 cochaperone. Nat. Genet. **37:** 1309–1311.

52. GOECKELER, J.L. et al. 2002. Overexpression of yeast Hsp110 homolog Sse1p suppresses *ydj1–151* thermosensitivity and restores Hsp90-dependent activity. Mol. Biol. Cell. **13:** 2760–2770.
53. SHANER, L. et al. 2004. The function of the yeast molecular chaperone Sse1 is mechanistically distinct from the closely related Hsp70 family. J. Biol. Chem. **279:** 21992–22001.
54. LIU, X.D., K.A. MORANO & D.J. THIELE. 1999. The yeast Hsp110 family member, Sse1, is an Hsp90 cochaperone. J. Biol. Chem. **274:** 26654–26660.
55. MCCLELLAN, A.J., M.D. SCOTT & J. FRYDMAN. 2005. Folding and quality control of the VHL tumor suppressor proceed through distinct chaperone pathways. Cell **121:** 739–748.
56. JOLLY, C. & R.I. MORIMOTO. 2000. Role of the heat shock response and molecular chaperones in oncogenesis and cell death. J. Natl. Cancer Inst. **92:** 1564–1572.
57. BEERE, H.M. & D.R. GREEN. 2001. Stress management - heat shock protein-70 and the regulation of apoptosis. Trends Cell. Biol. **11:** 6–10.
58. POCALY, M. et al. 2007. Overexpression of the heat-shock protein 70 is associated to imatinib resistance in chronic myeloid leukemia. Leukemia **21:** 93–101.
59. AGHDASSI, A. et al. 2007. Heat shock protein 70 increases tumorigenicity and inhibits apoptosis in pancreatic adenocarcinoma. Cancer Res. **67:** 616–625.
60. MUCHOWSKI, P.J. & J.L. WACKER. 2005. Modulation of neurodegeneration by molecular chaperones. Nat. Rev. Neurosci. **6:** 11–22.
61. NOVOSELOVA, T.V. et al. 2005. Treatment with extracellular Hsp70/Hsc70 protein can reduce polyglutamine toxicity and aggregation. J. Neurochem. **94:** 597–606.
62. WACKER, J.L. et al. 2004. Hsp70 and hsp40 attenuate formation of spherical and annular polyglutamine oligomers by partitioning monomer. Nat. Struct. Mol. Biol. **11:** 1215–1222.
63. COWAN, K.J., M.I. DIAMOND & W.J. WELCH. 2003. Polyglutamine protein aggregation and toxicity are linked to the cellular stress response. Hum. Mol. Genet. **12:** 1377–1391.
64. EVANS, C.G., S. WISEN & J.E. GESTWICKI. 2006. Heat shock proteins 70 and 90 inhibit early stages of amyloid beta-(1–42) aggregation *in vitro*. J. Biol. Chem. **281:** 33182–33191.
65. LOOVERS, H.M., E. GUINAN & G.W. JONES. 2007. Importance of the Hsp70 ATPase domain in yeast prion propagation. Genetics **175:** 621–630.

Heat Shock Factors at a Crossroad between Stress and Development

MALIN ÅKERFELT,[a] DIANE TROUILLET,[b] VALÉRIE MEZGER,[b] AND LEA SISTONEN[a]

[a]*Turku Centre for Biotechnology and Department of Biology, Åbo Akademi University, 20520 Turku, Finland*

[b]*Biologie Moléculaire du Stress, Centre National de la Recherche Scientifique (CNRS) UMR8541, Ecole Normale Supérieure, 75005 Paris, France*

> ABSTRACT: Organisms must be able to sense and respond rapidly to changes in their environment in order to maintain homeostasis and survive. Induction of heat shock proteins (Hsps) is a common cellular defense mechanism for promoting survival in response to various stress stimuli. Heat shock factors (HSFs) are transcriptional regulators of Hsps, which function as molecular chaperones in protecting cells against proteotoxic damage. Mammals have three different HSFs that have been considered functionally distinct: HSF1 is essential for the heat shock response and is also required for developmental processes, whereas HSF2 and HSF4 are important for differentiation and development. Specifically, HSF2 is involved in corticogenesis and spermatogenesis, and HSF4 is needed for maintenance of sensory organs, such as the lens and the olfactory epithelium. Recent evidence, however, suggests a functional interplay between HSF1 and HSF2 in the regulation of *Hsp* expression under stress conditions. In lens formation, HSF1 and HSF4 have been shown to have opposite effects on gene expression. In this chapter, we present the different roles of the mammalian HSFs as regulators of cellular stress and developmental processes. We highlight the interaction between different HSFs and discuss the discoveries of novel target genes in addition to the classical *Hsp*s.
>
> KEYWORDS: heat shock factor; heat shock response; corticogenesis; spermatogenesis; transcription

THE MAMMALIAN HEAT SHOCK FACTOR FAMILY

The eukaryotic heat shock response (HSR) is mediated by a positive control element, the heat shock element (HSE), which is present in multiple copies upstream of the *Hsp* genes. The first evidence for a factor that could interact with

Address for correspondence: Lea Sistonen, Ph.D., Turku Centre for Biotechnology, P.O. Box 123, FI-20521 Turku, Finland. Voice: +358-2-333-8028; fax: +358-2-333-8000.
  lea.sistonen@btk.fi

the HSE originated from studies of protein–DNA interactions in *Drosophila* cell nuclei.[1] An activator protein, named heat shock factor (HSF), was identified to specifically bind to the HSE and regulate the *Hsp* expression on stress stimulation. Since then efforts from a large number of investigators have shown that the HSR is conserved in all organisms from yeast to plants and animals. In yeast, fruit fly, and nematode, only a single HSF exists, whereas in vertebrates and plants, the HSF family consists of several members.[2,3] HSF1 and HSF2 exist in all vertebrates, while HSF3 is specific for avian species and HSF4 for mammals.[4] HSF1 was originally identified as the transcriptional regulator of the HSR and has been most extensively investigated in mammals. HSF1 is activated in response to elevated temperatures, exposure to oxidants, heavy metals, and bacterial or viral infections, and genetic studies indicate that no other HSF is able to compensate for HSF1 in the HSR.[5–7] HSF2 is known to be involved in development and differentiation-related processes such as spermatogenesis and corticogenesis in mice and hemin-mediated differentiation of human K562 erythroleukemia cells.[8–14] No stress-related functions have been shown for HSF4, but its importance in lens formation and maintenance of olfactory epithelium has been well documented.[15–17]

HSF activation is a multistep process, including trimerization, localization to the nucleus, and binding to DNA. Several inducible posttranslational modifications, such as phosphorylation and sumoylation, are involved in regulation of the transactivation capacity of HSF1.[18–21] Upon activation, HSF1 undergoes a transition from monomer to trimer,[22,23] whereas HSF2 undergoes a transition from dimer to trimer.[9] Similar to most transcriptional regulators, HSFs are composed of different functional domains, of which the DNA-binding domain (DBD) is best preserved (FIG. 1.).[21,23] HSFs bind to DNA where each DBD recognizes the HSE in the major groove of the double helix.[23] HSEs are highly conserved, consisting of multiple inverted repeats of the pentameric sequence nGAAn.[24] The promoters of HSF target genes can also have more than one HSE, thereby allowing simultaneous binding of multiple HSFs. HSF binding to an HSE occurs in a cooperative manner, where binding of one HSF trimer facilitates the binding of the next.[25] Trinklein and colleagues confirmed the finding of Xiao and Lis, identifying guanines to be the most conserved nucleotides within the HSEs.[26,27] In addition to typical HSEs, binding of HSF1 to a discontinuous type of HSE was recently observed *in vitro*.[28] These results suggest that both the nucleotides and the spacing of the repeated units are critical determinants for recognition by HSFs and transcriptional activation.

HSFs AS CELLULAR STRESS REGULATORS

HSF1 is the bona fide stress-responsive prototype in mammals. *Hsf1* knockout mouse models have demonstrated that HSF1 is required as a transcriptional

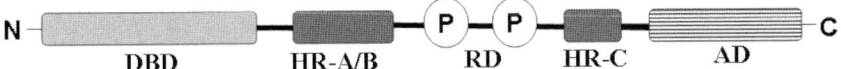

FIGURE 1. The functional domains of HSF1. DBD: DNA-binding domain, HR-A/B and HR-C: hydrophobic heptad repeats, RD: regulatory domain, AD: activation domain, P: posttranslational modifications (PTMs). Note that PTMs occur also within other functional domains than RD.

TABLE 1. Mammalian HSF target genes identified *in vivo*

DNA-binding factor	Target gene promoter	Reference
HSF1 and HSF2	*Hsp*s	27
HSF1 and HSF2	*Hsp70.1* and *Hsp25.1*	14
HSF1 and HSF2	*Clu*	32
HSF1	*Il-6*	34
HSF4	*Crygf*	16
HSF1 and HSF4	*Fgf7*	16
HSF1 and HSF4	*Lif*	17
HSF2	*p35*	13

activator of *Hsp* genes during the HSR. Moreover, HSF1 is critical for maintaining cellular integrity during stress, because cells from $Hsf1^{-/-}$ mice lack the ability to develop thermotolerance.[5–7,29] In contrast, the role of HSF2 in the HSR was not revealed until very recently.[14] *Hsf2*-null mice are viable and have no gross defects in the HSR, but several reports have proposed that HSF2 could contribute in the transcriptional regulation through interplay with HSF1.[10–12,14,30–32]

Functional Interaction between HSF1 and HSF2

Using chromatin immunoprecipitation (ChIP), binding of both HSF1 and HSF2 was detected on *Hsp* promoters upon heat shock and hemin treatment (TABLE 1).[14,27] Further studies using both knockdown and knockout strategies elucidated that during stress, HSF2 is recruited to the *Hsp70* promoter only in the presence of HSF1 and that this cooperation requires an intact HSF1 DBD.[14] Importantly, gene expression analyses showed that HSF2 is able to modulate the HSF1-mediated inducible expression of *Hsp*s, and reintroduction of HSF2 into $Hsf2^{-/-}$ fibroblasts potentiated the expression of major *Hsp* genes. In addition, individual targets were differently regulated, depending on stimuli. These findings indicate that HSF2, in contrast to the previous model, actively participates in the transcriptional regulation of the HSR.[14] HSF2 has also been reported to regulate the chromatin structure of the *Hsp70* promoter during mitosis,[33] and it will be interesting to find out whether HSF1, possibly through interplay with HSF2, would participate in bookmarking. HSF1 and HSF2 were also found to interact during proteasome inhibition, and both factors bound to the *clusterin (Clu)* promoter in such stress conditions.[32] Because the HSE located on the *Clu* promoter contains only three pentamers, HSF1 and HSF2 binding as a heterocomplex is an intriguing possibility.

Apart from target gene promoters, HSF1 concentrates rapidly upon stress into nuclear stress bodies (nSBs). nSBs form on specific chromosomal loci, mainly q12 of human chromosome 9, where HSF1 binds to a subclass of satellite III repeats.[35,36] Stress-inducible HSF1-dependent transcription of the satellite III repeats, originating from the 9q12 locus, has been shown to produce noncoding RNA molecules, whose functions remain to be established.[37] Intriguingly, HSF2 was also found to localize in nSBs in HeLa cells exposed to heat stress.[38] Upon heat shock, HSF1 and HSF2 colocalize in the nSBs,[39] and an interaction between HSF1 and HSF2 during both control and heat shock treatment has been detected,[31,39] suggesting a possible functional interplay between the two transcription factors. Further studies are required to determine whether HSF2 functions as a modulator of HSF1-mediated transactivation of other targets than *Hsps*, including satellite III repeats, and whether competition between HSFs is a common phenomenon in the regulation of their target gene expression.

HSF1 in the Immune Response

Hsf1 knockout mice display a significantly impaired T cell–dependent B cell response.[34] Transcriptional profiling of $Hsf1^{-/-}$ fibroblasts has revealed that HSF1, in addition to *Hsp* genes, regulates immunologically important genes. In mouse spleen cells, HSF1 was found to directly bind to the *Il-6* gene (TABLE 1),[34] coding for a proinflammatory cytokine secreted by T cells to stimulate an immune response and is required for B cell differentiation.[40] In response to immunization with sheep red blood cells, the *Hsf1*-deficient mice showed 50% lower production of immunoglobulins, especially IgG2a. These results unraveled a novel molecular link between HSF1 and a gene related to immune response and inflammation.[34]

The nematode *C. elegans* has evolved an immune system that is excellent for studying the effect of elevated temperatures upon immunity. On heat shock the worms become more resistant to bacterial pathogens. The enhanced resistance requires HSF1-mediated activation of Hsp90 and small Hsps, effectors of the immune protection.[41] In addition, the HSF1 defense pathway interacts with the insulin/IGF-1 signaling pathway, including FOXO transcription factor DAF-16 and its upstream receptor DAF-2 that are known to affect aging and immunity in *C. elegans*.[41–43] These findings indicate that HSF1 has multiple ways of regulating the immune system.

HSFs AS DEVELOPMENTAL REGULATORS

A developmental role for HSFs was introduced when the *Drosophila* HSF was found to be required for early larval development and oogenesis.[44] Surprisingly, these developmental effects were not mediated by *Hsp* gene expression, which is consistent with the subsequent studies showing that basal *Hsp* expression during mouse embryonic development is not affected by the lack of HSF1.[6] Gene inactivation studies in mice have revealed functions beyond the HSR and demonstrated roles in embryonic development, reproduction, cortical lamination, lens development, and maintenance of olfactory epithelium.[10,11,13,16,17,30,45] A major challenge is to establish the genes that are directly controlled by HSFs and, most importantly, the processes where these gene products play a key role.

HSF1: A Maternal Factor

Mice lacking HSF1 can survive to adulthood but they exhibit multiple defects, including placental insufficiency, prenatal lethality, growth retardation, and female infertility.[6] In developing $Hsf1^{-/-}$ embryos, no extensive defects were evident and no changes were detected in the expression of *Hsp70*, which is the earliest sign of zygotic genome activation. In contrast, an abnormal architecture of the placenta was observed at E11.5, suggesting that the prenatal lethality was due to failure in the extra-embryonic tissue. No fertilized oocytes developed past the zygotic stage when $Hsf1^{-/-}$ females were mated with wild-type males. These results demonstrate that HSF1 is a maternal factor, essential for early postfertilization development.[46] Disturbed control by maternal HSF1 during oogenesis or in the initiation phase of embryogenesis could therefore be associated with infertility in mammals.

HSF1 and HSF4 Interplay in Sensory Organs

Little was known about the physiological function of HSF4 before a genetic study by Bu and coworkers showed that inherited cataract in certain Chinese and Danish families was associated with a mutation in the DBD of HSF4.[47] The phenotype of *Hsf4*-null mice supports an important role for HSF4 in lens formation; although *Hsf4* knockout mice displayed normal lens development during embryogenesis, abnormalities in the lens appeared soon after birth and the mice developed cataract by 6 weeks of age.[16,45] $Hsf4^{-/-}$ lens fiber cells were abnormal, containing inclusion-like structures, probably due to a reduction in the expression of γ-*crystallin* gene family members. HSF4 was found to directly bind and regulate the γ*F-crystallin (Crygf)* gene.[16] Binding of HSF1 and HSF4 to the *Fgf7* promoter showed opposite effects on gene expression, that is, repression by HSF4 and activation by HSF1, providing evidence for a competition between these two HSFs during mouse lens development (TABLE 1).[16] This finding is the first example of an interplay between two different mammalian HSFs in development.

During early postnatal period the $Hsf1^{-/-}$ mice display severe atrophy of the olfactory epithelium, increased cell death of olfactory sensory neurons and increased expression of the *Lif* gene.[17] In contrast, this phenotype is alleviated to some extent in the $Hsf4^{-/-}$ olfactory epithelium. A similar interplay between HSF1 and HSF4 as detected during lens formation, also occurs on the *Lif* promoter in the olfactory epithelium (TABLE 1).[16,17] HSF1 and HSF4 are required for the maintenance of different sensory organs, the lens and the olfactory epithelium, specifically when these organs are exposed to environmental stimuli for the first time after birth.[16,17] The increased sensitivity of these organs may be partly due to the altered expression of *Crygf*, *Fgf7*, and *Lif*. In addition, decreased levels of *Hsp25*, *Hsp70*, and *Hsp90* were observed in $Hsf1^{-/-}$ olfactory epithelium, and *Hsf4*-deficient lens fiber cells had compromised expression of *Hsp25*. The results suggest that the preservation of the protein homeostasis by Hsps could be an important determinant in sensory organ maintenance.[16,17] Although these gene inactivation studies mainly focused on the cooperative and competitive roles of HSF1 and HSF4, it is also possible that HSF2 might have a function in the neuronal part of retinal formation, owing to similar expression patterns.[48]

HSFs in Brain Development

During rodent brain development, HSF2 is highly expressed in the neuroepithelium, with nuclear localization in the developing neural tube, and HSF2 DNA-binding activity can be detected in the cortex, striatum, olfactory bulbs, and mesencephalon before birth (Y. Chang's unpublished results).[10,11,13,49–51] In the mouse, HSF2 is expressed in the proliferative neuronal progenitors of the ventricular zone. In addition, HSF2 expression is detected in the cortical plate, when the most superficial layers of cortex are being established.[10,50] HSF2 expression and activity profiles implicate a major role for HSF2 as a transcriptional regulator in the development of the fore- and midbrain, and possibly also in the cerebellum.

Hsf2 inactivation studies have been performed by three different laboratories.[10,11,30] In all three cases, *Hsf2*-null mice did not display any overt morphological abnormalities. While one laboratory did not observe any brain phenotype in adult mice,[30] embryonic brain defects were reported by the two others groups.[10,11] Adult brains displayed enlarged ventricles and reduction of hippocampus and striatum, as well as in the width of the cortex. In addition, prominent abnormalities in the central nervous system (CNS), with collapse of the ventricular systems and hemorrhages in cerebral regions at early stages, were detected.[10,11] HSF2 was also found to be involved in later brain development, in the migration phase of newborn cortical neurons.[13] When migrating, cortical neurons receive migration inputs, such as Reelin secreted from Cajal-Retzius cells, and the neurons benefit from architectural guides provided by radial glia cell fibers, which extend all the way from the ventricular zone to the marginal zone.[52] In the absence of HSF2, a reduced number of radial glia and Cajal-Retzius cells, together with disturbances in the Reelin signaling cascade, were observed.[13] Moreover, the expression of p35, which is an activator of cyclin-dependent kinase 5 (Cdk5) and essential for radial migration,[52] was found to be dependent on the amount of HSF2.[13] As demonstrated in vivo by ChIP experiments, HSF2 directly bound to the promoter of p35, which was thereby identified as the first HSF2 target gene in development (TABLE 1).[13] In the light of present knowledge, HSF2 could function as a fine-tuner of gene expression, required for correct neuronal positioning in superficial layers in the developing cortex. The role of HSF2 in cortical development is unlikely to be restricted only to the late phase of migration, as HSF2 is also expressed at high levels in the cells of the neuroepithelium and neuronal progenitors in the ventricular zone (D. Trouillet's unpublished results).[10,11] It is plausible that HSF2 participates in the regulation of neuronal proliferation, which is well in line with defects observed in the early CNS of Hsf2-null mice.[11]

Although no substantial data on the role of HSF1 in brain development is currently available, HSF1 has been implicated in the maintenance of the postnatal brain under nonstressed conditions.[53] In accordance with Hsf2-null mice, Hsf1 disruption resulted in enlarged ventricles. Moreover, astrogliosis and neurodegeneration occurred in specific areas.[53] Interestingly, the expression levels of Hsp27 and αB-crystallin, which protect cells against stress and apoptosis, were decreased in Hsf1 knockout brain regions. Because $Hsf1^{-/-}$ embryonic brains are still normal at E18.5, the abnormalities probably originate from a later stage in the perinatal and postnatal development.[6]

HSFs in Spermatogenesis

HSFs have been found to be involved in the regulation of gametogenesis in both genders.[10–12,46,54,55] In males, experimental evidence reveals a critical function for both HSF1 and HSF2 in germ cell production. A constitutively active form of HSF1 caused disruption of spermatogenesis and death of pachytene spermatocytes.[54] Although $Hsf1^{-/-}$ mice are fertile and exhibit normal spermatogenesis, decreased heat-induced elimination of the pachytene spermatocytes was observed, which is an opposite effect to that detected in HSF1-overexpressing mice.[55] In general, mutations affecting spermatogenesis result in apoptosis at the pachytene stage,[56] and HSF1 is activated at this specific stage, which could be a marker for accumulation of damaged proteins and a signal to induce cell death.[55]

Hsf2 deficiency resulted in reduced size of testes, increased apoptosis, and decreased sperm count.[10,11] Severe disruption and vacuolization of the seminiferous tubules were observed, reflecting the absence of differentiating spermatocytes and spermatids. At the late pachytene stage, up to 90% of spermatocytes were dead. Furthermore, in the $Hsf2^{-/-}$ pachytene spermatocytes, the synaptonemal complex, which forms an axis of paired chromosomes, was often disorganized, showing an abnormal loop-like structure between pairs of homologous chromosomes.[10] Disruption of both *Hsf1* and *Hsf2* caused a more severe phenotype associated with male sterility and a potentiation of the phenotype seen in $Hsf2^{-/-}$ mice, suggesting that transcriptional activity of both factors is required for normal spermatogenesis.[12] Global expression analyses in the testes of double knockout mice demonstrated changes in expression patterns of genes involved in spermatogenesis.[12] Together, these observations strongly suggest that the activities of HSF1 and HSF2 are tightly intertwined during spermatogenesis. Identification of the direct target genes is a prerequisite for understanding the physiological functions of HSFs in testis.

FUTURE PERSPECTIVES

Previously, HSFs were identified solely as regulators of *Hsp*s, whereas now there is unambiguous evidence for HSFs' having a great variety of target genes (TABLE 1). In *S. cerevisiae* and *Drosophila,* about 3% of the genomic loci were identified as targets for HSF on heat stress.[57,58] The existence of multiple HSFs in higher eukaryotes with different expression patterns, suggests that they may have functions that are triggered by distinct stimuli, leading to activation of specific target genes. The *Hsf* knockout mice have rendered the possibility of identifying novel targets. However, a challenging genomewide ChIP-microarray approach to investigate *in vivo* targets of mammalian HSFs could uncover entirely novel gene clusters, pathways and functions for these transcription factors. This approach would most certainly broaden the current view of HSFs.

The functional relationship between different HSFs, both in cell stress and in developmental processes, is of great interest, and a novel dimension of the cooperation between HSFs is emerging. Synergy of DNA-binding activities among different transcription factors offers an efficient way to control gene expression in a cell- and stimulus-specific manner. By interacting with distinct partners and responding to both stress and developmental stimuli, HSFs could orchestrate differential gene regulation. It will be intriguing to elucidate whether HSF-mediated regulation depends on the activity of individual trimers, or whether homo- or heterotrimer formation is a common theme in HSF-mediated transcription (FIG. 2.). Obvious questions for future studies are the stoichiometry between HSF1 and HSF2 in a possible complex and the mechanism by which the factors interact with each other. Given the slightly

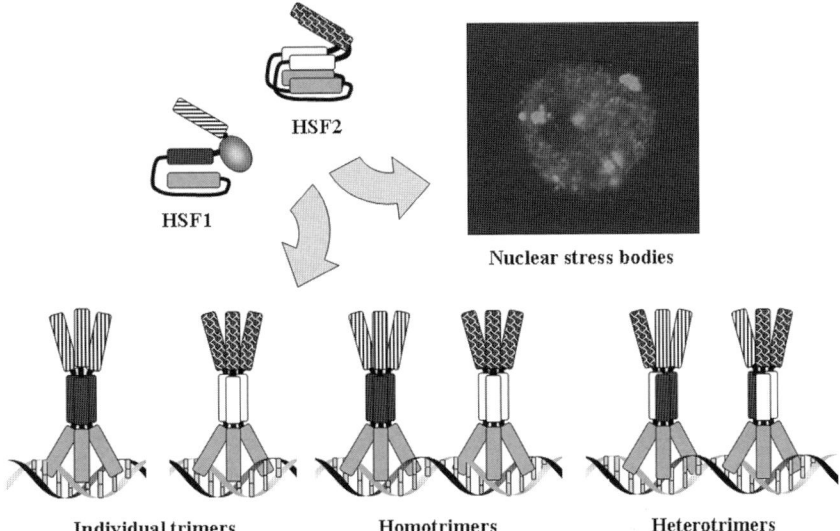

FIGURE 2. Activation and complex formation of HSF1 and HSF2 on the DNA. Inactive HSF1 is kept as a monomer, whereas HSF2 is a dimer. Upon activation HSFs trimerize and translocate to their target gene promoters and nuclear stress bodies (nSBs) mainly formed on locus 9q12 (confocal microscopy image is a courtesy of Anton Sandqvist). The possible composition of HSF trimers, either individual, or combined homo- and heterotrimers, on the DNA is displayed in the figure.

different binding preferences of HSF1 and HSF2,[2] the composition of the HSE on the target promoter could direct the formation of a specific heterocomplex. Sequence variations of the HSE, in a specific chromatin environment determined by histone modifications, could be an efficient way of regulating the DNA-binding ability of HSFs.

It is not exactly known how the cells sense stress. The correlation between longevity and stress resistance suggests that the ability to sense and respond to environmental challenges is important for the regulation of lifespan. Results from several groups indicate a direct role for HSF1 in the regulation of lifespan.[42,43,59] Interestingly, downregulation of HSF1 leads to both decreased lifespan and an accelerated aging phenotype in *C. elegans*. Recent discoveries demonstrate that a mutation conferring longevity also delays polyQ aggregation and toxicity, suggesting a link between the regulation of aging and aging-related diseases. Inactivation of *Daf-16*, *Hsf*, or small *Hsps*, accelerates the aggregation of polyQ expansion proteins in *C. elegans*.[42,43] Correspondingly, human diploid fibroblasts show attenuated heat-inducible HSF1 DNA-binding activity and a decrease in Hsps on aging.[60] These findings support a model where HSF1 is a key molecule for coupling the regulation of lifespan with the ability of cells to sense stress. Many pathologies in humans are associated with stress, age, and expression of misfolded proteins, and several HSF-targeted therapeutic strategies have already been proposed. Small molecular regulators of the HSF activity have been identified and will be valuable tools for discovering novel therapies.[61] The functional interplay between different mammalian HSFs emphasize that great consideration is required when planning future HSF-targeted therapies.

ACKNOWLEDGMENTS

We apologize to our colleagues whose original work could only be cited indirectly due to space limitation. Members of our laboratories are acknowledged for valuable comments on the manuscript. Our own work is supported by the Academy of Finland, Sigrid Jusélius Foundation, the Finnish Life and Pension Insurance Companies, the Finnish Cancer Organizations, Åbo Akademi University, the Association pour la Recherche contre le Cancer (ARC #3609, #3997), and the Agence Nationale pour la Recherche (ANR Neurosciences). M. Åkerfelt is supported by the Turku Graduate School for Biomedical Sciences (TuBS) and D. Trouillet by the French Ministry of Research and Technology and by ARC.

REFERENCES

1. WU, C. 1984. Activating protein factor binds in vitro to upstream control sequences in heat shock gene chromatin. Nature **311:** 81–84.
2. PIRKKALA, L., P. NYKÄNEN & L. SISTONEN. 2001. Roles of the heat shock transcription factors in regulation of the heat shock response and beyond. FASEB J. **15:** 1118–1131.
3. NOVER, L., K. BHARTI, P. DORING, et al. 2001. Arabidopsis and the heat stress transcription factor world: how many heat stress transcription factors do we need? Cell Stress Chap. **6:** 177–189.
4. NAKAI, A. 1999. New aspects in the vertebrate heat shock factor system: Hsf3 and Hsf4. Cell Stress Chap. **4:** 86–93.
5. MCMILLAN, D.R., X. XIAO, L. SHAO, et al. 1998. Targeted disruption of heat shock transcription factor 1 abolishes thermotolerance and protection against heat-inducible apoptosis. J. Biol. Chem. **273:** 7523–7528.
6. XIAO, X., X. ZUO, A.A. DAVIS, et al. 1999. HSF1 is required for extra-embryonic development, postnatal growth and protection during inflammatory responses in mice. EMBO J. **18:** 5943–5952.
7. ZHANG, Y., L. HUANG, J. ZHANG, et al. 2002. Targeted disruption of hsf1 leads to lack of thermotolerance and defines tissue-specific regulation for stress-inducible hsp molecular chaperones. J. Cell. Biochem. **86:** 376–393.
8. SISTONEN, L., K.D. SARGE, B. PHILLIPS, et al. 1992. Activation of heat shock factor 2 during hemin-induced differentiation of human erythroleukemia cells. Mol. Cell. Biol. **12:** 4104–4111.

9. SISTONEN, L., K.D. SARGE & R.I. MORIMOTO. 1994. Human heat shock factors 1 and 2 are differentially activated and can synergistically induce hsp70 gene transcription. Mol. Cell. Biol. **14:** 2087–2099.
10. KALLIO, M., Y. CHANG, M. MANUEL, et al. 2002. Brain abnormalities, defective meiotic chromosome synapsis and female subfertility in HSF2 null mice. EMBO J. **21:** 2591–2601.
11. WANG, G., J. ZHANG, D. MOSKOPHIDIS, et al. 2003. Targeted disruption of the heat shock transcription factor (hsf)-2 gene results in increased embryonic lethality, neuronal defects, and reduced spermatogenesis. Genesis **36:** 48–61.
12. WANG, G., Z. YING, X. JIN, et al. 2004. Essential requirement for both hsf1 and hsf2 transcriptional activity in spermatogenesis and male fertility. Genesis **38:** 66–80.
13. CHANG, Y., P. ÖSTLING, M. ÅKERFELT, et al. 2006. Role of heat-shock factor 2 in cerebral cortex formation and as a regulator of p35 expression. Genes Dev. **20:** 836–847.
14. ÖSTLING, P., J.K. BJÖK, P. ROOS-MATTJUS, et al. 2007. Heat shock factor 2 (HSF2) contributes to inducible expression of hsp genes through interplay with HSF1. J. Biol. Chem. **282:** 7077–7086.
15. NAKAI, A., M. TANABE, Y. KAWAZOE, et al. 1997. HSF4, a new member of the human heat shock factor family which lacks properties of a transcriptional activator. Mol. Cell. Biol. **17:** 469–481.
16. FUJIMOTO, M., H. IZU, K. SEKI, et al. 2004. HSF4 is required for normal cell growth and differentiation during mouse lens development. EMBO J. **23:** 4297–4306.
17. TAKAKI, E., M. FUJIMOTO, K. SUGAHARA, et al. 2006. Maintenance of olfactory neurogenesis requires HSF1, a major heat shock transcription factor in mice. J. Biol. Chem. **281:** 4931–4937.
18. HOLMBERG, C.I., S.E.F. TRAN, J.E. ERIKSSON, et al. 2002. Multisite phosphorylation provides sophisticated regulation of transcription factors. Trends Biochem. Sci. **27:** 619–627.
19. HIETAKANGAS, V., J.K. AHLSKOG, A.M. JAKOBSSON, et al. 2003. Phosphorylation of serine 303 is a prerequisite for the stress-inducible SUMO modification of heat shock factor 1. Mol. Cell. Biol. **23:** 2953–2968.
20. HIETAKANGAS, V., J. ANCKAR, H.A. BLOMSTER, et al. 2006. PDSM, a motif for phosphorylation-dependent SUMO modification. Proc. Natl. Acad. Sci. USA **103:** 45–50.
21. ANCKAR, J. & L. SISTONEN. 2007. Heat shock factor 1 as a coordinator of stress and developmental pathways. Adv. Exp. Med. Biol. **594:** 78–88.
22. SORGER, P.K. & H.C.M. NELSON. 1989. Trimerization of a yeast transcriptional activator via a coiled-coil motif. Cell **59:** 807–813.
23. WU, C. 1995. Heat shock transcription factors: structure and regulation. Annu. Rev. Cell Dev. Biol. **11:** 441–469.
24. AMIN, J., J. ANANTHAN & R. VOELLMY. 1988. Key features of heat shock regulatory elements. Mol. Cell. Biol. **8:** 3761–3769.
25. XIAO, H., O. PERISIC & J.T. LIS. 1991. Cooperative binding of *Drosophila* heat shock factor to arrays of a conserved 5 bp unit. Cell **64:** 585–593.
26. XIAO, H. & J.T. LIS. 1988. Germline transformation used to define key features of heat-shock response elements. Science **239:** 1139–1142.

27. TRINKLEIN, N.D., W.C. CHEN, R.E. KINGSTON, et al. 2004. Transcriptional regulation and binding of heat shock factor 1 and heat shock factor 2 to 32 human heat shock genes during thermal stress and differentiation. Cell Stress Chap. **9:** 21–28.
28. SAKURAI, H. & Y. TAKEMORI. 2007. Interaction between heat shock transcription factors (HSFs) and divergent binding sequences: different binding specificities of yeast HSFs and human HSF1. J. Biol. Chem. **282:** 13334–13341.
29. PIRKKALA, L., T.-P. ALASTALO, X. ZUO, et al. 2000. Disruption of heat shock factor 1 reveals an essential role in the ubiquitin proteolytic pathway. Mol. Cell. Biol. **20:** 2670–2675.
30. MCMILLAN, D.R., E. CHRISTIANS, M. FORSTER, et al. 2002. Heat shock transcription factor 2 is not essential for embryonic development, fertility, or adult cognitive and psychomotor function in mice. Mol. Cell. Biol. **22:** 8005–8014.
31. HE, H., F. SONCIN, N. GRAMMATIKAKIS, et al. 2003. Elevated expression of heat shock factor (HSF) 2A stimulates HSF1-induced transcription during stress. J. Biol. Chem. **278:** 35465–35475.
32. LOISON, F., L. DEBURE, P. NIZARD, et al. 2005. Up-regulation of the clusterin gene after proteotoxic stress: implication of HSF1/HSF2 heterocomplexes. Biochem. J. **395:** 223–231.
33. XING, H., D.C. WILKERSON, C.N. MAYHEW, et al. 2005. Mechanism of hsp70i gene bookmarking. Science **307:** 421–423.
34. INOUYE, S., H. IZU, E. TAKAKI, et al. 2004. Impaired IgG production in mice deficient for heat shock transcription factor 1. J. Biol. Chem. **279:** 38701–38709.
35. JOLLY, C., L. KONECNY, D.L. GRADY, et al. 2002. In vivo binding of active heat shock transcription factor 1 to human chromosome 9 heterochromatin during stress. J. Cell Biol. **156:** 775.
36. DENEGRI, M., D. MORALLI, M. ROCCHI, et al. 2002. Human chromosomes 9, 12, and 15 contain the nucleation sites of stress-induced nuclear bodies. Mol. Biol. Cell **13:** 2069–2079.
37. BIAMONTI, G. 2004. Nuclear stress bodies: a heterochromatin affair? Nat. Rev. Mol. Cell Biol. **5:** 493–498.
38. SHELDON, L.A. & R.E. KINGSTON. 1993. Hydrophobic coiled-coil domains regulate the subcellular localization of human heat shock factor 2. Genes Dev. **7:** 1549–1558.
39. ALASTALO, T.-P., M. HELLESUO, A. SANDQVIST, et al. 2003. Formation of nuclear stress granules involves HSF2 and coincides with the nucleolar localization of Hsp70. J. Cell Sci. **116:** 3557–3570.
40. HEINRICH, P.C., I. BEHRMANN, S. HAAN, et al. 2003. Principles of interleukin (IL)-6-type cytokine signalling and its regulation. Biochem. J. **374:** 1–20.
41. SINGH, V. & A. ABALLAY. 2006. Heat-shock transcription factor (HSF)-1 pathway required for *Caenorhabditis elegans* immunity. Proc. Natl. Acad. Sci. USA **103:** 13092–13097.
42. HSU, A.L., C.T. MURPHY & C. KENYON. 2003. Regulation of aging and age-related disease by DAF-16 and heat-shock factor. Science **300:** 1142–1145.
43. MORLEY, J.F. & R.I. MORIMOTO. 2004. Regulation of longevity in *Caenorhabditis elegans* by heat shock factor and molecular chaperones. Mol. Biol. Cell **15:** 657–664.
44. JEDLICKA, P., M.A. MORTIN & C. WU. 1997. Multiple functions of *Drosophila* heat shock transcription factor in vivo. EMBO J. **16:** 2452–2462.

45. MIN, J.N., Y. ZHANG, D. MOSKOPHIDIS, et al. 2004. Unique contribution of heat shock transcription factor 4 in ocular lens development and fiber cell differentiation. Genesis **40:** 205–217.
46. CHRISTIANS, E., A.A. DAVIS, S.D. THOMAS, et al. 2000. Maternal effect of Hsf1 on reproductive success. Nature **407:** 693–694.
47. BU, L., Y. JIN, Y. SHI, et al. 2002. Mutant DNA-binding domain of HSF4 is associated with autosomal dominant lamellar and marner cataract. Nat. Genet. **31:** 276–278.
48. KWONG, J.M., M. LALEZARY, J.K. NGUYEN, et al. 2006. Co-expression of heat shock transcription factors 1 and 2 in rat retinal ganglion cells. Neurosci. Lett. **405:** 191–195.
49. WALSH, D., Z. LI, Y. WU, et al. 1997. Heat shock and the role of the HSPs during neural plate induction in early mammalian CNS and brain development. Cell Mol. Life Sci. **53:** 198–211.
50. RALLU, M., M. LOONES, Y. LALLEMAND, et al. 1997. Function and regulation of heat shock factor 2 during mouse embryogenesis. Proc. Natl. Acad. Sci. USA **94:** 2392–2397.
51. MIN, J.N., M.Y. HAN, S.S. LEE, et al. 2000. Regulation of rat heat shock factor 2 expression during the early organogenic phase of embryogenesis. Biochim. Biophys. Acta **1494:** 256–262.
52. AYALA, R., T. SHU & L.H. TSAI. 2007. Trekking across the brain: the journey of neuronal migration. Cell **128:** 29–43.
53. SANTOS, S.D. & M.J. SARAIVA. 2004. Enlarged ventricles, astrogliosis and neurodegeneration in heat shock factor 1 null mouse brain. Neuroscience **126:** 657–663.
54. NAKAI, A., M. SUZUKI & M. TANABE. 2000. Arrest of spermatogenesis in mice expressing an active heat shock transcription factor 1. EMBO J. **19:** 1545–1554.
55. IZU, H., S. INOUYE, M. FUJIMOTO, et al. 2004. Heat shock transcription factor 1 is involved in quality-control mechanisms in male germ cells. Biol. Reprod. **70:** 18–24.
56. DE ROOIJ, D.G. & P. DE BOER. 2003. Specific arrests of spermatogenesis in genetically modified and mutant mice. Cytogenet. Genome Res. **103:** 267–276.
57. HAHN, J.S., Z. HU, D.J. THIELE, et al. 2004. Genome-wide analysis of the biology of stress responses through heat shock transcription factor. Mol. Cell. Biol. **24:** 5249–5256.
58. BIRCH-MACHIN, I., S. GAO, D. HUEN, et al. 2005. Genomic analysis of heat-shock factor targets in *Drosophila*. Genome Biol. **6:** R63.
59. GARIGAN, D., A.L. HSU, A.G. FRASER, et al. 2002. Genetic analysis of tissue aging in *Caenorhabditis elegans*: a role for heat-shock factor and bacterial proliferation. Genetics **161:** 1101–1112.
60. LEE, Y.K., D. MANALO & A.Y. LIU. 1996. Heat shock response, heat shock transcription factor and cell aging. Biol. Signals **5:** 180–191.
61. WESTERHEIDE, S.D. & R.I. MORIMOTO. 2005. Heat shock response modulators as therapeutic tools for diseases of protein conformation. J. Biol. Chem. **280:** 33097–33100.

Extracellular Heat Shock Proteins in Cell Signaling and Immunity

STUART K. CALDERWOOD, SALAMATU S. MAMBULA, AND PHILLIP J. GRAY, JR.

Department of Radiation Oncology, Beth Israel Deaconess Medical Center, Harvard Medical School, Boston, Massachusetts, USA

ABSTRACT: Extracellular stress proteins including heat shock proteins (Hsps) and glucose-regulated proteins (Grps) are emerging as important mediators of intercellular signaling and transport. Release of such proteins from cells is triggered by physical trauma and behavioral stress as well as exposure to immunological "danger signals." Stress protein release occurs both through physiological secretion mechanisms and during cell death by necrosis. After release into the extracellular fluid, Hsp or Grp may then bind to the surfaces of adjacent cells and initiate signal transduction cascades as well as the transport of cargo molecules, such as antigenic peptides. In addition, Hsp60 and Hsp70 are able to enter the bloodstream and may possess the ability to act at distant sites in the body. Many of the effects of extracellular stress proteins are mediated through cell-surface receptors. Such receptors include toll-like receptors (TLRs) 2 and 4, CD40, CD91, CCR5, and members of the scavenger receptor family, such as LOX-1 and SREC-1. The possession of a wide range of receptors for the Hsp and Grp family permits binding to a diverse range of cells and the performance of complex multicellular functions particularly in immune cells and neurons.

KEYWORDS: extracellular; heat; shock; protein; secretion; receptor

INTRODUCTION

Mammalian stress proteins appear to have been derived from prokaryotic ancestors that evolved to solve problems in protein folding.[1] Eukaryotic descendents of these proteins retain these properties and tend the tertiary and quaternary structures of a large group of intracellular proteins.[2] The properties of these proteins in stress involve the direct maintenance of protein structure (chaperoning) as well as the regulation of death pathways. In the heat shock response, for instance, elevated temperatures trigger massive synthesis

---

Address for correspondence: Stuart K. Calderwood, Ph.D., Department of Radiation Oncology, BIDMC, Harvard Medical School, Rm553B, 21–27 Burlington Ave., Boston, MA 02215. Voice: 617-632-0628; fax: 617-632-0365.
scalderw@bidmc.harvard.edu

TABLE 1. Intracellular and extracellular properties of Hsp and Grp

Stress Protein	Protein Function	
	Intracellular	Extracellular
Hsp27	Chaperone antideath	Anti-inflammatory
Hsp60	Chaperonin	Proinflammatory
Hsp70	Chaperone antideath	Immunoregulatory proinflammatory neuronal survival
Hsp90	Chaperone cell regulation	Proimmune prometastatic
Hsp110	Chaperone co-chaperone	Proimmune
Grp78	ER chaperone	Anti-inflammatory
Grp94	ER chaperone	Immunoregulatory
Grp170	ER chaperone	Proimmune

of heat shock proteins (Hsps) that fold heat-denatured proteins and block caspase-dependent apoptosis, permitting repair and thwarting death.[3] Extreme exposure to heat overwhelms the capacity of intracellular stress proteins and triggers death.[3] In addition to the intracellular response, stress also triggers the release of proteins into the extracellular spaces.[4-6] Indeed, stress proteins, such as Hsp 27, 60, 70, 90, and 110 and glucose-regulated proteins (Grps) 78, 94, 170, and calreticulin, are released from cells in a variety of circumstances and interact with adjacent cells or in some cases enter the bloodstream (TABLE 1). We will discuss here mechanisms by which such proteins are released from cells and the cell-surface structures involved in recognizing them through autocrine loops, by adjacent cells or at distant sites.

A variety of cell types secrete stress proteins, including neuronal cells, monocytes, macrophages, B cells, and tumor cells of epithelial origin.[7-9] This suggests that stress protein release is a fairly widespread phenomenon and may be implicated in a number of physiological or pathological events. Furthermore, it appears that some cell types may be adapted for specialized secretion of stress proteins into the blood stream and Hsp70 is released from brown fat tissue into the circulation in response to behavioral stress.[10] In addition, Hsps are released from cells undergoing necrosis after extremes of heat stress or other toxic treatments.[11,12] Extracellular Hsp60 and Hsp70 may indeed be physiological alarm signals for cell trauma. Just as Hsp70 release is common to multiple cell types, the ability to bind stress proteins is also shared by many cell types including many cells of the hemopoietic lineage, neuronal cells, vascular, and other epithelial cells.[13-15]

EXTRACELLULAR STRESS PROTEINS AND NEURONAL CELLS

The first reports of stress protein release involved Hsp70 secretion released from neuronal cells along with Hsp110 and actin.[16] Indeed, further studies

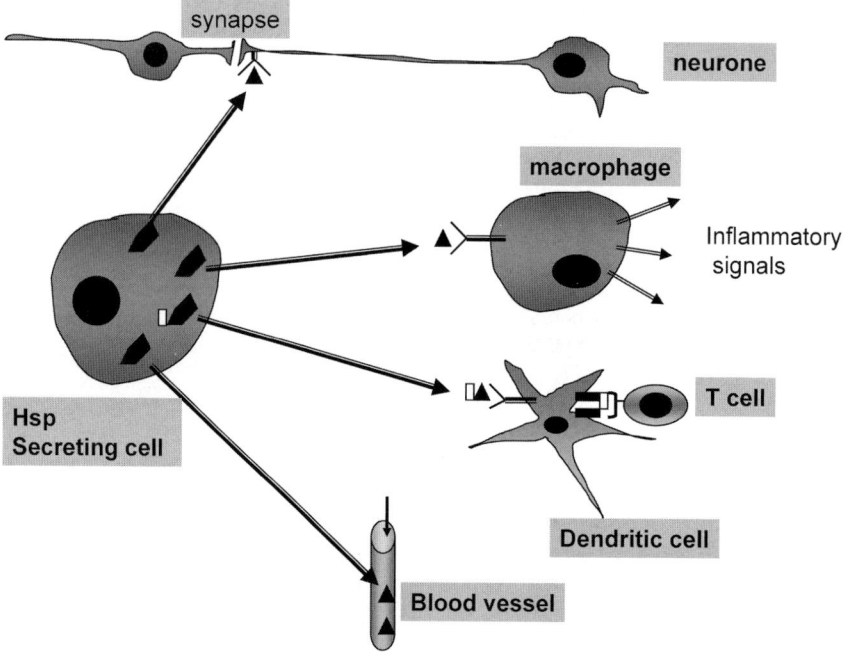

FIGURE 1. Hsp are released from cells and may interact with a wide range of target cells. Hsp may be released by active secretion mechanisms or from cells undergoing necrosis. The resulting extracellular Hsp (triangles) may then interact with neuronal cells, monocytes, or macrophages or enter the circulation. Hsp may also be released conjugated to antigenic peptides (white rectangles) and Hsp–peptide complexes are taken up by antigen-presenting cells (APCs), such as dendritic cells. Such peptides may then be transferred to major histocompatibility complex (MHC) class I molecules through the process known as *cross-presentation,* and such MHC I-peptide complexes can be recognized by CD8+ T lymphocytes leading to T cell activation.

indicate that Hsp70 can be released from glial cells and then taken up by adjacent neurons.[4,17] The rationale for this altruistic behavior between different neuronal cells appears to be that many neurons have a deficit in terms of Hsp expression due to inadequate Hsp gene transcription [18] (FIG. 1). The neuronal cytoplasm in the region of the synapse may thus have a deficit in Hsp levels due to a twofold cause: decreased *de novo* synthesis of Hsp that characterizes such cells and the long axonal distances along which the Hsp must be transported to reach the synapse.[18] One solution to this problem might be to synthesize Hsp in glial cells adjacent to the synapse, release them into the extracellular fluid for uptake by neuronal receptors.[17,19] The functions of such donated Hsps are not known but probably include increasing the chaperoning power of the neuron and protection from programmed cell death, which can accompany neurotransmission.[20]

There are also reports indicating the Hsp release into circulation following behavioral stress or the trauma associated with extreme exercise.[10,21] Hsp70 levels increase in the circulation of experimental animals and human subjects after these physiological stresses although the cellular source and physiological functions of such Hsp70 are not clear.[10,21] It has been suggested that brown fat tissue is the source of much of such Hsp70.[22]

IMMUNE AND INFLAMMATORY EFFECTS OF EXTRACELLULAR HSP

Extracellular stress proteins of the Hsp and Grp families have powerful effects on the immune response.[23–25] These stress proteins interact with the immune response in a number of contexts. During exposure to many pathogens, prokaryotic Hsps are released at high levels and are dominant antigens in the immunological responses to such pathogens.[26,27] Mammalian cells express endogenous stress proteins to high levels after trauma or exposure to bacteria or bacterial proteins.[28] Such stress proteins can be proinflammatory and lead to cytokine transcription and release.[29,30] In addition, stress proteins can act as stimulants of the adaptive immune response through their ability to bind antigenic peptides during antigen processing.[31] When such stress protein–peptide complexes are released from dead and dying cells they bind to receptors on antigen-processing cells and antigens can be delivered to major histocompatibility complex (MHC) class I molecules on the surfaces of such cells through a process known as antigen cross-presentation (FIG. 1).[32,33]

Such interactions form the basis for molecular chaperone-based anticancer vaccines. Hsp–peptide complexes extracted from tumors can stimulate a specific CD8+ T cell-mediated immune response in the tumor-bearing host.[32,33] The potency of such vaccines has been ascribed to the ability of stress proteins to stimulate both the innate and adaptive arms of the antitumor immune response (FIG. 1).[24]

Stress proteins can also be anti-inflammatory and such interactions are noted specifically in inflammatory diseases. Diseases such as RA can be triggered by cross-reactive T cells that recognize common epitopes in mammalian and highly immunogenic prokaryotic Hsp.[34,35] It was speculated that the close degree of conservation in the sequences of these stress proteins might trigger an autoimmune response to mammalian stress proteins. Interestingly, however, application of the corresponding mammalian Hsp suppresses the proinflammatory responses to bacterial Hsp epitopes and leads to remission of inflammatory diseases.[36] Hsps can thus be both profoundly immunostimulatory or immunosuppressive, depending on context.[34,37]

MECHANISMS OF HSP70 RELEASE

Hsp70 is not secreted by the classical pathway; its sequence encodes no secretion leader signal and inhibitors, such as brefeldin A, which antagonize transport through the ER-Golgi system do not inhibit its release. However, a number of non-canonical pathways for release of "leaderless" proteins exist. In some cases release of leaderless proteins involves cell lysis and this may occur both in pathological conditions that give rise to necrosis or in physiologically regulated release of cytokines.[38] Hsp70, for instance, is released when tumor cells undergo necrotic death, presumably when cell membranes become compromised.[11] It has also been suggested that Hsp70 may be released and enter the blood stream under a number of pathological conditions that lead to widespread cell death.[21] In addition, a second pathway involves release of intracellular proteins in secretary vesicles.[39] This mechanism has been observed in macrophages stimulated to release Il-1β by lipopolysaccharide (LPS) and extracellular ATP.[39] Stress proteins, such as Hsp27, Hsp70, Hsc70, and Hsp90, can apparently be released within the lumen of "exosomes" through such a pathway when B cells, for instance, are exposed to heat shock.[8] The postulation of vesicles or exosomes as a source of extracellular Hsp70 requires that these structures should rupture or lyse on entering the extracellular microenvironment.[39] A third secretion pathway involves the entry of the leaderless protein into secretary lysosomal endosomes, migration of these organelles to the cell surface, and release of the contents of the endolysosome into the extracellular space.[40] Again, this pathway has been observed in Il-1β secretion by macrophages stimulated with LPS and ATP.[41,42] Indeed, Hsp70 has recently been shown to be secreted from tumor cells and macrophages by this pathway.[6] Fever range conditions have been shown to induce Hsp70 to enter endolysosomes and be secreted in an extracellular ATP-dependent manner.[6] Study of these processes is still in its infancy and further studies are required to determine the favored pathways for Hsp70 release by neuronal and immune cells.

HSP BINDING AND INTERNALIZATION

Stress proteins interact with a range of receptors on target cells (FIG. 2). These include the oxidized LDL-binding protein CD91/LRP found on antigen-presenting cells (APCs) and other cell types.[43,44] It was in fact proposed that CD91 is the common receptor for all immunogenic Hsp, including Hsp 60, 70, Gp96, and calreticulin.[43] However, its role as a direct high-affinity Hsp binder is still not clear. Theriault *et al.* examined the ability of Hsp70 in free solution to bind cells with, or without CD91 expression and observe minimal differences.[15] In addition, overexpression of substrate-binding domains of CD91 in cells deficient for Hsp70 binding does not restore Hsp70 binding.[15]

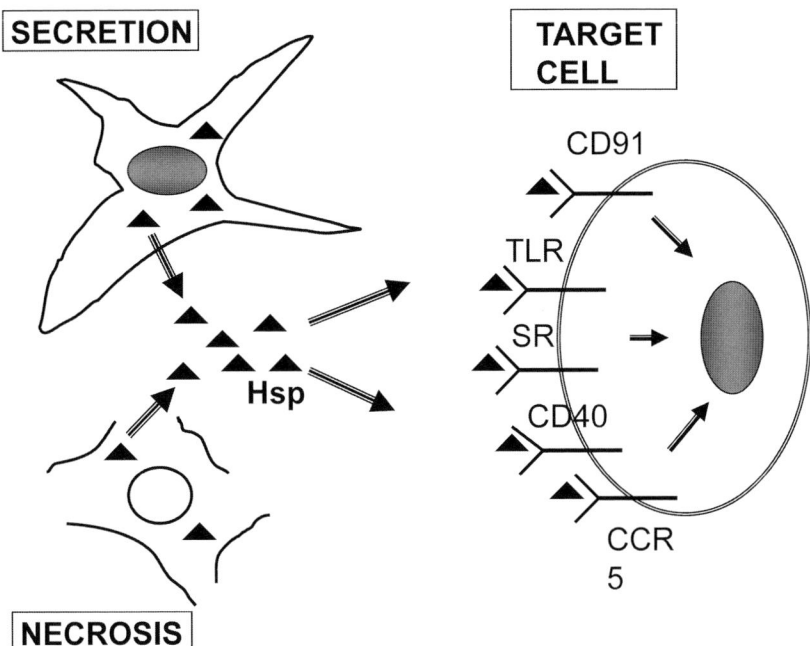

FIGURE 2. Release of stress proteins from cells and binding to cell-surface receptors on the donor cells, adjacent cells, or distant cells. Hsp or Grp reach the extracellular microenvironment through active secretion or passive release after necrosis. The stress proteins can be recognized by an array of receptors on the target cell. Cells may have one or a number of receptors, which may be arrayed in parallel as here or hierarchically with binding to one receptor influencing another.

Hsp70 binding may thus involve low-affinity interactions or be indirect. In addition, CD40 can function as a Hsp70 receptor.[45] CD40 is a member of the *tumor necrosis factor* receptor family and plays a major role in APC maturation through binding to its counter receptor on activated T cells (CD40L).[46] The studies of Lehrner *et al.* showed that mycobacterial Hsp70 can bind avidly to CD40, a finding that may have major implications suggesting the ability to activate APC and cause the release of CC cytokines CCL3, CCL4, and CCL5.[45] It has also been reported that human Hsp70 binds to CD40.[47] However, subsequent studies indicated that when CD40 is overexpressed in previously null cells, they fail to acquire Hsp70-binding capacity.[15] The exact role of CD40 as a direct binder for mammalian Hsp70 is thus still in the balance. However, Millar *et al.* showed an essential role for CD40 in Hsp70 stimulation of APCs indicating its importance at one of the stages in the pathways of antigen presentation and T cell stimulation.[48] As major *pattern recognition receptors* (PRR), the toll-like receptors (TLRs) sparked the interest of a number of groups and instigated investigations of TLR as Hsp receptors. The TLR couple exposure of APC to prokaryotic cell-derived danger signals, such as bacterial LPS, lipopeptides, and CpG DNA, to intracellular signal transduction and transcription pathways that include the NF-κB and interferon response factor pathways.[49] There are now at least 11 members of the TLR, most of which have not been tested as Hsp receptors.[49] However, at least two TLR members function as Hsp receptors and can couple the binding of Hsp60, Hsp70, and gp96 to NF-κB activity.[50–55] In addition, the cell-surface protein CD14 that couples LPS exposure to TLR4 activation is also required for Hsp70 induction of the cytokines TNF-α, IL-1β, and IL-6.[50] Again, however, some studies suggest that these interactions between Hsp70 and TLR are not likely to be exerted through the direct binding of Hsp70 to CD14, TLR2, or TLR4, as null cells that stably express CD14, TLR2, or TLR4 do not bind avidly to Hsp70.[15] Thus, low-affinity interactions may be involved in TLR activation by Hsp. Alternatively, an indirect mechanism, involving Hsp binding to a primary receptor that secondarily activates TLR signaling may mediate the activation of Toll-like receptors. However, previous studies have shown that TLR activation by Hsp70 requires the internalization of the Hsp70; simple experiments using TLR gene overexpression in Hsp null cell lines may thus be inadequate to assess direct Hsp–TLR binding.[56] Another receptor that can directly bind to mycobacterial Hsp70 is the chemokine receptor CCR5; this finding may have considerable significance for signaling cascades induced by Hsp.[57]

Scavenger receptors (SRs) constitute another family of PRR. The SRs are receptors for chemically modified forms of lipoproteins including oxidized and acetylated low-density lipoproteins (oxLDL and acLDL). The SR family is subdivided into eight different subclasses (A to H) and many receptors belonging to this family are expressed on the surface of APC.[58,59] It has been shown that Hsp70 can interact with at least three members of the SR family including LOX-1, SREC-1, and FEEL-1/CLEVER-1.[15,60,61] Hsp70 can both be bound at high affinity by these SR and internalized.[12,60] Both Hsp90 and Hsp60 can also bind to LOX-1.[60] In addition, Gp96/Grp94 and calreticulin show significant affinity to Scavenger Receptor-A (SR-A) and SREC-1 and are internalized by this receptor.[62] Extracellular calreticulin uptake can also be mediated by SREC-1 but apparently not by LOX-1.[62,63] Clearly, members of the SR, which are expressed widely in a range of cell types may play important roles in stress protein binding and internalization.[60,61] Besides its functional relationship to other SR family members, LOX-1 shares structural homology to type V c-type lectin family members Dectin-1, NKG2D, and CD94.[64] Previous studies have demonstrated that another type V c-type lectin CD94 binds to Hsp70 but not to type II c-type lectin DC-SIGN.[15,65] Although, of the c-type lectins, Dectin-1 shares the greatest structural homology with LOX-1, Hsp70 binding was not detected in CHO overexpressing Dectin-1. However, a significant affinity was seen between Hsp70 and C type lectins found in NK cells, such as NKG2D and to a lesser extent the heterodimer CD94/NKG2A. Binding to these c-type lectins was selective in that other family members overexpressed in CHO cells, including DC-SIGN, CLEC-1, and CLEC-2 showed minimal Hsp70 association (J. Theriault & S.K. Calderwood, unpublished). Thus, a number of c-type lectins can serve as receptors for Hsp70 although the conserved *c-type lectin domain* was not the sole determinant required for Hsp70 interaction, as indicated by Hsp70 binding to unrelated SR, such as SREC-1 that do not contain such a domain. C-type lectins may play a significant role in the interaction of Hsp70-expressing tumor cells with NK cells.[66]

This perplexingly large group of receptors may reflect the large and heterogeneous group of stress proteins often with radically different cellular effects. Indeed, we have found that even between quite closely related members of the Hsp70 family there are differences in interactions with individual SR members (J. Theriault & S. K. Calderwood, unpublished). In addition, stress proteins likely recognize different receptors on different cell types.[15] The multiplicity of receptors may also indicate specialization for individual functions: receptors, such as the TLR, CD40, and CCR5, may be adapted for transmembrane signaling while CD91 and SR may play more important roles in internalization of Hsp (FIG. 2).

CONCLUSIONS

Secretion of Hsps and Grps thus extends the reach of the stress response into the extracellular microenvironment. Such stress proteins escape the plasma membrane through both active secretion and passive loss from necrotic cells. Once in the extracellular milieu, the stress proteins increase stress resistance after binding to stress-sensitive recipient cells, such as neurons, signal tissue destruction, and danger to inflammatory cells and aid in immunosurveillance by transporting intracellular peptides to distant immune cells.

ACKNOWLEDGMENTS

We acknowledge the support of the Department of Radiation Oncology at Beth Israel Deaconess Medical Center, Boston. We thank out colleagues Kishiko Ogawa and M. Abdul Khaleque for sharing their thoughts and Rong Zhong for managing the laboratory. These studies were supported by grants 5RO1CA047407 and 3RO1CA094397.

REFERENCES

1. MAYER, M.P. & B. BUKAU. 2005. Hsp70 chaperones: cellular functions and molecular mechanism. Cell. Mol. Life Sci. **62:** 670–684.
2. BUKAU, B., J. WEISSMAN & A. HORWICH. 2006. Molecular chaperones and protein quality control. Cell **125:** 443–451.

3. BEERE, H.M. 2004. 'The stress of dying': the role of heat shock proteins in the regulation of apoptosis. J. Cell Sci. **117:** 2641–2651.
4. TYTELL, M., S.G. GREENBERG & R.J. LASEK. 1986. Heat shock-like protein is transferred from glia to axon. Brain Res. **363:** 161–164.
5. HIGHTOWER, L.E. & P.T. GUIDON, Jr. 1989. Selective release from cultured mammalian cells of heat-shock (stress) proteins that resemble glia-axon transfer proteins. J. Cell. Physiol. **138:** 257–266.
6. MAMBULA, S.S. & S.K. CALDERWOOD. 2006. Heat shock protein 70 is secreted from tumor cells by a nonclassical pathway involving lysosomal endosomes. J. Immunol. **177:** 7849–7857.
7. ROBINSON, M.B. *et al.* 2005. Extracellular heat shock protein 70: a critical component for motoneuron survival. J. Neurosci. **25:** 9735–9745.
8. CLAYTON, A. *et al.* 2005. Induction of heat shock proteins in B-cell exosomes. J. Cell. Sci. **118:** 3631–3638.
9. DAVIES, E.L. *et al.* 2006. Heat shock proteins form part of a danger signal cascade in response to lipopolysaccharide and GroEL. Clin. Exp. Immunol. **145:** 183–189.
10. CAMPISI, J. & M. FLESHNER. 2003. Role of extracellular HSP72 in acute stress-induced potentiation of innate immunity in active rats. J. Appl. Physiol. **94:** 43–52.
11. MAMBULA, S.S. & S.K. CALDERWOOD. 2006. Heat induced release of Hsp70 from prostate carcinoma cells involves both active secretion and passive release from necrotic cells. Int. J. Hyperthermia **22:** 575–585.
12. TODRYK, S. *et al.* 1999. Heat shock protein 70 induced during tumor cell killing induces Th1 cytokines and targets immature dendritic cell precursors to enhance antigen uptake. J. Immunol. **163:** 1398–1408.
13. SINGH-JASUJA, H. *et al.* 2000. The heat shock protein gp96 induces maturation of dendritic cells and down-regulation of its receptor. Eur. J. Immunol. **30:** 2211–2215.
14. SRIVASTAVA, P. 2002. Interaction of heat shock proteins with peptides and antigen presenting cells: chaperoning of the innate and adaptive immune responses. Annu. Rev. Immunol. **20:** 395–425.
15. THERIAULT, J.R. *et al.* 2005. Extracellular HSP70 binding to surface receptors present on antigen presenting cells and endothelial/epithelial cells. FEBS Lett. **579:** 1951–1960.
16. HIGHTOWER, L.E. & P.T. GUIDON. 1989. Selective release from cultured cells of heat shock (stress) proteins that resemble glia-axon proteins. J. Cell. Physiol. **135:** 257–266.
17. GUZHOVA, I. *et al.* 2001. *In vitro* studies show that Hsp70 can be released by glia and that exogenous Hsp70 can enhance neuronal stress tolerance. Brain Res. **914:** 66–73.
18. TONKISS, J. & S.K. CALDERWOOD. 2005. Regulation of heat shock gene transcription in neuronal cells. Int. J. Hyperthermia **21:** 433–444.
19. CHEN, S. & I.R. BROWN. 2007. Translocation of constitutively expressed heat shock protein Hsc70 to synapse-enriched areas of the cerebral cortex after hyperthermic stress. J. Neurosci. Res. **85:** 402–499.
20. CALDERWOOD, S.K. 2005. Evolving connections between molecular chaperones and neuronal function. Int. J. Hyperthermia **21:** 375–378.
21. POCKLEY, A.G. 2002. Heat shock proteins, inflammation, and cardiovascular disease. Circulation **105:** 1012–1017.

22. FLESHNER, M. & J.D. JOHNSON. 2005. Endogenous extra-cellular heat shock protein 72: releasing signal(s) and function. Int. J. Hyperthermia **21:** 457–471.
23. SRIVASTAVA, P.K. 2000. Heat shock protein-based novel immunotherapies. Drug News Perspect. **13:** 517–522.
24. SRIVASTAVA, P.K. & R.J. AMATO. 2001. Heat shock proteins: the 'Swiss Army Knife' vaccines against cancers and infectious agents. Vaccine **19:** 2590–2597.
25. CALDERWOOD, S.K., J.R. THERIAULT & J. GONG. 2005. Message In A Bottle: Role of the 70 kilodalton heat shock protein family in anti-tumor immunity. Eur. J. Immunol. **35:** 2518–2527.
26. BURNIE, J.P. *et al*. 2006. Fungal heat-shock proteins in human disease. FEMS Microbiol. Rev. **30:** 53–88.
27. YOUNG, R.A. & T.J. ELLIOTT. 1989. Stress proteins, infection, and immune surveillance. Cell **59:** 5–8.
28. HUNTER-LAVIN, C. *et al*. 2004. Hsp70 release from peripheral blood mononuclear cells. Biochem. Biophys. Res. Commun. **324:** 511–517.
29. ASEA, A. *et al*. 2000. HSP70 stimulates cytokine production through a CD14-dependant pathway, demonstrating its dual role as a chaperone and cytokine. Nat. Med. **6:** 435–442.
30. ASEA, A. *et al*. 2002. Novel signal transduction pathway utilized by extracellular HSP70: role of toll-like receptor (TLR) 2 and TLR4. J. Biol. Chem. **277:** 15028–15034.
31. NOESSNER, E. *et al*. 2002. Tumor-derived heat shock protein 70 peptide complexes are cross-presented by human dendritic cells. J. Immunol. **169:** 5424–5432.
32. ARNOLD-SCHILD, D. *et al*. 1999. Cutting edge: receptor-mediated endocytosis of heat shock proteins by professional antigen-presenting cells. J. Immunol. **162:** 3757–3760.
33. SINGH-JASUJA, H. *et al*. 2000. Cross-presentation of glycoprotein 96-associated antigens on major histocompatibility complex class I molecules requires receptor-mediated endocytosis. J. Exp. Med. **191:** 1965–1974.
34. VAN EDEN, W., R. VAN DER ZEE & B. PRAKKEN. 2005. Heat-shock proteins induce T-cell regulation of chronic inflammation. Nat. Rev. Immunol. **5:** 318–330.
35. HAUET-BROERE, F. *et al*. 2006. Heat shock proteins induce T cell regulation of chronic inflammation. Ann. Rheum. Dis. **65**(Suppl 3): iii65–iii68.
36. KINGSTON, A.E. *et al*. 1996. A 71-kD heat shock protein (hsp) from Mycobacterium tuberculosis has modulatory effects on experimental rat arthritis. Clin. Exp. Immunol. **103:** 77–82.
37. DANIELS, G.A. *et al*. 2004. A simple method to cure established tumors by inflammatory killing of normal cells. Nat. Biotechnol. **22:** 1125–1132.
38. WEWERS, M.D. 2004. IL-1beta: an endosomal exit. Proc. Natl. Acad. Sci. USA **101:** 10241–10242.
39. MACKENZIE, A. *et al*. 2001. Rapid secretion of interleukin-1beta by microvesicle shedding. Immunity **15:** 825–835.
40. BARALDI, P.G., F. DI VIRGILIO & R. ROMAGNOLI. 2004. Agonists and antagonists acting at P2'7 receptor. Curr. Top. Med. Chem. **4:** 1707–1717.
41. ANDREI, C. *et al*. 1999. The secretory route of the leaderless protein interleukin 1beta involves exocytosis of endolysosome-related vesicles. Mol. Biol. Cell **10:** 1463–1475.
42. ANDREI, C. *et al*. 2004. Phospholipases C and A2 control lysosome-mediated IL-1 beta secretion: Implications for inflammatory processes. Proc. Natl. Acad. Sci. USA **101:** 9745–9750.

43. BASU, S. *et al.* 2001. CD91 is a common receptor for heat shock proteins gp96, hsp90, hsp70, and calreticulin. Immunity **14:** 303–313.
44. BINDER, R.J., D.K. HAN & P.K. SRIVASTAVA. 2000. CD91: a receptor for heat shock protein gp96. Nat. Immunol. **1:** 151–155.
45. WANG, Y. *et al.* 2001. CD40 is a cellular receptor mediating mycobacterial heat shock protein 70 stimulation of CC-chemokines. Immunity **15:** 971–983.
46. MACKEY, M.F., R.J. BARTH, Jr. & R.J. NOELLE. 1998. The role of CD40/CD154 interactions in the priming, differentiation, and effector function of helper and cytotoxic T cells. J. Leukoc. Biol. **63:** 418–428.
47. BECKER, T., F.U. HARTL & F. WIELAND. 2002. CD40, an extracellular receptor for binding and uptake of Hsp70-peptide complexes. J. Cell. Biol. **158:** 1277–1285.
48. MILLAR, D.G. *et al.* 2003. Hsp70 promotes antigen-presenting cell function and converts T-cell tolerance to autoimmunity *in vivo*. Nat. Med. **9:** 1469–1476.
49. TAKEDA, K., T. KAISHO & S. AKIRA. 2003. Toll-like receptors. Annu. Rev. Immunol. **21:** 335–376.
50. ASEA, A. *et al.* 2000. HSP70 peptidembearing and peptide-negative preparations act as chaperokines. Cell Stress Chaperones **5:** 425–431.
51. ASEA, A. *et al.* 2002. Novel signal transduction pathway utilized by extracellular HSP70: role of toll-like receptor (TLR) 2 and TLR4. J. Biol. Chem. **277:** 15028–15034.
52. VABULAS, R.M. *et al.* 2001. Endocytosed HSP60s use toll-like receptor 2 (TLR2) and TLR4 to activate the toll/interleukin-1 receptor signaling pathway in innate immune cells. J. Biol. Chem. **276:** 31332–31339.
53. VABULAS, R.M. *et al.* 2002. HSP70 as endogenous stimulus of the Toll/interleukin-1 receptor signal pathway. J. Biol. Chem. **277:** 15107–15112.
54. QUINTANA, F.J. *et al.* 2004. Inhibition of adjuvant-induced arthritis by DNA vaccination with the 70-kd or the 90-kd human heat-shock protein: immune cross-regulation with the 60-kd heat-shock protein. Arthritis Rheum. **50:** 3712–3720.
55. QUINTANA, F.J. & I.R. COHEN. 2005. Heat Shock Proteins Regulate Inflammation by Both Molecular and Network Cross-Reactivity. Cambridge University Press. Cambridge.
56. VABULAS, R.M. & H. WAGNER. 2005. Toll-Like Receptor-Dependent Activation of Antigen-Presenting Cells by Hsp60, Gp96 and Hsp70. Cambridge University Press. Cambridge.
57. FLOTO, R.A. *et al.* 2006. Dendritic cell stimulation by mycobacterial Hsp70 is mediated through CCR5. Science **314:** 454–458.
58. MURPHY, J.E. *et al.* 2005. Biochemistry and cell biology of mammalian scavenger receptors. Atherosclerosis **182:** 1–15.
59. ADACHI, H. & M. TSUJIMOTO. 2006. Endothelial scavenger receptors. Prog. Lipid Res. **45:** 379–404.
60. DELNESTE, Y. *et al.* 2002. Involvement of LOX-1 in dendritic cell-mediated antigen cross-presentation. Immunity **17:** 353–362.
61. THÉRIAULT, J.R., H. ADACHI & S.K. CALDERWOOD. 2006. Role of scavenger receptors in the binding and internalization of heat shock protein 70. J. Immunol. **177:**8604–8611.
62. BERWIN, B. *et al.* 2003. Scavenger receptor-A mediates gp96/GRP94 and calreticulin internalization by antigen-presenting cells. EMBO J. **22:** 6127–6136.
63. BERWIN, B. *et al.* 2004. SREC-I, a type F scavenger receptor, is an endocytic receptor for calreticulin. J. Biol. Chem. **279:** 51250–51257.

64. ZELENSKY, A.N. & J.E. GREADY. 2005. The C-type lectin-like domain superfamily. FEBS J. **272:** 6179–6217.
65. GROSS, C. *et al*. 2003. Interaction of heat shock protein 70 peptide with NK cells involves the NK receptor CD94. Biol. Chem. **384:** 267–279.
66. MULTHOFF, G. 2002. Activation of natural killer cells by heat shock protein 70. Int. J. Hyperthermia **18:** 576–585.

Membrane Regulation of the Stress Response from Prokaryotic Models to Mammalian Cells

LASZLO VIGH,[a] HITOSHI NAKAMOTO,[b] JACQUES LANDRY,[c] ANTONIO GOMEZ-MUNOZ,[d] JOHN L. HARWOOD,[e] AND IBOLYA HORVATH[a]

[a]*Institute of Biochemistry, Biological Research Center of the Hungarian Academy of Sciences, Szeged, Hungary*

[b]*Department of Biochemistry and Molecular Biology, Saitama University, Saitama, Japan*

[c]*Cancer Research Centre, Laval University, Quebec, Canada*

[d]*Department of Biochemistry and Molecular Biology, University of the Basque Country, Bilbao, Spain*

[e]*School of Biosciences, Cardiff University, Cardiff, United Kingdom*

ABSTRACT: "Membrane regulation" of stress responses in various systems is widely studied. In poikilotherms, membrane rigidification could be the first reaction to cold perception: reducing membrane fluidity of membranes at physiological temperatures is coupled with enhanced cold inducibility of a number of genes, including desaturases (see J.L. Harwood's article in this Proceedings volume). A similar role of changes in membrane physical state in heat (oxidative stress, etc.) sensing- and signaling gained support recently from prokaryotes to mammalian cells. Stress-induced remodeling of membrane lipids could influence generation, transduction, and deactivation of stress signals, either through global effects on the fluidity of the membrane matrix, or by specific interactions of boundary (or raft) lipids with receptor proteins, lipases, ion channels, etc. Our data point to membranes not only as *targets* of stress, but also as *sensors* in activating a stress response.

KEYWORDS: membrane sensor; membrane fluidity; membrane-associated heat shock proteins; sphingomyelin pathway; ceramide; Hsp coinducers; membrane microdomains

Address for correspondence: Laszlo Vigh, Ph.D., D.Sc., Institute of Biochemistry, Biology Research Centre, Hungarian Academy of Sciences,Temesvari krt 62, H-6726, Szeged, Hungary. Voice/fax: +36-62-432-048.
vigh@brc.hu

INTRODUCTION

Subtle alterations in the membranes of yeast and cyanobacteria have been shown to transform environmental signals into the transcriptional activation of heat shock protein (Hsp) genes. In addition, stress protein molecular chaperones appeared to be "Janus-faced," since Hsps are involved not only in the protein folding but also in the maintenance of membrane integrity. The latter role of Hsps in quality control of thylakoid membrane-associated proteins will be discussed.

In most cells, prior to the massive accumulation of Hsps, heat shock rapidly activates many distinct membrane-initiated signaling pathways. The overall balance of these pathways and their interplay are primordial determinants of cell fates. Many stress-induced and lipid-mediated signaling pathways are known to exist and contribute to cellular stress responses: one of those mechanisms is the so-called sphingomyelin pathway, which is discussed here in detail.

Severe heat shock of mammalian cells causes protein damage, but mild, fever-type heat does not. Yet, small changes in temperature can alter the microdomain organization and/or lipid molecular species of membranes. Aging or pathophysiological conditions can also be linked to the development of subtle membrane changes or "membrane-defects." The importance of the operation of membrane-based and membrane-controlled "molecular switches" can be especially important under various disease states. The common feature of certain nonproteotoxic membrane fluidizers and more importantly HSP coinducer drug candidates is, that they are all capable to cause a distinct reorganization of membrane microdomains necessary and apparently sufficient to refine the expression of distinct *hsp* genes.

MOLECULAR CHAPERONES IN QUALITY CONTROL OF MEMBRANES AND MEMBRANE- ASSOCIATED PROTEINS IN PROKARYOTES

Molecular chaperones play important roles in protein quality control. They also associate with membranes although they do not contain transmembrane domains or signal sequences. Such a unique face of molecular chaperones, that is, related to the quality control of membrane and membrane-associated proteins, is well studied in cyanobacterial models. Cyanobacteria are photoautotrophic prokaryotes that are phylogenetically and physiologically related to the chloroplast of photosynthetic eukaryotes. In contrast to heterotrophic organisms, such as *E. coli*, they have layers of green membranes called thylakoid membranes where photosynthesis takes place. Thylakoid membranes possess membrane-embedded protein complexes, such as photosystem II as well as peripheral soluble protein complexes, such as phycobilisomes. Both

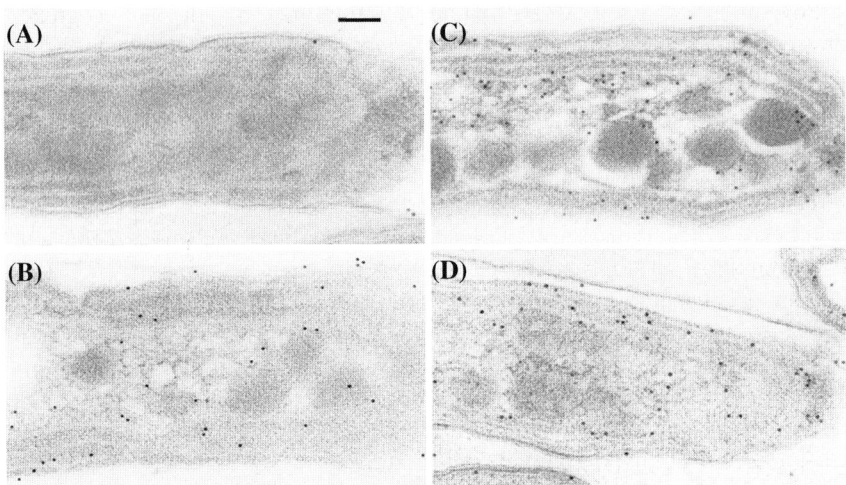

FIGURE 1. Immunocytochemical localization of small Hsp in small Hsp-expressing cyanobacterial cells after heat treatment. (**A**) A reference strain at 30°C devoid of small Hsp (negative control). (**B**) Small Hsp-expressing cells at 30°C. Gold particles are distributed at thylakoid membranes, carboxysomes, and cytoplasm. (**C**) Small Hsp-expressing cells at 45°C for 15 min. Many gold particles are in the cytoplasmic area. (**D**) Small Hsp-expressing cells at 45°C for 60 min. A majority of gold particles are associated with thylakoid membranes. Bar indicates 0.1 μm. (Reprinted by permission of the Federation of the European Biochemical Societies from (Ultrastructural stability under high temperature or intensive light stress conferred by a small heat shock protein in cyanobacteria), by (Nitta K., N. Suzuki, D. Honma, Y. Kaneko & Nakamoto) FEBS Letters, 579, 1235–1242, 2005).

photosystem II and phycobilisomes are thermolabile elements of the thylakoid membrane.

Small Hsp (Hsp17), GroEL (Hsp60), DnaK (Hsp70), and HtpG (Hsp90) have been shown to associate with thyakoid membranes.[1-4] Among them, the small Hsp has been studied most extensively in terms of cellular localization and physiological relevance of the association. Immunocytochemical studies directly demonstrate that small Hsp is located on thylakoid membranes in the cyanobacterium *Synechococcus* sp. PCC7942 (Ref. 5, see FIG.3). Interestingly, the main localization of small Hsp shifts transiently to the cytoplasm after heat shock. The liberated small Hsp may interact with thermolabile cytoplasmic proteins. Direct evidence for the physiological relevance of small Hsp–thylakoid association was presented by genetic studies. Inactivation of a small Hsp gene resulted in a reduced activity of photosynthetic oxygen evolution in heat-stressed cells,[6] while constitutive expression of a small Hsp in cyanobacterial cells increased thermostability of the photosystem II electron transport system and light-harvesting phycocyanins, the major component for phycobilisomes.[7]

Phycobilisomes are deassembled during heat shock. Small Hsp recognizes phycocyanins in deassembled phycobilisomes under severe stress, as probably a major cellular target in order to protect them from irreversible denaturation.[8] Constitutive expression of a small Hsp stabilizes subcellular structures, such as thylakoid membranes under elevated temperature or intensive light stress.[5] These results are consistent with *in vitro* studies that showed that small Hsp possesses an ability to stabilize the lipid phase of membranes.[9] Thus, small Hsp possesses not only protein-protective activity located either in cytosol or in membranes, but also an ability to stabilize the lipid phase of membranes *in vivo*.

HEAT SHOCK-INDUCED SIGNALING PATHWAYS IN MAMMALIAN CELLS

Heat shock, like many other stresses, induces specific and highly regulated signaling cascades that promote cellular homeostasis. The triggering events responsible for activating these pathways are hitherto unclear. Besides protein denaturation, chromatin structure perturbation, changes in cellular redox state, etc., however, the membrane alteration is a potential initiating factor. The three major mitogen-activated protein kinases (MAPK)—extracellular signal-regulated kinase (ERK1/2), stress-activated protein kinase 1 (SAPK1)-c-Jun N-terminal kinase, and SAPK2-p38—together with protein kinase B (PKB/Akt) are the most notable of these HS-stimulated pathways. Their activation occurs rapidly and sooner than the transcriptional upregulation of heat shock proteins, which ultimately generates a transient state of extreme resistance against subsequent thermal stress. The direct link of these early signaling pathways to cellular death or survival mechanisms suggests that they contribute importantly to the HS response. Some of them may counteract early noxious effects of heat, while others may bolster key apoptosis events.

Different lines of evidence suggest a model in which the initial heat shock signal would originate from the cellular membrane and involve, for instance, receptors or receptor-like molecules.[10] This concept is not unique to thermal stress since activation of the ERK and JNK pathways by UV light and osmotic shock also involves agonist-independent receptor activation.[11–13] HS-induced reorganization of membrane hyperstructures might also accelerate the clustering of receptors into lipid rafts, thereby changing the activation threshold of certain receptors leading to agonist-independent activation. The ERK pathway is already known to be heat-triggered by the agonist−independent activation of a receptor, probably EGFR.[14] PKB is commonly activated via PI3K downstream of membrane receptors.[15] Moreover, the HS desensitization of the p38 pathway is reminiscent of the growth factor receptor desensitization that follows their stimulation. Interestingly, clathrin, one of the major components of the machinery involved in the internalization and desensitization of these receptors, is tyrosine phosphorylated in response to HS and H_2O_2 (Dorion, S and Landry J, unpublished results). Under oxidative stress, this phosphorylation seems to affect positively the endocytosis of certain molecules, such as transferrin.[16] Moreover, the phosphorylation of clathrin by the tyrosine kinase Src has already been shown to favor the uptake of the agonist-stimulated EGFR.[17] The possibility that heat might activate some specific membrane receptor is further accredited by the discovery of a receptor for noxious heat stimuli in nociceptive neurons.[18,19] In addition to facilitating receptor activation, HS-induced changes in the fluidity of the bilayer lipid membrane might alter membrane tension and stretching. Membrane stretching leads to the activation of the MAPK pathways in a variety of cell types.[10,20,21] Furthermore, in yeast, the cell wall mechanosensor Wsc1 is required for HS activation of a specific stress kinase pathway composed of PKC1 and Mpk1 and is a critical determinant for the induction of thermotolerance.[22,23] Yeast strains defective in Wsc1 or any of the proteins of the downstream HS-activated pathway are characterized by increased cell lysis at high temperatures, a phenotype very similar to that observed in HSF1 mutant strains.[24,25] It is noted, that the mechanisms of the activation and regulation of the heat shock-sensitive signaling pathways have been reviewed recently.[10,26]

The plasma membrane may still contribute in another way to the early HS-induced cell responses. HS induces the production of certain lipid signaling molecules derived from sphingolipids, such as ceramide. As discussed below, these lipids play pivotal role in the stimulation of signaling pathways under a variety of stress conditions.[10,27]

SIGNALING OF STRESS VIA CERAMIDE AND ITS METABOLITES

Cellular stress has been defined as the threat of damage to macromolecules and membranes. Since many lipids, enzymes, and signaling pathways contribute to the cellular stress response, it is necessary to identify the key players that are located at major nodes within the stress response network. Many types of stresses, including UV or ionizing radiation, oxidative stress, chemotherapeutic drugs, or starvation, cause DNA or protein damage. This can result in growth arrest, apoptosis, or inflammatory responses.[28-32] One of the mechanisms involved in these actions is the sphingomyelin pathway. Ceramide, the central molecule in this pathway, is an important second messenger that engages different downstream effectors depending on the concomitant activation of other second messengers and the activity of enzymes that convert ceramide to other related metabolites, such as sphingosine, sphingosine 1-phosphate (S1P), or ceramide 1-phosphate (C1P)[33] (FIG. 2). While ceramide is proapoptotic and can induce cell cycle arrest, S1P or C1P are antiapoptotic and have mitogenic properties.[34,35] Ceramide and C1P can be interconverted

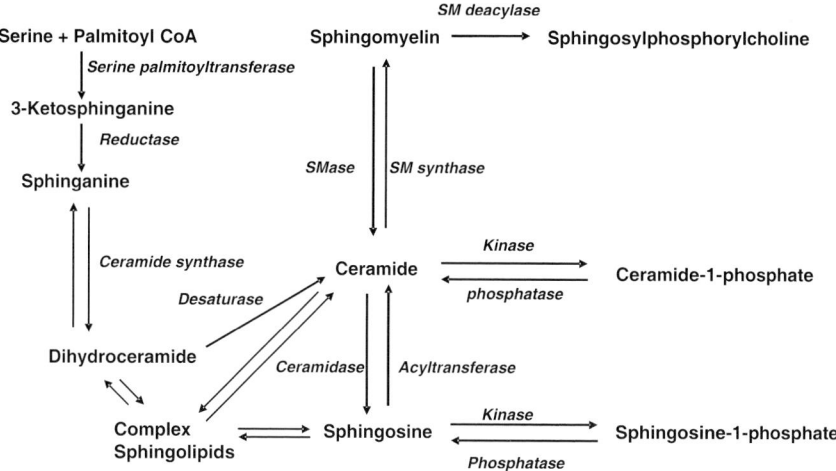

FIGURE 2. Formation of bioactive sphingolipids in mammalian cells. Ceramide can be produced through degradation of sphingomyelin (SM) by sphingomyelinases (SMase), or by *de novo* synthesis by the concerted action of serine palmitoyltransferase and dihydroceramide synthase. It can also be generated through metabolism of complex sphingolipids. Ceramide can be metabolized to ceramide-1-phosphate by ceramide kinase. The reverse reaction is catalyzed by ceramide-1-phosphate phosphatase, or by lipid phosphate phosphatases (LPPs). Alternatively, ceramide can be degraded by ceramidases to form sphingosine, which can, in turn, be phosphorylated to sphingosine-1-phosphate by sphingosine kinases. The reverse reaction is catalyzed by sphingosine-1-phosphate phosphatases, or by lipid phosphate phosphatases. Sphingomyelin *N*-deacylase generates sphingosylphosphorylcholine.

in cells by kinase and phosphatase activities (FIG. 2). An appropriate balance between the levels of these metabolites is crucial for cell and tissue homeostasis. Switching this balance toward accumulation of one or the other can result in metabolic dysfunction or disease. Therefore, the activity of the enzymes that are involved in C1P and ceramide metabolism must be efficiently coordinated to ensure normal cell functioning. At least in primary macrophages, the mechanisms whereby C1P and S1P block apoptosis involve inhibition of acid sphingomyelinase (A-SMase) activity[36] and stimulation of the phosphatidylinositol 3-kinase (PI3K)/protein kinase B (Akt)/nuclear factor-kappa B (NF-κB) pathway.[37] Although both C1P and S1P are mitogenic and antiapoptotic, their mechanisms of action are not always identical. Specifically, S1P is a potent activator of phospholipase D (PLD), induces calcium mobilization, and can alter the levels of cAMP in different cell types (reviewed in Ref. 34). However, C1P does not activate PLD, and does not alter calcium or cAMP levels. Of note, although S1P and C1P can both inhibit A-SMase, C1P acts directly on this enzyme activity,[35] whereas the inhibition of A-SMase by S1P is indirect and requires cell integrity.[38] The physiological relevance of the prosurvival effect of C1P is underscored by the demonstration that the intracellular levels of C1P are substantially decreased in apoptotic cells. This could release A-SMase from inhibition, thereby triggering ceramide generation and apoptotic cell death.[36] Although the antiapoptotic effects of S1P and C1P involve activation the PI3K/PKB pathway with no intervention of mitogen-activated protein kinases (MAPK), the C1P- or S1P-induced cell proliferation also implicates activation of the MAPKs ERK1/2 and upregulation of *c-myc* and cyclin D1.

SPECIFIC MEMBRANE CHANGES CAN REFINE THE STRESS PROTEIN EXPRESSION

As illustrated above, a large number of membrane-associated stress sensing and signaling mechanisms operate in mammalian cells that are capable to trigger both apoptotic and nonapoptotic cell death programs and/or survival.[10,26] Stress-induced alterations in the lipid phase of membranes could affect these stress signaling pathways either through global effects on the membrane's physical state, or via specific interactions of lipids with proteins.[10,39,40] Moreover, individual lipid species at specific membrane locations (like at "rafts") can recruit a unique set of proteins, forming specified signaling platforms involved in a particular signal transduction pathway. Distribution of lipid species is directed by proteins, and, conversely, lipids influence the distribution of proteins within membranes.[10] Taken together, more than 1,000 membrane lipids can play important roles in regulating stress-triggered cell signalings.[10,40,41]

Modification of membrane properties can also be linked to the operation of another category of signaling pathways, which ultimately affects the preexisting level and profile of cellular HSPs.[41,42] As reviewed recently, stress-induced membrane changes may accelerate the clustering of certain growth factor receptors and enhance their autophosphorylation.[10,41] As described typically for ErbB activation in tumor cells, this step leads to a downstream signal cascade with concomitant upregulation of Hsp synthesis.[41] Several membrane-intercalating compounds (benzyl alcohol, or bimoclomol) or disease states (cancer, diabetes) that alter specifically membrane lipid properties can simultaneously modulate the activation of various protein kinases (including PKC, PKA, MAPKs) implicated in refining the stress protein response (see Refs. 41–44). There exists a known connection also between *hsp* gene activation and Ras signaling pathways. Lipid second messenger arachidonic acid (AA), together with its metabolic products, such as prostaglandins, has been reported to be potent Hsp inducers. Heat (or other) stress can alter the physical state of certain membrane phospholipids in a way, that this change then makes these phospholipids available as substrates for phospholipases resulting in the release of free fatty acids, like AA (see Refs. 10, 41). A cholesterol derivative steryl glucoside is rapidly induced in cells upon exposure to stress, and it is followed by activation of a certain protein kinases and induction of heat shock proteins.[45] Another lipid, sphingosine 1-phosphate (S1P) was shown to regulate the induction of Hsp27 by a p38 MAPK-dependent mechanism pathway [see Ref. 10]. Taken together, modification of membrane properties by various means can alter the level of cellular Hsps.

A recent study revealed, that the same level of membrane hyperfluidization of melanoma cells with the closely analogous, nonproteotoxic membrane fluidizers is not necessarily and automatically coupled with the generation of membrane-based stress protein-inducing signal.[43,44] Benzyl alcohol (BA) at a concentration that activates heat shock genes exerts a profound effect on the melting of raft-like cholesterol–sphingomyelin domains both *in vitro* and *in vivo*. On the contrary, the close analog phenetyl alcohol at a concentration equipotent with BA in membrane fluidization has no such effect and is unable to activate Hsp synthesis. It is suggested that, apart from membrane hyperfluidization in the deep hydrophobic region, a distinct reorganization of cholesterol-rich microdomains may also be required for the generation and transmission of stress signals to activate *hsp* genes.[45]

As reviewed elsewhere, pharmacological activation (or attenuation) of the HSP response with lipid-interacting drug molecules may be a successful future therapeutic approach for the treatment of most various diseases (like diabetes or cancer).[10,39,41,46–48]

In conclusion, while many inducers of the heat shock response may function through a protein unfolding mechanism, the above observations clearly suggest that some inducers may work through a distinct, membrane remodeling-based mechanism.[10,39,41] A membrane sensor model could explain specific HSP expression patterns with different stressors.[39,41] This model predicts that plasma membrane, which is the barrier to the external environment and well suited for sensing thermal stress in mammalian cells, acts as an important regulatory interface. Subtle alterations in its lipid phase (characteristic "membrane defects" in disease states as reviewed in Ref. 41) may influence membrane-initiated stress-signaling processes via causing changes in the distribution and size of membrane microdomains, membrane thickness, or curvature. Real-time monitoring of the changes of geometry, topology, and dynamics of membranes and combination of that with lipidomics and computational methods may allow the identification of novel stress-sensory membrane domains.[49] In addition, the lipid-selective association of a subpopulation of heat shock proteins with membranes, leading to altered molecular order and polymorphic characteristics[9,49–52] may regulate further the membrane-controlled expression of heat shock genes.[39,41]

ACKNOWLEDGMENTS

For A G-M was supported by "Ministerio de Educación y Ciencia" (Grant BFU2006-13689), Madrid, Spain, and Universidad del Pais Vasco, UPV/EHU

(Grant 9/UPV 00042-310-15852/2004), Bilbao, Spain. Work by LV and IH was supported by grants from the Agency for Research Fund Management and Research Exploitation (RET OMFB 0067/2005) and from Marie Curie Host Fellowship MTKI-CT-2004-003091.

REFERENCES

1. HORVÁTH, I., A. GLATZ, V. VARVASOVSZKI, et al. 1998. Membrane physical state controls the signaling mechanism of the heat shock response in *Synechocystis* PCC 6803: Identification of *hsp17* as a 'fluidity gene'. Proc. Natl. Acad. Sci. USA **95**: 3513–3518.
2. LEHEL, C., H. WADA, E. KOVÁCS, et al. 1992. Plant Mol. Biol. **18**: 327–336.
3. NIMURA, K., H. YOSHIKAWA & H. TAKAHASHI. 1996. DnaK3, one of the three DnaK proteins of cyanobacterium *Synechococcus* sp. PCC7942, is quantitatively detected in the thylakoid membrane. Biochem. Biophys. Res. Commun. **229**: 334–340.
4. WATANABE, S., T. KOBAYASHI, M. SAITO, et al. 2007. Studies on the role of HtpG in the tetrapyrrole biosynthesis pathway of the cyanobacterium *Synechococcus elongatus* PCC 7942. Biochem. Biophys. Res. Commun. **352**: 36–41.
5. NITTA, K., N. SUZUKI, D. HONMA, et al. 2005. Ultrastructural stability under high temperature or intensive light stress conferred by a small heat shock protein in cyanobacteria. FEBS Lett. **579**: 1235–1242.
6. LEE, S., H.A. OWEN, D.J. PROCHASKA, et al. 2000. HSP16.6 is involved in the development of thermotolerance and thylakoid stability in the unicellular cyanobacterium, *Synechocystis* sp. PCC 6803. Curr. Microbiol. **40**: 283–287.
7. NAKAMOTO, H., N. SUZUKI & S.K. ROY. 2000. Constitutive expression of a small heat-shock protein confers cellular thermotolerance and thermal protection to the photosynthetic apparatus in cyanobacteria. FEBS Lett. **483**: 169–174.
8. NAKAMOTO, H. & D. HONMA. 2006. Interaction of a small heat shock protein with light-harvesting cyanobacterial phycocyanins under stress conditions. FEBS Lett. **580**: 3029–3034.
9. TÖRÖK, Z., P. GOLOUBINOFF, I. HORVÁTH, et al. 2001. *Synechocystis* HSP17 is an amphitropic protein that stabilizes heat-stressed membranes and binds denatured proteins for subsequent chaperone-mediated refolding. Proc. Natl. Acad. Sci. USA **98**: 3098–3103.
10. VIGH, L., P.V. ESCRIBA, A. SONNLEITNER, et al. 2005. The significance of lipid composition for membrane activity: new concepts and ways of assessing function. Prog. Lipid Res. **44**: 303–344.
11. SACHSENMAIER, C., A. RADLER-POHL, R. ZINCK, et al. 1994. Involvement of growth factor receptors in the mammalian UVC response. Cell **78**: 963–972.
12. COFFER, P.J., B.M. BURGERING, M.P. PEPPELENBOSCH, et al. 1994. UV activation of receptor tyrosine kinase activity. Oncogene **11**: 561–569.
13. ROSETTE, C. & M. KARIN. 1996. Ultraviolet light and osmotic stress: activation of the JNK cascade through multiple growth factor and cytokine receptors. Science **274**: 1194–1197.
14. LIN, R.Z., Z.W. HU, J.H. CHIN, et al. 1997. Heat shock activates c-Src tyrosine kinases and phosphatidylinositol 3-kinase in NIH3T3 fibroblasts. J. Biol. Chem. **272**: 31196–31202.

15. BURGERING, B.M. & P.J. COFFER. 1995. Protein kinase B (c-Akt) in phosphatidylinositol-3-OH kinase signal transduction. Nature **376:** 599–602.
16. IHARA, Y., C. YASUOKA, K. KAGEYAMA, *et al.* 2002. Tyrosine phosphorylation of clathrin heavy chain under oxidative stress. Biochem. Biophys. Res. Commun. **297:** 353–360.
17. WILDE, A., E.C. BEATTIE, L. LEM, *et al.* 1999. EGF receptor signaling stimulates SRC kinase phosphorylation of clathrin, influencing clathrin redistribution and EGF uptake. Cell **96:** 677–687.
18. CATERINA, M.J., M.A. SCHUMACHER, M. TOMINAGA, *et al.* 1997. The capsaicin receptor: a heat-activated ion channel in the pain pathway. Nature **389:** 816–824.
19. VOETS, T., G. DROOGMANS, U. WISSENBACH, *et al.* 2004. The principle of temperature-dependent gating in cold- and heat-sensitive TRP channels. Nature **430:** 748–754.
20. LI, C. & Q. XU. 2000. Mechanical stress-initiated signal transductions in vascular smooth muscle cells. Cell Signal **12:** 435–445.
21. SADOSHIMA, J. & S. IZUMO. 1993. Mechanical stretch rapidly activates multiple signal transduction pathways in cardiac myocytes: potential involvement of an autocrine/paracrine mechanism. EMBO J. **12:** 1681–1692.
22. GRAY, J.V., J.P. OGAS, Y. KAMADA, *et al.* 1997. A role for the Pkc1 MAP kinase pathway of Saccharomyces cerevisiae in bud emergence and identification of a putative upstream regulator. EMBO J. **16:** 4924–4937.
23. VERNA, J., A. LODDER, K. LEE, *et al.* 1997. A family of genes required for maintenance of cell wall integrity and for the stress response in Saccharomyces cerevisiae. Proc. Natl. Acad. Sci. USA **94:** 13804–13809.
24. KAMADA, Y., U.S. JUNG, J. PIOTROWSKI, *et al.* 1995. The protein kinase C-activated MAP kinase pathway of Saccharomyces cerevisiae mediates a novel aspect of the heat shock response. Genes Dev. **9:** 1559–1571.
25. IMAZU, H. & H. SAKURAI. 2005. Saccharomyces cerevisiae heat shock transcription factor regulates cell wall remodeling in response to heat shock. Eukaryot. Cell **4:** 1050–1056.
26. NADEAU, S.I. & J. LANDRY. 2007. Mechanism of the activation and regulation of the heat shock-sensitive signaling pathways. *In* Molecular Aspects of the Stress Protein Response: Chaperones, Membranes and Networks. P. Csermely & L. Vigh, Eds.: 100–113. *Series: Advances in Experimental Medicine and Biology,* Vol 594. Springer.
27. JENKINS, G.M. 2003. The emerging role for sphingolipids in the eukaryotic heat shock response. Cell. Mol. Life Sci. **60:** 701–710.
28. MERRILL, A.H.J. & D.D. JONES. 1990. An update of the enzymology and regulation of sphingomyelin metabolism. Biochim. Biophys. Acta **1044:** 1–12.
29. KOLESNICK, R.N. 1987. 1,2-Diacylglycerols but not phorbol esters stimulate sphingomyelin hydrolysis in GH_3 pituitary cells. J. Biol. Chem. **262:** 16759–16762.
30. SPIEGEL, S. & A.H. MERRILL JR. 1996. Sphingolipid metabolism and cell growth regulation. FASEB J. **10:** 1388–1397.
31. HANNUN, Y.A. & L.M. OBEID. 1995. Ceramide: an intracellular signal for apoptosis. Trends Biochem. Sci. **20:** 73–79.
32. HANNUN, Y.A. 1996. Functions of ceramide in coordinating cellular responses to stress. Science **274:** 1855–1859.
33. GÓMEZ-MUÑOZ, A. 2006. Sphingomyelinases and the regulation of cell death and survival. Curr. Enz. Inhibit. **2:** 125–134.

34. GÓMEZ-MUÑOZ, A. 1998. Modulation of cell signalling by ceramides. Biochim. Biophys. Acta **1391:** 92–109.
35. GÓMEZ-MUÑOZ, A. 2006. Ceramide 1-phosphate/ceramide, a switch between life and death. Biochim. Biophys. Acta **1758:** 2049–2056.
36. GÓMEZ-MUÑOZ, A., J.Y. KONG, B. SALH, *et al.* 2004. Ceramide-1-phosphate blocks apoptosis through inhibition of acid sphingomyelinase in macrophages. J. Lipid Res. **45:** 99–105.
37. GÓMEZ-MUÑOZ, A., J.Y. KONG, K. PARHAR, *et al.* 2005. Ceramide-1-phosphate promotes cell survival through activation of the phosphatidylinositol 3-kinase/protein kinase B pathway. FEBS Lett. **579:** 3744–3750.
38. GÓMEZ-MUÑOZ, A., J.Y. KONG, B. SALH & U.P. STEINBRECHER. 2003. Sphingosine-1-phosphate inhibits sphingomyelinase and blocks apoptosis in macrophages. FEBS Lett. **539:** 56–60.
39. VIGH, L., B. MARESCA & J. HARWOOD. 1998. Does the membrane's physical state control the expression of heat shock and other genes? Trends Biochem. Sci. **23:** 369–374.
40. VEREB, G., J. SZÖLLÖSI, J. MATKÓ, *et al.* 2003. Dynamic, yet structured: the cell membrane three decades after the Singer-Nicolson model. Proc. Natl. Acad. Sci. USA **100:** 8053–8058.
41. VIGH, L., I. HORVATH, B. MARESCA & J.L. HARWOOD. 2007. Can the stress response be controlled by "membrane-lipid herapy"? Trends Biochem. Sci. **32:** 357–363.
42. SOTI, C., E. NAGY, Z. GIRICZ, *et al.* 2005. Heat shock proteins as emerging therapeutic targets. Brit. J. Pharmacol. **146:** 769–780.
43. BALOGH, G., I. HORVÁTH, E. NAGY, *et al.* 2005. The hyperfluidisation of mammalian cell membranes acts as a signal to initiate the heat shock protein response. FEBS J. **272:** 6077–6086.
44. NAGY, E., Z. BALOGI, I. GOMBOS, *et al.* 2007. Hyperfluidization-coupled membrane microdomain reorganization is linked to activation of the heat shock response in a murine melanoma cell line. Proc. Natl. Acad. Sci. USA **104:** 7945–7950.
45. KUNIMOTO, S., W. MUROFUSHI, I. YAMATSU, *et al.* 2003. Cholesteryl glucoside-induced protection against gastric ulcer. Cell Struct. Funct. **28:** 179–186.
46. VIGH, L., N.P. LITERÁTI, I. HORVÁTH, *et al.* 1997. Bimoclomol: a nontoxic, hydroxylamine derivative with stress protein-inducing activity and cytoprotective effects. Nat. Med. **3:** 1150–1154.
47. TÖRÖK, Z., N.M. TSVETKOVA, G. BALOGH, *et al.* 2003. Heat shock protein co-inducers with no effect on protein denaturation specifically modulate the membrane lipid phase. Proc. Natl. Acad. Sci. USA **100:** 3131–3136.
48. KIEARAN, D., B. KALMAR, J.R. DICK, *et al.* 2004. Treatment with arimoclomol, a coinducer of heat shock proteins, delays disease progression in ALS mice. Nat. Med. **4:** 402–405.
49. VIGH, L., Z. TÖRÖK, G. BALOGH, *et al.* 2007. Membrane associated stress response: a theoretical and practical approach. *In* Molecular Aspects of the Stress Protein Response: Chaperones, Membranes and Networks. P. Csermely & L. Vigh, Eds.: 114–131. *Series: Advances in Experimental Medicine and Biology,* Vol 594. Springer.
50. TÖRÖK, Z., I. HORVÁTH, P. GOLOUBINOFF, *et al.* 1997. Evidence for a lipochaperonin: association of active protein-folding GroESL oligomers with lipids can

stabilize membranes under heat shock conditions. Proc. Natl. Acad. Sci. USA **94:** 2192–2197.
51. TSVETKOVA, N.M., I. HORVÁTH, Z. TÖRÖK, *et al*. 2002. Small heat-shock proteins regulate membrane lipid polyporphism. Proc. Natl. Acad. Sci. USA **99:** 13504–13509.
52. NAKAMOTO, H. & L. VIGH. 2007. The small heat shock proteins and their clients. Cell. Mol. Life. Sci. **3:** 294–306.

Temperature Stress

Reacting and Adapting: Lessons from Poikilotherms

JOHN L. HARWOOD

School of Biosciences, Cardiff University, Cardiff, United Kingdom

ABSTRACT: *Acanthamoeba castellanii* (*A. castellanii*) is a common soil- or water-borne protozoon that feeds on bacteria by phagocytosis. *A. castellanii* can grow between 4 and 32°C and has to adapt quickly to chilling in order to survive. We have identified a Δ12-fatty acid desaturase as key to low temperature adaptation. The activity of this enzyme is mainly increased through gene expression and new protein synthesis. Interestingly, the activity can also be altered independently by dissolved oxygen levels. In addition, we have identified a gene for the Δ12-desaturase, which, when expressed in yeast, catalyses Δ15-desaturation also. Moreover, it is also capable of producing very unusual n-1 polyunsaturated products.

KEYWORDS: *Acanthamoeba castellanii*; eukaryotic temperature adaptation; fatty acid Δ12-desaturase; bifunctional enzyme

INTRODUCTION

Temperature adaptation is a primary characteristic of all poikilotherms that need the ability to survive in a range of environmental temperatures. The need to adapt is governed principally by the need to maintain membrane function. In the case of lower environmental temperatures, alterations in membrane lipid order are found while higher temperatures may cause protein unfolding and denaturation.

When a poikilotherm is exposed to lower environmental temperatures then it is likely that membrane lipid order will increase ("fluidity" decrease). Since membrane lipids are present as a controlled mixture of lipid classes and also, to an extent, of molecular species then lowering growth temperature can cause one or more of these species to be below its transition temperature (Tc). When an appreciable amount of lipid is below its Tc, then phase separation can occur

---

Address for correspondence: John L. Harwood, School of Biosciences, Cardiff University, Cardiff CF10 3US, UK. Voice: +44-0-29-2087-4108; fax: +44-0-29-2087-4117.

Harwood@Cardiff.ac.uk

and cause loss of membrane function. To obviate this problem, poikilotherms have evolved a series of mechanisms to restore membrane lipid fluidity.[1,2] Of the tactics adopted by different organisms, increase in *cis*-unsaturation of membrane lipids is the most prevalent.

USING A SOIL PROTOZOON AS A MODEL

Acanthamoeba castellanii (*A. castellanii*) is a common soil protozoon, which is found in most soil or freshwater habitats in temperate regions. Because of its abundance, *A. castellanii* plays a key role in ecosystems by controlling bacterial numbers. In order to feed, *Acanthamoeba* needs to use phagocytosis, which, of course, requires a fully functional plasma membrane. The organism grows naturally within a temperature range of 4–32°C, below which it encysts.

We have been studying *A. castellanii* for some time, ever since we defined the first adaptive change in response to a lowering of environmental temperature as induction of a fatty acid $\Delta 12$-desaturase. (See Ref. 3) This enzyme was, itself, quite unusual for an animal system because, in general, the production of polyunsaturated fatty acids in which the second double bond is toward the methyl end (e.g., linoleic acid) is considered the preserve of oxygen-evolving photosynthetic organisms (plants, algae, cyanobacteria). So the immediate response of *Acanthamoeba castellanii* to lowered temperature was induction of the $\Delta 12$-desaturase that converted oleate to linoleate. Subsequent changes in fatty acid composition involved elongation and further desaturation reactions (TABLE 1).

Because of the unusual nature of the enzyme reaction for an animal we studied its characteristics in more detail. The desaturase used an intact phospholipid as substrate with preference for the *sn*-2 position of phosphatidylcholine. Activity was mainly increased by elevated transcription, although there was some evidence that preexisting desaturase protein could be activated. A sequence of events following chilling ensured – detection of lowered temperature, increase in desaturase activity, increased membrane unsaturation and, then, fluidity and, finally, recommenced phagocytosis (TABLE 2).[4]

TEMPERATURE IS NOT THE ONLY SIGNAL TO INDUCE DESATURASE ACTIVITY

While studying changes in *A. castellanii* in response to chilling we noticed that fatty acid composition was also altered in stationary phase compared to the logarithmic growth phase. Moreover, this meant that the organism was not responsive to chilling for early log. phase cultures—since the fatty acids in the membranes were already sufficiently unsaturated for fluidity to be unaffected by moderate temperature lowering. (Moreover, the $\Delta 12$-desaturase was

TABLE 1. Sequential changes in fatty acid composition in *A. castellanii* on lowering environmental temperature[10]

Time (h)	Fatty acid composition (%)								
	14:0	16:0	18:0	18:1	n6–18:2	n6–20:2	n6–20:3	n6–20:4	n3–20:5
0	9 ± 5	7 ± tr	3 ± 1	33 ± 1	7 ± 2	14 ± 2	3 ± 1	11 ± 2	1 ± tr
2	6 ± 1	8 ± 1	3 ± tr	37 ± 1	12 ± 1	12 ± 1	4 ± tr	13 ± 1	1 ± tr
7	7 ± 1	8 ± 1	4 ± 1	26 ± tr	20 ± 1	9 ± 1	5 ± tr	16 ± 3	1 ± 1
24	7 ± tr	4 ± 1	6 ± 1	20 ± 1	24 ± 2	9 ± 1	8 ± 2	23 ± 1	4 ± 1

Fatty acids are abbreviated with the number before the colon showing the number of carbons and the figure after giving the number of double bonds. Only major fatty acids are listed. The data are for *A. castellanii* harvested in mid-log growth-phase. Means ± SD ($n = 3$). tr. = trace (<0.5).

TABLE 2. Adaptive changes in *A. castellanii* on chilling

1. Lowered environmental temperature perceived.
2. Fatty acid delta12-desaturase activity increased.
3. Membrane unsaturation increased.
4. Lipid order reduced.
5. Phagocytosis and growth recommence.

already present at high activity). This made it clear that chilling alone was not enough to raise desaturase activity and pointed to a mechanism that was more related to physiological need. One such mechanism would be to use membrane lipid order as the sensor, as proposed for cyanobacteria by Murata, Vigh, and colleagues.[5]

However, one other environmental factor, which changes at the same time as temperature is oxygen availability—because oxygen is more soluble in aqueous solutions at low temperatures. Moreover, oxygen is required for fatty acid desaturation in most organisms, including amoebae. Therefore, we explored whether oxygen availability could influence desaturase activity.

Unexpectedly, we found that activity was influenced by the dissolved oxygen level and, moreover, oxygen could affect desaturase expression *independent* of temperature. Careful control experiments also showed that the temperature induction of the $\Delta 12$-desaturase could also be independent of oxygen.[6]

Thus, fatty acid $\Delta 12$-desaturase activity would be altered by two factors—oxygen and temperature. The next question was whether these two factors induced the same enzyme. We do not have a definitive answer to this question yet but enzymatic characteristics of the oxygen-induced activity are certainly similar to those of the chilling-induced enzyme (TABLE 3). Both enzymes use phosphatidylcholine as a substrate with oleate being converted to linoleate at the sn-2 position. Of the three proteins involved in the desaturase complex, it

TABLE 3. Comparison of Δ12-desaturase properties following oxygen or temperature manipulation

	Temperature	Oxygen-induced
Main mechanism	Transcription	Transcription
Preferred substrate	Phosphatidylcholine	Phosphatidylcholine
Position of oleate desat.	sn-2 preferred	sn-2 preferred
Desaturase component changed	Not measured	Cyanide-sens. protein
Desaturation position measured	Not determined	Either n-6 or from existing n-9 bond

is the terminal (cyanide-sensitive) desaturase protein that is altered rather than the oxidoreductase or cytochrome b5.[7]

THE DESATURASE GENE SHOWS UNUSUAL PROPERTIES

In order to examine further the mechanism of desaturase induction and details of the enzyme(s) concerned, we needed to identify the gene(s) involved. Using known sequences for the conserved "histidine box" motifs for desaturases, we screened ESTs for likely candidates. Only one sequence was identified and this was used to obtain a full-length cDNA, which was then used for heterologous expression in *S. cerevisiae*.[8] The expressed desaturase was shown to have Δ12-activity and was able to desaturate both palmitoleate and oleate to their dienoic products. It was likely that intact phospholipids were again the substrates.[8]

However, there were two unexpected features of the *Acanthamoeba* desaturase when expressed in yeast. First, it was bifunctional and also contained Δ15-activity. This meant that it could also generate α-linolenate as a product. Moreover, because it was active at the 16C level it could convert palmitoleic acid successively first to Δ9,12 hexadecadienoate and then to Δ9,12,15- hexadecatrienoate.[8] n-1 Polyunsaturated fatty acids had been reported before in nature but this was the first time that an enzyme had been described that could synthesize them. Since the writing of our paper, other bifunctional fatty acid desaturases and enzymes capable of generating n-1 polyunsaturated fatty acids have been reported. For the *A. castellanii* gene identified, it would be interesting to determine whether its expression can be induced by both oxygen and temperature or, if selectively induced, what other fatty acid desaturases are still to be uncovered in the organism's genome.

Small amounts of n-3 eicosapentaenoic acid are present in *A. castellanii* and the isolated desaturase gene is clearly capable of coding for a desaturase to produce both the linoleate and α-linolenate precursors for the n-6 and n-3 pathways, respectively (FIG. 1.). Further metabolism of these polyunsaturated fatty acids uses the so-called alternative pathway and a further two desaturases.[9]

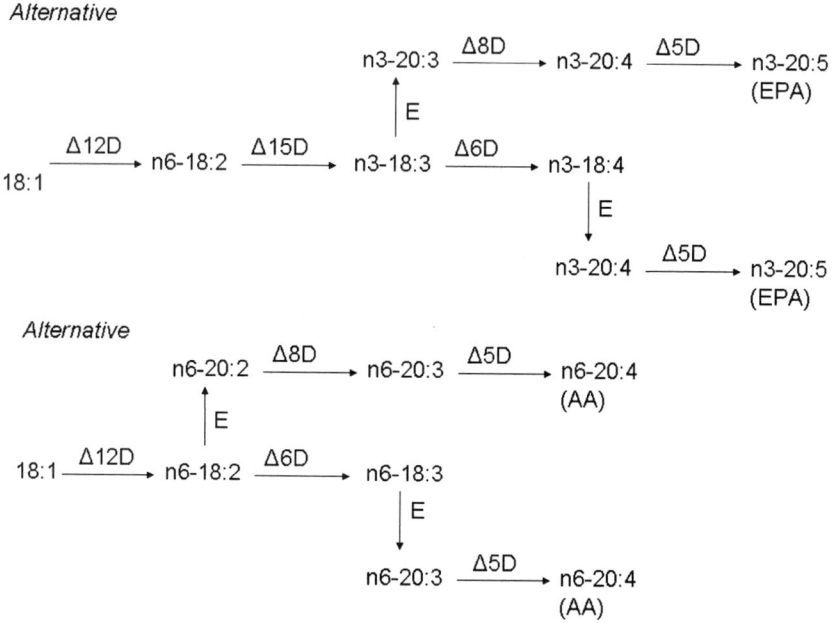

FIGURE 1. Potential pathways for the biosynthesis of unsaturated fatty acids in *Acanthamoeba castellanii*. The "alternative" pathways for the metabolism of linoleate and α-linoleate are supported by the isolation of a C_{20} Δ8-desaturase gene[9] and the presence of appropriate, characterized intermediates in *A. castellanii*.[10]

CONCLUSIONS AND FUTURE QUESTIONS

A. castellanii has proven to be a very useful and responsive eukaryotic system with which to study temperature and adaptation. Our studies have revealed that, as might be expected, eukaryotes share some features with prokaryotes but are distinct in other ways.

But many important questions remain to be answered. Among these are how many fatty acid desaturases are there in *A. castellanii* ? How many are induced by temperature and how many by oxygen separately or as well ? Do the properties of the desaturases control the lower limit at which *A. castellanii* can grow? Finally, and most interesting, how is temperature perceived and desaturase activity controlled? Judging from the unexpected findings that the last decade has produced, future research in this area is not only important but is sure to be interesting!

REFERENCES

1. GURR, M.I., J.L. HARWOOD & K.N. FRAYN. 2002. Lipid Biochemistry. Fifth edition. Blackwell Scientific. Oxford, UK.

2. GUSCHINA, I.A. & J.L. HARWOOD. 2006. Mechanisms of temperature adaptation in poikilotherms. FEBS Lett. **580:** 5477–5483.
3. JONES, A.L., D. LLOYD & J.L. HARWOOD. 1993. Rapid induction of microsomal $\Delta 12(\omega 6)$-desaturase activity in chilled *Acanthamoeba castellanii*. Biochem. J. **296:** 183–188.
4. AVERY, S.V., D. LLOYD & J.L. HARWOOD. 1995. Temperature-dependent changes in plasma membrane lipid order and the phagocytotic activity of the amoeba *Acanthamoeba castellanii* are closely correlated. Biochem. J. **312:** 811–816.
5. VIGH, L., D.A. LOS, I. HORVATH & N. MURATA. 1993. The primary signal in the biological perception of temperature: Pd-catalysed hydrogenation of membrane lipids stimulated the expression of the desA gene in *Synechocystis* PCC6803. Proc. Natl. Acad. Sci. USA **90:** 9090–9094.
6. THOMAS, K., A.J. RUTTER, M. SULLER, *et al.* 1998. Oxygen induces fatty acid (n-6) desaturation independently of temperature in *Acanthamoeba castellanii*. FEBS Lett. **425:** 171–174.
7. RUTTER, A.J., K.L. THOMAS, D. HERBERT, *et al.* 2002. Oxygen induction of a novel n-6 desaturase in the soil protozoon, *Acanthamoeba castellanii*. Biochem. J. **368:** 57–67.
8. SAYANOVA, O., R. HASLAM, I. GUSCHINA, *et al.* 2006. A bifunctional $\Delta 12$, $\Delta 15$-desaturase from *Acanthamoeba castellanii* directs the synthesis of highly unusual n-1 series unsaturated fatty acids. J. Biol. Chem. **281:** 36533–36541.
9. SAYANOVA, O., R. HASLAM, B. QI, *et al.* 2006. The alternative pathway C20 $\Delta 8$-desaturase from the non-photosynthetic organism *Acanthamoeba castellanii* is an atypical cytochrome b5-fusion desaturase. FEBS Lett. **580:** 1946–1952.
10. JONES, A.L., N.L. PRUITT, D. LLOYD & J.L. HARWOOD. 1991. Temperature-induced changes in the synthesis of unsaturated fatty acids by *Acanthamoeba castellanii*. J. Protozool. **38:** 532–536.

Endoplasmic Reticulum Stress

GÁBOR BÁNHEGYI,[a] PETER BAUMEISTER,[b] ANGELO BENEDETTI,[c] DEZHENG DONG,[b] YONG FU,[b] AMY S. LEE,[b] JIANZE LI,[b] CHANGHUI MAO,[b] ÉVA MARGITTAI,[a] MIN NI,[b] WULF PASCHEN,[d] SIMONA PICCIRELLA,[c] SILVIA SENESI,[c] ROBERTO SITIA,[e] MIAO WANG,[b] AND WEI YANG[d]

[a]*Institute of Medical Chemistry, Molecular Biology and Pathobiochemistry, Semmelweis University of Budapest, Hungary*

[b]*Department of Biochemistry and Molecular Biology and the USC/Norris Comprehensive Cancer Center, Keck School of Medicine of the University of Southern California, Los Angeles, California, USA*

[c]*Department of Pathophysiology, Experimental Medicine and Public Health, University of Siena, Italy*

[d]*Multidisciplinary Neuroprotection Laboratories, Departments of Anesthesiology and Neurobiology, Duke University Medical Center, Durham, North Carolina, USA*

[e]*Università Vita-Salute San Raffaele, DiBiT Istituto Scientifico San Raffaele, Milano, Italy*

ABSTRACT: Stress is the imbalance of homeostasis, which can be sensed even at the subcellular level. The stress-sensing capability of various organelles including the endoplasmic reticulum (ER) has been described. It has become evident that acute or prolonged ER stress plays an important role in many human diseases; especially those involving organs/tissues specialized in protein secretion. This article summarizes the emerging role of ER stress in diverse human pathophysiological conditions such as carcinogenesis and tumor progression, cerebral ischemia, plasma cell maturation and apoptosis, obesity, insulin resistance, and type 2 diabetes. Certain components of the ER stress response machinery are identified as biomarkers of the diseases or as possible targets for therapeutic intervention.

KEYWORDS: endoplasmic reticulum; stress; GRP78; cerebral ischemia; plasma cell; glucocorticoids; metabolic syndrome; obesity

Address for correspondence: Gábor Bánhegyi, Institute of Medical Chemistry, Molecular Biology and Pathobiochemistry, Semmelweis University, Puskin utca 9, 1088, Budapest, Hungary. Voice: +36-1-266-2755-4061; fax: +36-1-266-2615.
banhegyi@puskin.sote.hu

INTRODUCTION

The efficient functioning of the endoplasmic reticulum (ER) is essential for proper cellular activities and survival. Discrepancies between the demand for and the capacities of the ER functions lead to ER stress.[1] In most cases, conditions that interfere with ER functions cause the accumulation and aggregation of unfolded proteins. ER transmembrane receptors detect the onset of ER stress and initiate the unfolded-protein response (UPR) to restore normal ER function.[2] Mammalian cells are equipped with multiple adaptive mechanisms to react to the perturbations of the ER homeostasis. Two major signaling branches lead to the general downregulation of translation and to the increased expression of proteins required for the restoration of normal ER functioning. If the stress *nec arte nec marte* can be surmounted, that is, the adaptive response fails, apoptotic cell death ensues.[3]

This article resumes four important aspects of ER stress and their relation to the pathomechanism of important human diseases. GRP78, the major sensor of ER stress, seems to be a promising target in neuroprotection and cancer therapy, as well as a marker of tumor progression. Modification of ER stress response can help neurons withstand the stressful conditions during cerebral ischemia. A Janus-faced ER stress is shown in plasma cells, which controls both immunoglobulin production and lifespan. Finally, the possible role of glucocorticoids as an autocrine signal of ER stress is discussed.

ER STRESS INDUCTION OF UPR REGULATOR GRP78: ROLE IN DEVELOPMENT AND DISEASE

The ER is an essential cellular compartment for protein synthesis and maturation. It also functions as a Ca^{2+} storage organelle and resource of calcium signals. The perturbation of ER functions, such as disruption of Ca^{2+} homeostasis, inhibition of protein glycosylation or disulfide bond formation, hypoxia, and virus or bacteria infection, can result in accumulation of unfolded or misfolded proteins, which leads to ER stress.[4,5] Mammalian cells have evolved multiple adaptive pathways referred to as the UPR that allow them to respond to perturbations in ER homeostasis.[6] One major prosurvival mechanism is mediated by molecular ER chaperone GRP78/BiP.[7,8] In addition to assisting in the formation, folding, assembly, and quality control of newly synthesized proteins to preserve cellular homeostasis, GRP78 is also critical for the quality control of proteins processed in the ER and for regulation of ER signaling in response to ER stress (Fig. 1).

Despite these advances, the physiologic function of GRP78/BiP is just emerging. Gene knockout technology allows definitive tests for the requirement and role of GRP78 during embryonic development and pathogenesis. This leads to the discovery that complete depletion of GRP78 results in

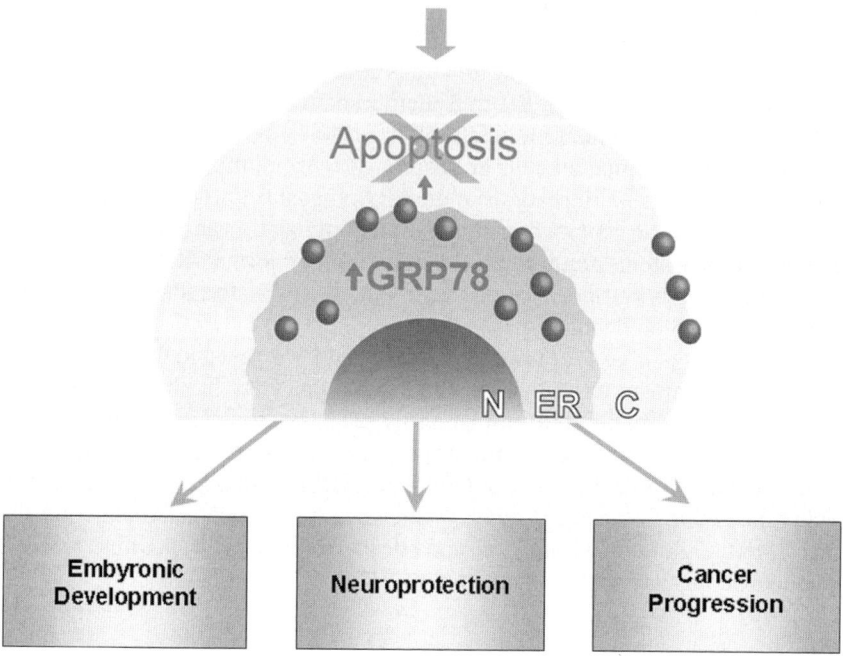

FIGURE 1. Implications of ER stress induction of GRP78 in cell and cancer biology. A variety of physiological stress conditions that target the ER leads to the induction of GRP78 and the UPR. Examples of ER stress conditions include glucose starvation, severe hypoxia, and efflux of the ER Ca^{2+} store. These result in the production and accumulation of underglycosylated and malfolded proteins in the ER. On transcriptional activation, GRP78 level is elevated. The majority of GRP78 resides in the lumen of the ER; however, a subpopulation of GRP78 exists as ER transmembrane protein with the N-terminal half in the cytosol. There are reports of GRP78 on the cell surface of certain cancer cells, cells in some pathological states and growth-stimulated endothelial cells. Cell surface GRP78 is not detected in normal organs. GRP78 is able to block apoptosis at multiple steps and plays important and essential roles in embryonic development, neuroprotection, and cancer progression. Abbreviations: N, nucleus; ER, endoplasmic reticulum; C, cytoplasm.

lethality in 3.5-day-old embryos due to the failure of embryo periimplantation.[9] The $Grp78^{-/-}$ embryos cannot hatch from the zona pellucida *in vitro*, fail to grow in culture, and exhibit proliferation defects and a massive increase of apoptosis in the inner cell mass, which are precursors of embryonic stem cells. These findings show that GRP78 is essential for embryonic cell growth and pluripotent cell survival, as its function cannot be compensated by other ER chaperones or folding enzymes which are present in the *Grp78* knockout embryos. To probe the transcriptional activation of the *Grp78* during embryonic development, transgenic mouse lines bearing a *lacZ* reporter gene driven by 3 kilobases of the rat *Grp78* promoter or a mutant promoter with deletion of the ER stress-response elements were constructed.[10,11] In addition, heterozygous *Grp78* mice with 50% level of wild-type GRP78 in all tissues were evaluated for developmental abnormalities as well as susceptibility to diseases.[9] Collectively, the results show that 50% of GRP78 is sufficient to maintain cellular homeostasis during development, and are consistent with the view that compared to normal tissues and organs, GRP78 is more critically needed in cells undergoing physiological or pathological stress, as exemplified by protection of vulnerable neuronal cells and allowing cancer cells to evade the host defense system and cancer therapy.[12,13]

The protective role of GRP78 in neurodegeneration is supported by the studies of woozy mutant mice and Marinesco–Sjögren (MS) syndrome patients.[14–16] SIL1 stimulates the release of ADP from GRP78, activating the ATPase cycle to promote the binding and folding of the substrate proteins. Using mouse models deficient in GRP78 in specific regions of the brain, we provide direct evidence that GRP78 is essential for the survival of specific brain cells and brain function. Furthermore, these experimental systems can also be used to evaluate the status of the UPR signaling pathways when GRP78 is totally depleted in specific organs and provide mechanistic explanations why GRP78 deficiency leads to the phenotypes observed.

Owing to hypoxic conditions and glucose deprivation caused by poor vascularization, the microenvironment of tumors represents physiological ER stress and the UPR is activated for the survival of tumor cells.[17,18] GRP78 is induced in a wide variety of human cancers, correlating with tumor progression, metastasis, and drug resistance of both proliferating and dormant cancer cells.[19–22] Previous studies showed that fibrosarcoma cells where GRP78 expression was suppressed by antisense were either unable to form tumors or quickly regressed.[12] By breeding mouse models deficient in GRP78 with mouse models of breast and prostate cancer, the role of GRP78 in tumor progression and metastasis is directly examined. Our results showed that GRP78 is a major contributor to tumor progression such that partial reduction of GRP78 results in a prolonged latency period as well as substantial reduction in tumor size. The underlying mechanisms for these observations are also being explored.

In summary, because of its central role in preserving cellular homeostasis, GRP78 is a novel target for therapeutic intervention for organ preservation, neuroprotection, and cancer therapy.[5,19–22] For human cancer, GRP78 may also be useful as a novel biomarker for tumor behavior and responsiveness to therapy.[22–24] These translational applications warrant vigorous testing in preclinical and clinical settings.

CEREBRAL ISCHEMIA/STROKE IMPAIRS ER FUNCTION

When cells are exposed to a severe form of stress, stress responses are activated that are highly conserved from yeast to mammalian cells. These stress responses comprise activation of two major signal transduction pathways, resulting in shutdown of translation and activation of the expression of genes coding for proteins that are required to restore function. Stress-induced shutdown of translation is triggered by phosphorylation of the alpha subunit of the eukaryotic initiation factor 2 (eIF2α), phosphorylated eIF2α being an inhibitor of the initiation step of protein synthesis. The pattern of stress responses varies depending on which subcellular compartment is hit by the respective stress. Detailed analysis of the stress response pattern will therefore permit identification of those subcellulars exhibiting stress-induced impairment of function.

Cerebral ischemia is a severe form of metabolic stress that interferes with most biochemical and molecular biology pathways. Like other severe forms of stress, transient cerebral ischemia causes a shutdown of translation triggered by phosphorylation of eIF2α, and it activates the expression of various stress genes.[25,26] Four different kinases have been identified that specifically phosphorylate eIF2α: the heme-regulated inhibitor that plays a central role in the regulation of protein synthesis by heme iron, the general control nonderepressing 2 kinase (GCN2) that is activated under conditions of nutritional deprivation, the double-stranded RNA-activated kinase (PKR) that is induced by various signals including interferon, double-stranded RNA and cytokines, and the PKR-like ER kinase (PERK) that is specifically activated under experimental conditions associated with impaired ER function. The eIF2α kinases, which are activated after ischemia and responsible for ischemia-induced phosphorylation of eIF2α, have been investigated in detail, and it has been convincingly demonstrated that it is PERK and not one of the other eIF2α kinases that is activated in models of transient cerebral ischemia.[27] This implies that ischemia impairs ER function.[28]

Three ER transmembrane proteins are involved in the ER stress response: PERK, the inositol-requiring enzyme 1 (IRE1) and the activating transcription factor 6 (ATF6). In the physiological state, when ER-resident protein folding and processing capacity matches the ER load of newly synthesized proteins, the ER chaperon GRP78 is bound to these proteins, thus blocking activation. Under conditions of ER stress when unfolded proteins accumulate in the ER lumen, GRP78 dissociates from PERK, IRE1, and ATF6 and binds to the unfolded proteins to assist in refolding. This is the warning signal that triggers activation of the stress-sensing proteins. Activated IRE1 is turned into an endonuclease that cuts out a sequence of 26 bases from the coding region of xbp1 mRNA. This causes a frame-shift and the processed message is translated into a new protein that functions as a transcription factor activating expression of genes coding for proteins involved in the ER resident folding and processing reactions.

Activated ATF6 translocates to the Golgi compartment, where it is split by proteases into the active form that functions as a transcription factor.

The effect of transient cerebral ischemia on activation of the ER stress response has been studied in detail. PERK has been found to be activated after ischemia and levels of GRP78 bound to PERK to be considerably reduced.[27,29] Furthermore, preconditioning has been shown to be associated with less pronounced activation of PERK and less marked dissociation of GRP78 from PERK and phosphorylation of eIF2α, in parallel with a several-fold rise in GRP78 protein levels.[29] This suggests that activation of the ER stress response may play a major role in the preconditioning process. Furthermore, transient cerebral ischemia activates splicing of *xbp1* mRNA.[30] It has also been shown that transient cerebral ischemia induces expression of various ER stress genes, including *grp78*, *grp94*, *erp72*, *gadd34*, *gadd153*, and that it activates caspase-12, a specific indicator of ER stress.[31] Despite demonstration of the induction of these unequivocal indicators of ER stress by transient ischemia, efforts to illustrate activation of IRE1 and ATF6 after transient cerebral ischemia have not been successful.[32]

What are the possible mechanisms underlying ischemia-induced impairment of ER function? The observation that the ischemia-induced ER stress response is markedly less pronounced in animals overexpressing SOD1 strongly suggests that superoxide radicals play a role in this pathological process.[33] Furthermore, evidence has been presented suggesting that nitric oxide, excessively produced after ischemia, may contribute to ER stress: ischemia-induced activation of PERK and phosphorylation of eIF2α is completely blocked in animals with targeted deletion of the endothelial or neuronal nitric oxide synthase gene.[34] Furthermore, exposure of primary neuronal cell cultures to NO suppressed ER calcium pump activity, depleted ER calcium stores, and suppressed protein synthesis.[35] The notion of a role of NO in ischemia-induced ER stress is supported by the observation that ER calcium stores have been found to be depleted after ischemia and that recovery of ER calcium homeostasis was observed only in animals pretreated with neuroprotective levels of a NO synthase inhibitor.[36]

Various avenues can be envisaged for therapeutic intervention.[31] First, strategies could be developed to make cells more resistant to conditions associated with ER dysfunction, for example, by preconditioning. Second, because oxidative stress is believed to contribute to ischemia-induced ER stress, oxygen radical scavenger could be used to block the pathological process. Third, strategies could be developed to restore ER function by stimulating refolding of unfolded proteins.

ER stress-induced shutdown of translation is believed to be a protective response, as it blocks the new synthesis of proteins that cannot be correctly folded. On the other hand, stress-induced gene expression requires programmed recovery from translational repression, a process induced by GADD34.[37] Transient cerebral ischemia activates transcription of the *gadd34* gene. In the resistant cortex, but not in the vulnerable hippocampal CA1 subfield, the rise in *gadd34* mRNA is followed by an increase in GADD34 protein levels.[38] Forced postischemic induction of *gadd34* translation may help vulnerable neurons to restore protein synthesis and thus to translate stress messages into the specific proteins needed to help cells withstand the stressful conditions associated with transient cerebral ischemia.

MOLECULAR MECHANISMS CONTROLLING PLASMA CELL BIOGENESIS AND LIFESPAN

Plasma cells are shortlived professional secretory cells, each of them capable of releasing several thousands of antibodies per second. Their differentiation from B lymphocytes entails the spectacular enlargement of the ER, finalized to sustain massive Ig production.[39] The production of spliced XBP-1, a transcription factor controlled by the ER stress sensors ATF6 and IRE1 is essential for plasma cell differentiation. These observations led to propose that abundant Ig production elicits a UPR that in turn drives differentiation. In addition, because a prolonged UPR can lead to apoptosis, the intense synthetic activity of plasma cells might link lifespan control to Ig secretion, thus limiting antibody responses. Several lines of evidence indicate that both the biogenesis and death of plasma cells rely on additional mechanisms. First, the PERK-CHOP subpathway does not seem to be critical for plasma cell activity (S. Masciarelli *et al.*, unpublished results). Second, a proteomics analysis performed in a suitable model of B cell differentiation revealed that many ER resident proteins, whose synthesis is controlled by UPR elements in most cell types, appear well before the accumulation of IgM subunits. This approach also revealed that waves of functionally related proteins appear sequentially during terminal B cell differentiation. Thus, proteins involved in amino acid or lipid synthesis peak at day 2, whereas ER resident proteins and molecules responsible in redox homeostasis increase linearly for the entire duration of the experiment. IgM biosynthesis and secretion become preponderant after day 2. Interestingly, a number of proteins appear concomitantly with IgM, suggesting a specific role in the manufacturing, quality control, and/or transport of antibodies.[40] This approach led to the identification of novel ER folding factors, whose functional role is under investigation. In addition, the temporal expression pattern emerged as a powerful tool to classify groups of proteins sharing a functional role.

Surprisingly, the development of such an efficient factory is matched by a decrease in proteasomes. In striking correlation with impaired protein degradation, polyubiquitinated proteins accumulate, death-inducing proteins are stabilized, and apoptotic sensitivity to proteasome inhibitors ensues, prior to spontaneous apoptosis.[41] Attesting to a cause–effect relationship between proteasomal overload and death, similar events can be recapitulated by overexpressing an orphan μ chain, a proteasome substrate deriving from the ER, in HeLa cells.

Additional elements could make plasma cells susceptible to apoptosis. For instance, cells secreting large amounts of Ig molecules could undergo metabolic or redox imbalances. In their late differentiation phases, before the apoptotic program is activated, primary plasma cells produce large amounts of IgM (about 10^3/cell/sec). It follows that up to 10^5 disulfides are formed per second in each plasma cell, solely to sustain IgM production. In agreement with an increased need for oxidative power, both Ero1α and Ero1β are upregulated during plasma cell differentiation (M. Otsu et al., unpublished results).[42,43] In addition to redox imbalances, amino acid supply or energy production may become limiting during terminal plasma cell differentiation.

In summary, shortlived plasma cells display a redundant array of pathways that may cause their death.[44] In particular, although proteasomal insufficiency profoundly perturbs protein homeostasis and lowers the apoptotic threshold, toxic byproducts of intensive Ig synthesis may accumulate as secretion proceeds, in an exponentially less favorable energetic and metabolic cellular environment, so as to provide the coup the grace that shuts off the life program of Ig factories.[45]

The proteasome load-versus-capacity model summarized above may offer a sensible strategy to end the immune response. A decrease in proteasomes when the cell increases the rate of protein synthesis seems at first a paradox. What is the teleology of such a program? By engineering a bottleneck in an otherwise efficient quality control strategy devoted to Ig folding and assembly, plasma cells may have evolved a mechanism to count the work accomplished, and predispose themselves to die. A selective accumulation of certain protein species, like the pro-apoptotic Bcl-2 relatives Bax and Bim, or the NF-kB counteractor IkBα, may facilitate death, additional stress being imposed by oxidative stress, UPR, ER calcium release, etc. These processes have been implicated in inducing cell death in a variety of systems. Collectively, these stressful conditions may link plasma cell death to antibody production, providing a molecular counter for secreted molecules, as well as an explanation for the peculiar sensitivity of myeloma cells toward proteasome inhibitors.

GLUCOCORTICOIDS AND ER STRESS

Glucocorticoids are main actors in the pathomechanism of stress. In the original context formulated by Hans Selye in his stress theory, glucocorticoids are produced in the adrenal cortex upon the activation of the hypothalamic–pituitary–adrenal axis.[46] Their increased production mediates alarm reactions in acute stress, facilitating metabolic alterations in the general adaptation syndrome allowing the individual to attempt countermeasures, such as the fight-or-flight response. Recent observations show that active glucocorticoids can be also formed from their inactive counterparts in various tissues, including liver and adipose tissue. This prereceptorial activation takes place in the lumen of the ER[47] and depends on the activity of the glucose-6-phosphate transporter (G6PT)–hexose-6-phosphate dehydrogenase (H6PD)–11β-hydroxysteroid dehydrogenase type 1 (11βHSD1) axis.

Cortisone reductase activity of 11βHSD1 is strictly dependent on the redox state of the luminal pyridine nucleotide pool of the ER. Although *in vitro* 11βHSD1 catalyzes the reversible conversion of cortisone and cortisol,[48] under *in vivo* conditions this reaction is predominantly shifted toward cortisone reduction,[49] indicating a high luminal [NADPH]/[NADP$^+$] ratio. Although their *in vivo* redox state and concentration have not been fully elucidated, this observation clearly shows that reduced pyridine nucleotides must be present in the ER lumen. In fact, ER-derived microsomal vesicles contain a significant amount of pyridine nucleotides.[50] Reduced NAD(P)H is the major component of this pool,[51] which presumably mirrors the *in vivo* situation. The high [NADPH]/[NADP$^+$] ratio is maintained by H6PD,[52] because the permeability of the ER membrane to pyridine nucleotides is negligible.[51] In accordance, the cortisone reductase activity is turned to dehydrogenase activity in H6PD knockout mice.[53] Therefore, the high luminal [NADPH]/[NADP$^+$] ratio is dependent on the H6PD activity and, consequently, on the transport of its substrate glucose-6-phosphate through the ER membrane.[54]

Interestingly, it seems that the NAD(P)H–NAD(P)$^+$ system is not connected by a direct enzymatic route to the thiol/disulfide systems in the ER lumen. In the cytosol, both NADPH and glutathione are present predominantly in the reduced state because of the enzymatic coupling of the two redox pairs. Glutathione disulfide is reduced at the expense of NADPH in a reaction catalyzed by glutathione reductase. In spite of the oxidizing environment in the ER lumen due to the high [GSSH]/[GSH] ratio, pyridine nucleotides can be present predominantly in reduced form. The different redox state of the two redox couples is ensured by their uncoupling, because glutathione reductase is hardly detectable in the lumen.[51] Consequently, alterations of the redox state of pyridine nucleotides do not influence the redox state of glutathione or secretory proteins; therefore, the "classic" UPR signaling pathways are presumably not activated. However, several observations indicate that conditions which can be linked to an altered redox state of luminal pyridine nucleotides are associated with ER stress.

Inhibition or downregulation of G6PT results in increased apoptosis in glioma cells[55,56] or neutrophil granulocytes.[57] The pro-apoptotic effect can be prevented by the addition of antioxidants in granulocytes.[57] In line with these observations, inborn deficiency of G6PT in glycogen-storage disease type 1b is characterized by severe neutropenia and neutrophil dysfunction.[58] These observations show that the high [NADPH]/[NADP$^+$] ratio is important not only for the substrate supply of luminal reductases (including 11βHSD1) but also because of the role NADPH plays as an antiapoptotic factor. Whether this phenomenon is related to an antioxidant function of NADPH or it can be attributed to other mechanisms needs to be clarified.

Although the *in vivo* redox state of the luminal NADPH/NADP$^+$ system is still unknown and there is no direct evidence of it being changed under pathological conditions, it can be hypothesized that the altered expression/activity of luminal NADP(H)-dependent oxidoreductases leads to a redox shift. For example, 11βHSD1 and H6PD expression and activity are changed in the experimental animal models of obesity, metabolic syndrome, and type 2 diabetes. These enzymes are expressed in the liver and in the adipose tissue at a high extent. Supporting this assumption, dietary or genetic obesity causes ER stress in adipose and liver tissues of mouse models. The stress, in turn, leads to the IRE-1-dependent activation of c-Jun N-terminal kinase and subsequent serine phosphorylation of insulin receptor substrate-1. Finally, insulin receptor signaling is suppressed as the outcome of the events. Deficiency of the Xbox-binding protein-1, a transcription factor that modulates the ER stress response, leads to insulin resistance in mice.[59] These findings demonstrate that the ER can be regarded as a sensor organelle for metabolic stress in obesity and the ER stress response is crucial for the development of further metabolic consequences, such as insulin resistance and type 2 diabetes.[60] These observations are confirmed by other independent studies that have demonstrated the close connection between ER function and insulin sensitivity.[61,62]

Because the ER is very sensitive to glucose, lipid, oxygen, and energy availability (FIG. 1), it could be regarded as an essential and ancient site of integration between nutrients. However, the mechanism of this integration is poorly understood. The cytosolic NADPH supply as an integrative tool of carbohydrate and lipid metabolism has been known for a long while; it seems likely that a similar mechanism can be operative in the ER lumen. High glucose-6-phosphate supply through the activation of H6PD and 11βHSD1 leads to increased cortisol production. This prereceptorial glucocorticoid activation is supposed to be responsible for the majority of metabolic changes seen in obesity, metabolic syndrome, and type 2 diabetes. These diseases are more common among socioeconomically disadvantaged individuals and are associated with lifestyle factors and chronic stress. It is known that stress promotes the intake of palatable food, and consumption of palatable foods may dampen psychological and physiological responses to stress. It has been recently suggested that the ER can function as a sensor for electron donors and acceptors, that is, nutrients and oxygen. Consequently, the high food input provoked by the stress of life together with the minimal physical activity, lead to a nutrient (electron) overload in the ER. The evolving redox imbalance in the lumen may cause ER stress and consequent signaling events. Furthermore, reductive effects favor the increased prereceptorial glucocorticoid activation. In conclusion, sensing of metabolic alterations through the altered activity of certain luminal oxidoreductases could change the luminal redox state, which leads to glucocorticoid activation and autocrine signaling. The chronic exogenous stress alleviated by food intake through this mechanism will be converted to a glucocorticoid response initiated by the autonomous sensing of (nutrient) stress at a cellular level instead of the central neuroendocrine mechanism.

REFERENCES

1. MARCINIAK, S.J. & D. RON. 2006. Endoplasmic reticulum stress signaling in disease. Physiol. Rev. **86:** 1133–1149.
2. WU, J. & R.J. KAUFMAN. 2006. From acute ER stress to physiological roles of the unfolded protein response. Cell Death Differ. **13:** 374–384.
3. SZEGEZDI, E. *et al.* 2006. Mediators of endoplasmic reticulum stress-induced apoptosis. EMBO Rep. **7:** 880–885.
4. KAUFMAN, R.J. 1999. Stress signaling from the lumen of the endoplasmic reticulum: coordination of gene transcriptional and translational controls. Genes Dev. **13:** 1211–1233.
5. LEE, A.S. 2001. The glucose-regulated proteins: stress induction and clinical applications. Trends Biochem. Sci. **26:** 504–510.
6. RUTKOWSKI, D.T. & R.J. KAUFMAN. 2004. A trip to the ER: coping with stress. Trends Cell Biol. **14:** 20–28.
7. HENDERSHOT, L.M. 2004. The ER function BiP is a master regulator of ER function. Mt. Sinai J. Med. **71:** 289–297.
8. LEE, A.S. 2005. The ER chaperone and signaling regulator GRP78/BiP as a monitor of endoplasmic reticulum stress. Methods **35:** 373–381.
9. LUO, S. *et al.* 2006. GRP78/BiP is required for cell proliferation and protecting the inner cell mass from apoptosis during early mouse embryonic development. Mol. Cell Biol. **26:** 5688–5697.
10. MAO, C. *et al.* 2004. Transgenic mouse models for monitoring endoplasmic reticulum stress *in vivo*. Nat. Med. **10:** 1013–1014.
11. MAO, C. *et al.* 2006. *In vivo* regulation of Grp78/BiP transcription in the embryonic heart: role of the endoplasmic reticulum stress response element and GATA-4. J. Biol. Chem. **281:** 8877–8887.
12. JAMORA, C., G. DENNERT & A.S. LEE. 1996. Inhibition of tumor progression by suppression of stress protein GRP78/BiP induction in fibrosarcoma B/C10ME. Proc. Natl. Acad. Sci. USA **93:** 7690–7694.
13. YU, Z. *et al.* 1999. The endoplasmic reticulum stress-responsive protein GRP78 protects neurons against excitotoxicity and apoptosis: suppression of oxidative stress and stabilization of calcium homeostasis. Exp. Neurol. **155:** 302–314.
14. ANTTONEN, A.K. *et al.* 2005. The gene disrupted in Marinesco-Sjogren syndrome encodes SIL1, an HSPA5 cochaperone. Nat. Genet. **37:** 1309–1311.
15. SENDEREK, J. *et al.* 2005. Mutations in SIL1 cause Marinesco-Sjogren syndrome, a cerebellar ataxia with cataract and myopathy. Nat. Genet. **37:** 1312–1314.
16. ZHAO, L. *et al.* 2005. Protein accumulation and neurodegeneration in the woozy mutant mouse is caused by disruption of SIL1, a cochaperone of BiP. Nat. Genet. **37:** 974–979.
17. FELDMAN, D.E., V. CHAUHAN & A.C. KOONG. 2005. The unfolded protein response: a novel component of the hypoxic stress response in tumors. Mol. Cancer Res. **3:** 597–605.

18. KOUMENIS, C. 2006. ER stress, hypoxia tolerance and tumor progression. Curr. Mol. Med. **6:** 55–69.
19. FU, Y. & A.S. LEE. 2006. Glucose regulated proteins in cancer progression, drug resistance and immunotherapy. Cancer Biol. Ther. **5:** 741–744.
20. LI, J. & A.S. LEE. 2006. Stress induction of GRP78/BiP and its role in cancer. Curr. Mol. Med. **6:** 45–54.
21. RANGANATHAN, A.C. *et al.* 2006. Functional coupling of p38-induced up-regulation of BiP and activation of RNA-dependent protein kinase-like endoplasmic reticulum kinase to drug resistance of dormant carcinoma cells. Cancer Res. **66:** 1702–1711.
22. LEE, A.S. 2007. GRP78 induction in cancer: therapeutic and prognostic implications. Cancer Res. **67:** 3496–3499.
23. LEE, E. *et al.* 2006. GRP78 as a novel predictor of responsiveness to chemotherapy in breast cancer. Cancer Res. **66:** 7849–7853.
24. POOTRAKUL, L. *et al.* 2006. Expression of stress response protein Grp78 is associated with the development of castration-resistant prostate cancer. Clin. Cancer Res. **12:** 5987–5993.
25. BURDA, J. *et al.* 1994. Phosphorylation of the α subunit of initiation factor 2 correlates with the inhibition of translation following transient cerebral ischemia in the rat. Biochem. J. **302:** 335–338.
26. MASSA, S.M., R.A. SWANSON & F.R. SHARP. 1996. The stress gene response in brain. Cereb. Brain Metabol. Rev. **8:** 95–158.
27. KUMAR, R. *et al.* 2001. Brain ischemia and reperfusion activates the eukaryotic initiation factor 2a kinase, PERK. J. Neurochem. **77:** 1418–1421.
28. PASCHEN, W. & J. DOUTHEIL. 1999. Disturbances of the functioning of endoplasmic reticulum: a key mechanism underlying neuronal cell injury? J. Cereb. Blood Flow Metabol. **19:** 1–18.
29. HAYASHI, T. *et al.* 2003. Induction of GRP78 by ischemic preconditioning reduces endoplasmic reticulum stress and prevents delayed neuronal cell death. J. Cereb. Blood Flow Metabol. **23:** 949–961.
30. PASCHEN, W. *et al.* 2003. Transient cerebral ischemia activates processing of xbp1 mRNA indicative of endoplasmic reticulum stress. J. Cereb. Blood Flow Metabol. **23:** 449–461.
31. PASCHEN, W. & T. MENGESDORF. 2005. Cellular abnormalities linked to endoplasmic reticulum dysfunction in cerebrovascular disease—therapeutic potential. Pharmacol. Ther. **108:** 362–375.
32. KUMAR, R. *et al.* 2003. Dysfunction of the unfolded protein response during global brain ischemia and reperfusion. J. Cereb. Blood Flow Metabol. **23:** 462–471.
33. HAYASHI, T. *et al.* 2003. Oxidative damage to the endoplasmic reticulum is implicated in ischemic neuronal death. J. Cereb. Blood Flow Metabol. **23:** 1117–1128.
34. DEGRACIA, D.J. & H.L. MONTIE. 2004. Cerebral ischemia and the unfolded protein response. J. Neurochem. **91:** 1–8.
35. DOUTHEIL, J. *et al.* 2000. Effect of nitric oxide on endoplasmic reticulum calcium homeostasis, protein synthesis and energy metabolism. Cell Calcium **27:** 107–115.
36. KOHNO, K. *et al.* 1997. Neuroprotective nitric oxide synthase inhibitor reduces intracellular calcium accumulation following transient global cerebral ischemia in gerbil. Neurosci. Lett. **275:** 17–20.
37. NOVOA, I. *et al.* 2003. Stress-induced gene expression requires programmed recovery from translational repression. EMBO J. **22:** 1180–1187.

38. PASCHEN, W. *et al.* 2004. GADD34 protein levels increase after transient ischemia in the cortex but not in the CA1 subfield: implications for post-ischemic recovery of protein synthesis in ischemia-resistant cells. J. Neurochem. **90:** 694–701.
39. MA, Y. & L.M. HENDERSHOT. 2003. The stressful road to antibody secretion. Nat. Immunol. **4:** 310–311.
40. SITIA, R. & I. BRAAKMAN. 2003. Quality control in the endoplasmic reticulum protein factory. Nature **426:** 891–894.
41. CENCI, S. *et al.* 2006. Progressively impaired proteasomal capacity during terminal plasma cell differentiation. EMBO J. **25:** 1104–1113.
42. VAN ANKEN, E. *et al.* 2003. Sequential waves of functionally related proteins are expressed when B cells prepare for antibody secretion. Immunity **18:** 243–253.
43. GROSS, E. *et al.* 2006. Generating disulfides enzymatically: reaction products and electron acceptors of the endoplasmic reticulum thiol oxidase Ero1p. Proc. Natl. Acad. Sci. USA **103:** 299–304.
44. BENHAM, A.M. & R. SITIA. 2006. The diversity of oxidative protein folding. Antioxid. Redox Signal **8:** 271–273.
45. HAYNES, C.M., E.A. TITUS & A.A. COOPER 2004. Degradation of misfolded proteins prevents ER-derived oxidative stress and cell death. Mol. Cell **15:** 767–776.
46. SELYE, H. 1950. Stress and the general adaptation syndrome. Br. Med. J. **1:** 1383–1392.
47. OZOLS, J. 1995. Lumenal orientation and post-translational modifications of the liver microsomal 11 beta-hydroxysteroid dehydrogenase. J. Biol. Chem. **270:** 2305–2312.
48. AGARWAL, A.K. *et al.* 1989. Cloning and expression of rat cDNA encoding corticosteroid 11 beta-dehydrogenase. J. Biol. Chem. **264:** 18939–18943.
49. ODERMATT, A. *et al.* 2006. Why is 11β-hydroxysteroid dehydrogenase type 1 facing the endoplasmic reticulum lumen? Physiological relevance of the membrane topology of 11β-HSD1. Mol. Cell Endocrinol. **248:** 15–23.
50. BUBLITZ, C. & C.A. LAWLER. 1987. The levels of nicotinamide nucleotides in liver microsomes and their possible significance to the function of hexose phosphate dehydrogenase. Biochem. J. **245:** 263–267.
51. PICCIRELLA, S. *et al.* 2006. Uncoupled redox systems in the lumen of the endoplasmic reticulum: pyridine nucleotides stay reduced in an oxidative environment. J. Biol. Chem. **281:** 4671–4677.
52. HEWITT, K.N., E.A. WALKER & P.M. STEWART. 2005. Minireview: hexose-6-phosphate dehydrogenase and redox control of 11β-hydroxysteroid dehydrogenase type 1 activity. Endocrinology **146:** 2539–2543.
53. LAVERY, G.G. *et al.* 2006. Hexose-6-phosphate dehydrogenase knockout mice lack 11β-hydroxysteroid dehydrogenase type 1-mediated glucocorticoid generation. J. Biol. Chem. **281:** 6546–6551.
54. BÁNHEGYI, G. *et al.* 2004. Cooperativity between 11β-hydroxysteroid dehydrogenase type 1 and hexose-6-phosphate dehydrogenase in the lumen of the endoplasmic reticulum. J. Biol. Chem. **279:** 27017–27021.
55. BELKAID, A. *et al.* 2006. Silencing of the human microsomal glucose-6-phosphate translocase induces glioma cell death: potential new anticancer target for curcumin. FEBS Lett. **580:** 3746–3752.
56. BELKAID, A. *et al.* 2006. The chemopreventive properties of chlorogenic acid reveal a potential new role for the microsomal glucose-6-phosphate translocase in brain tumor progression. Cancer Cell Int. **6:** 7.

57. LEUZZI, R. *et al.* 2003. Inhibition of microsomal glucose-6-phosphate transport in human neutrophils results in apoptosis: a potential explanation for neutrophil dysfunction in glycogen storage disease type 1b. Blood **101:** 2381–2387.
58. KUIJPERS, T.W. 2002. Clinical symptoms and neutropenia: the balance of neutrophil development, functional activity, and cell death. Eur. J. Pediatr. **161**(Suppl. 1): S75–S82.
59. ÖZCAN, U. *et al.* 2004. Endoplasmic reticulum stress links obesity, insulin action, and type 2 diabetes. Science **306:** 457–461.
60. HOTAMISLIGIL, G.S. 2006. Inflammation and metabolic disorders. Nature **444:** 860–867.
61. NAKATANI, Y. *et al.* 2005. Involvement of endoplasmic reticulum stress in insulin resistance and diabetes. J. Biol. Chem. **280:** 847–851.
62. OZAWA, K. *et al.* 2005. The endoplasmic reticulum chaperone improves insulin resistance in type 2 diabetes. Diabetes **54:** 657–663.

Chaperones and Proteases—Guardians of Protein Integrity in Eukaryotic Organelles

CLAUDIA LEIDHOLD[a,b] AND WOLFGANG VOOS[a]

[a]*Institut für Biochemie und Molekularbiologie, Universität Freiburg, Germany*
[b]*Fakultät für Biologie, Universität Freiburg, Germany*

ABSTRACT: Organelles like mitochondria, chloroplasts, or the endoplasmic reticulum are essential subcompartments of eukaryotic cells that fulfill important metabolic tasks. Organellar protein homeostasis is maintained by a combination of specific protein biogenesis processes and protein quality control (PQC) mechanisms that together guarantee the functional state of the organelle. According to their endosymbiontic origin, mitochondria and chloroplasts contain internal PQC systems that consist of a cooperative network of molecular chaperones and proteases. In contrast, the endoplasmic reticulum employs the main cytosolic degradation machinery, the proteasome, for the removal of damaged or misfolded proteins. Here we present and discuss recent experimental insights into the molecular mechanisms underlying organellar PQC processes.

KEYWORDS: mitochondria; chloroplast; endoplasmic reticulum; chaperone; protease

INTRODUCTION

Protein quality control (PQC) is a major biochemical process that contributes to cellular protein homeostasis and to the maintenance of cellular functions under normal and stress conditions. In general, PQC comprises a multitude of complex and interdependent biochemical reactions that contribute to the functional integrity of proteins, ranging from the support of protein (re)folding, the protection against aggregation, and the specific proteolytic removal of terminally damaged polypeptides. Although these processes are already important for cellular growth at normal conditions, environmental stress leads to an increased number of inactive, denatured, or damaged polypeptides that makes a functional PQC essential for the survival of the cell. A characteristic

Address for correspondence: Dr. Wolfgang Voos, Institut für Biochemie und Molekularbiologie, Universität Freiburg, Hermann-Herder-Str. 7, D-79104 Freiburg, Germany. Voice: +49-761-203-5280; fax: +49-761-203-5261.
wolfgang.voos@biochemie.uni-freiburg.de

feature of many human diseases is the accumulation of nonfunctional or misfolded proteins, indicating that PQC systems are defective or overwhelmed under these conditions. Understanding the mechanisms of PQC in pathological situations is therefore essential for the development of diagnostic and therapeutic procedures.

Cells have evolved a dedicated proteinaceous machinery, consisting of molecular chaperones and proteolytic enzymes that catalyze PQC reactions. In that respect, PQC is essentially a two-step process. First, damaged and/or misfolded polypeptides are recognized and bound by different types of molecular chaperones to be stabilized and potentially refolded. If this fails, terminally damaged proteins will be transferred to special, ATP-dependent proteases, where they are degraded as a second step. To avoid unspecific proteolysis, the active sites of the PQC proteases reside in the interior of large oligomeric protease complexes that are usually not accessible to substrate proteins. Hence, the degradation of damaged polypeptides requires specific biochemical mechanisms to target substrate polypeptides for proteolysis. In the majority of PQC reactions, chaperones and proteases cooperate closely at the functional level or even form stable oligomeric complexes. In certain cases, chaperone and proteolytic activities even reside on the same polypeptide chain.

The complex organization of eukaryotic cells leads to specific problems concerning PQC in subcellular compartments. The best-characterized eukaryotic PQC machinery, the ubiquitin–proteasome system resides in the cytoplasm and is largely responsible for the degradation of soluble cytosolic proteins.[1] However, phospholipid membranes that surround subcellular structures like the endoplasmic reticulum (ER), mitochondria or chloroplasts are by default impermeable for macromolecules such as proteins. Only two ways exist to overcome the problem of this physical separation between most organellar proteins and the major cellular protease: (a) either the organelles contain their own independent PQC components or (b) specific machineries exist that make organellar proteins accessible to degradation by the proteasome. Interestingly, both mechanisms have been adopted by eukaryotic cells. Organelles with an endosymbiontic origin like mitochondria and chloroplasts have retained at least some of their original PQC enzymes that are closely related to their bacterial ancestors.[2,3] In contrast, the ER couples an internal PQC system with the degradation of nonfunctional proteins by the proteasome in the cytosol.[4,5] To date, the major components of cellular PQC systems have been identified. However, important mechanistic questions remain largely unanswered: (a) How do PQC machineries distinguish between functional and damaged polypeptides? (b) How is the decision between repair (refolding) or proteolytic removal of damaged proteins made? and (c) What are the molecular mechanisms of the individual reactions in PQC pathways? These questions have to be answered for each single subcellular system in an eukaryotic cell.

MITOCHONDRIA

As essential eukaryotic organelles, mitochondria fulfill important metabolic functions, for example, TCA cycle, fatty acid metabolism, Fe/S-cluster assembly, and they are the main energy source of the cell. Maintaining mitochondrial function is therefore indispensable for the life of the cell. Malfunction of mitochondria has been shown to be involved in neurodegenerative diseases and aging processes.[6] Mitochondria also play an important role in the regulation of cell death by apoptosis. The functional and structural integrity of mitochondrial proteins is sustained by an endogenous PQC system (FIG. 1). At the frontline of damage control, chaperones of the Hsp60, Hsp70, and Hsp100 families assist in protein import, folding, prevention of aggregation, disaggregation, and degradation. If refolding to the active conformation fails, damaged proteins are handed over to proteases of the AAA family (ATPases associated with a variety of cellular activities). Proteases or peptidases have been identified in the mitochondrial matrix, the inner membrane, and the intermembrane space, indicating that all mitochondrial subcompartments possess their own conserved proteolytic system.

The majority of mitochondrial proteins are encoded in the nucleus and have to be imported posttranslationally.[7] Most of these proteins contain N-terminal targeting sequences that need to be cleaved off after import by one of various processing peptidases—mitochondrial processing peptidases (MPP), mitochondria intermediate peptidase (MIP), or inner membrane peptidase (IMP)—depending on the subcompartment they are addressed to.[8] Only unfolded proteins can be translocated through the membrane transport pores of mitochondria. Newly imported proteins are kept in a soluble form by binding to mitochondrial chaperones that assist in their folding process. Owing to multiple folding and refolding processes as a consequence of the import reaction, mitochondrial proteins have a relatively high intrinsic probability of being misfolded. Exceeding chaperone capacities under stress conditions induces additional denaturation events and leads to the accumulation of damaged polypeptide.

The Vanguard of Damage Control: Chaperones

A prominent role in the mitochondrial PQC system has been demonstrated so far for Hsp70 and Hsp100 family members. Mitochondrial Hsp70 (mtHsp70) binds with high affinity to hydrophobic segments within unfolded or denatured proteins. MtHsp70 is part of the ATP-dependent import motor (PAM) and binds to the newly arriving polypeptide chains in the matrix.[9] In addition, mtHsp70 plays a critical role in the subsequent folding of the imported protein to its active conformation. Under stress conditions, mtHsp70 as an abundant mitochondrial chaperone can either protect proteins from aggregation or make them accessible to degradation by the protease Pim1/Lon.[10]

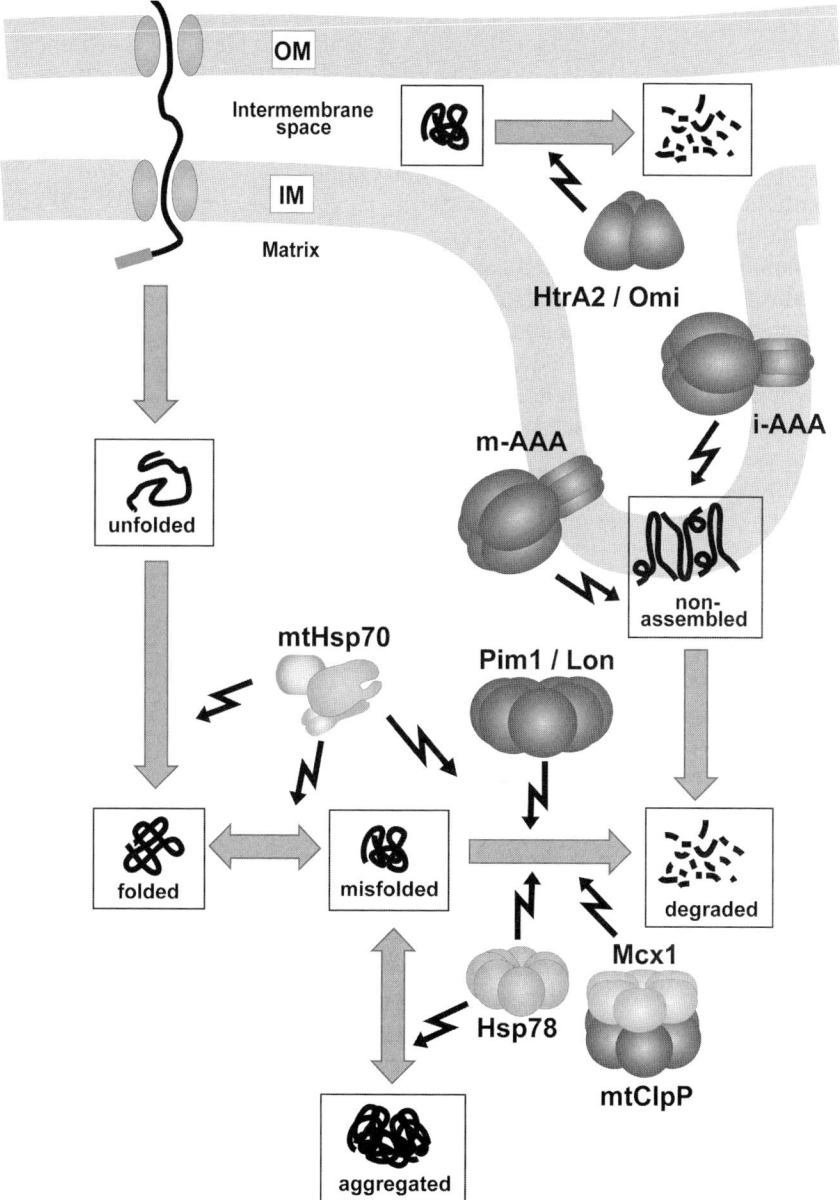

FIGURE 1. Chaperones and proteases in mitochondria as an example for organellar protein quality control (PQC). Imported mitochondrial polypeptides can acquire several conformational and functional states that in part lead to misfolding and aggregation. Shown are the identified chaperones (light gray) and proteases (dark gray) that control protein homeostasis (for details see text). OM = outer mitochondrial membrane; IM = inner mitochondrial membrane.

The bacterial ClpB is involved in disaggregation and conferment of thermotolerance.[11] Even though bacterial ClpB structure and function is well understood, less is known about the function of Hsp100 chaperones in mitochondria. The mitochondrial member of the Hsp100/ClpB family is Hsp78, which forms a homo-oligomeric complex in the matrix compartment.[12] Initially, Hsp78 has been shown to be necessary for adaptation of mitochondria to heat stress.[13] To confer thermotolerance, the main task of Hsp78 is to maintain mtHsp70 in a soluble state.[14] Cooperation between mtHsp70 and Hsp78 is necessary for the resolubilization of aggregated proteins under *in vivo* conditions. The function of Mcx1, a mitochondrial Hsp100/ClpX homolog has not been established so far.[15] Interestingly, cooperation of Hsp78 with the mitochondrial matrix protease Pim1 was reported to be required for the degradation of damaged polypeptides in the matrix compartment, a function that was not anticipated from the studies of its bacterial homolog.[16]

The Ultimate Destination: Mitochondrial Proteases

Protein degradation is an essential part of the mitochondrial PQC system but also necessary for general protein turnover in adaptation to different environmental conditions. The mitochondrial matrix as the main hydrophilic subcompartment contains two major proteases (Lon/Pim1 and ClpP). Lon is a highly conserved matrix protease of the AAA family.[17,18] Lon has been shown to be involved in three major functions in mitochondria: (a) ATP-dependent proteolysis, (b) chaperone-like function that facilitates protein complex assembly,[19] and (c) maintenance of mitochondrial DNA.[20] The proteolytic activity of Lon is tightly coupled to ATP hydrolysis. The serine protease Lon consists of two large protein domains, an N-terminal ATPase and a C-terminal protease domain, and was described as a ring-shaped soluble protein. In general, Lon is not only important under normal environmental conditions but also highly overexpressed after heat stress.[17] It is also important for the defense against oxidative stress because it was found to degrade oxidized proteins.[21] So far it is not clear what targets a protein for degradation by Lon. Lon was reported to degrade various imported model proteins that have to contain an unstructured N-terminal segment of certain length (<60 aa) because of its limited endogenous unfolding activity.[22,23] Human Lon (hLon) is expressed tissue specifically, and a reduced expression level of Lon results in caspase activation and apoptosis in fibroblasts.[24] Recent proteomic studies comparing wild-type and protease-deficient mitochondria in yeast revealed new insights in Pim1 substrate recognition.[25]

For ClpP, the other main protease in the mitochondrial matrix, no homolog was found in yeast. So far the functional aspects of eukaryotic ClpP are mostly unknown. Studies of the bacterial homolog showed that ClpP is the proteolytic subunit of a proteasome-like structure composed of two face-to-face assembled heptameric rings. The chaperone components ClpX and ClpA form hexamer ring structures on the top and at the bottom of ClpP and are described as regulatory factors exhibiting ATPase activity. Either ClpX or ClpA are necessary for substrate binding, unfolding, and translocation of the substrate into the proteolytic chamber formed by the ClpP subunits.[26] In mammals, not much is known about ClpP substrates; however, it is involved in stress response in mitochondria and, like Lon protease, it seems to have chaperone function.[27]

The mitochondrial inner membrane contains two membrane-integrated proteases that are facing their proteolytic domains to opposite sides of the inner membrane: the i-AAA protease into the intermembrane space and the m-AAA protease into the matrix.[28] The i-AAA protease complex is formed by six subunits of the protein Yme1, each containing one transmembrane domain. The ring-shaped i-AAA complex contains hydrophobic amino acid residues in the central pore that are probably the binding sites for membrane proteins exposing unfolded domains.[29] Substrate recognition is accomplished by two distinct substrate-binding sites: conserved helices at the N-terminus or an other helix–loop–helix structure at the C-terminal end of the proteolytic domain. The m-AAA protease is part of a large multisubunit complex composed of the Yta10 and Yta12 proteins in yeast and paraplegin in mammals. The m-AAA is highly homologous to the FtsH protease in bacteria.[30] The FtsH protease carries an ATPase and a proteolytic domain on one polypeptide chain, similar to the Lon protease. FtsH-type proteases form large ring-shaped hexameric protein complexes that are integrated into cellular membranes.[31] However, their proteolytic activities reside in the hydrophilic part of the complex. It is postulated that the proteolytic degradation of membrane proteins by proteases of this type encompasses an extraction step that is initiated at exposed soluble segments of substrate proteins. Inactivation of the human m-AAA protease causes hereditary spastic paraplegia (HSP). The molecular basis for HSP might be the missing processing of MrpL32 (a ribosomal protein) by the m-AAA.[32] The m-AAA together with a third protease of the inner membrane (rhomboid) was found to process newly imported Ccp1.[33] The mammalian homolog of the yeast rhomboid protease Pcp1 is PARL (presenilin-associated rhomboid-like protein). PARL was reported to be a regulatory link between PQC and morphology in mitochondria. Phosphorylation and cleavage of PARL can change mitochondrial morphology and induce fragmentation.[34]

Oma1, a recently found membrane-embedded metalloprotease with overlapping activities to m-AAA was described to degrade misfolded inner membrane proteins[35] and could be a novel component of the PQC system in the inner mitochondrial membrane.

Only little is known about the proteolysis system in the intermembrane space. Two oligopeptidases, Mop112/PreP and Prd1/neurolysin, and the serine protease HtrA2/Omi have been identified. Mop112 is a metallopeptidase of the pitrilysin family. Deletion of the peptidase in yeast leads to less peptide release from mitochondria. Its human homolog PreP was described as a peptidasome because it degrades unstructured peptides or presequences. Surprisingly, human PreP was found in the mitochondrial matrix.[36] Prd1 (saccharolysin or neurolysin) is another metallopeptidase with overlapping function to Mop112. Both are involved in degradation of peptide, debris from proteolysis of chambered proteases in the mitochondrial matrix and the inner membrane.[37] HtrA2/Omi is a serine protease of the IMS and homolog to the bacterial DegP. So far no protease substrates could be identified in eukaryotes. Interestingly it has been shown to be exported into the cytosol, where it seems to be involved in initiating apoptosis signaling.[38]

CHLOROPLASTS

Plant chloroplasts are endosymbiontic organelles derived from photosynthetic cyanobacteria. Not surprising, and similar to the situation in mitochondria, the PQC components of chloroplasts are highly similar to bacteria. Chloroplasts contain a full set of chaperones belonging to the Hsp100, Hsp70, and Hsp60 protein families.[39] All three types of chaperones have been implicated in protein import and subsequent folding reactions, because similar to mitochondria, chloroplast have only a limited protein synthesis capacity.[40] However, still many open questions remain about the specific roles of chaperones in chloroplast protein homeostasis. An additional level of complexity is generated by the presence of a chloroplast subcompartment, the thylakoids. These membrane stacks contain large hetero-oligomeric protein complexes, the photosystems I and II that are responsible for photosynthesis. Here, Hsp70 chaperones have been implicated in assembly and protection of photosystem II (PSII) components.[41] Plants also contain various members of the ClpB protein family that are mediators of protein stability. Mutants of the plastidic ClpB homolog have been shown to be essential in chloroplast development, indicating a major role not only under stress conditions but also in providing housekeeping functions.[42] The major PQC proteases characterized in chloroplasts belong to the ClpP, FtsH, and DegP families.[43] The recently completed genome of the model plant species *Arabidopsis thaliana* has revealed a much more complex situation in plants compared to metazoans or fungi. For each individual protease class, several homologs seems to be encoded by multiple genes.[44] The functional significance of these multiple homologs is still unknown. In particular, it is not known if the multitude of homologs reflects a heteromeric composition to the protease complexes or is due to tissue- and/or development-specific expression patterns.

Degradation of Thylakoid Membrane Proteins by the DegP/FtsH System

The biochemical characteristics of energy conversion by the photosystems in the thylakoid membranes can lead to an accumulation of oxidatively damaged components that have to be removed by proteases. The degradation of a photo-damaged D1 protein, a component of PSII, has been shown to be mediated by the chloroplast membrane protease FtsH in cooperation with homologs of the bacterial DegP. Two types of FtsH-like proteins have been identified in chloroplasts that are integrated into the thylakoid membrane system.[45,46] Proteins of the DegP family are Ser proteases that form homo-oligomeric protein complexes and usually exhibit an ATP-independent chaperone activity.[47] Interestingly, at higher temperatures the protein switches from a chaperone activity to a protease activity by making its proteolytic site accessible to external substrates. Chloroplasts contain several homologs of DegP in the thylakoid lumen, Deg1, Deg5, and Deg8; at least one, Deg2, in the stroma; and up to eleven homologs in other cellular localizations. Of the identified DegP homologs, the proteins Deg1 and Deg2 have been directly shown to be involved in PSII quality control.[48,49] Deg1 localizes to the thylakoid lumen while Deg2 resides on the stromal face of the thylakoid membrane. Owing to their localization, a two-step mechanism for the degradation of damaged PSII components has been proposed. First, single cleavages are introduced into hydrophilic domains at both sides of the membrane by Deg1 and Deg2[49,50] (and possibly by other homologs of the DegP family). Second, the complete proteolysis of the resulting fragments is mediated by the chloroplast FtsH protease.[48]

The Chloroplast ClpP Protease Complex

Similar to the related bacterial protease, the Clp protease in chloroplasts consists of a barrel-shaped ring system of the proteolytically active ClpP subunits that associates with a hexameric ring of ClpX or ClpA subunits that control substrate recognition and unfolding. Again, plants contain a higher number of Clp protein family members than other species. Six genes encoding homologs of the ClpP protease family have been identified. In addition, another four proteins with high homology to ClpP have been discovered that lack the catalytical triad typical for Ser proteases, named ClpR.[51] The functional significance of this special type of Clp protease subunits is not yet established. Most of these Clp family members localize to the chloroplast stroma where they are able to form hetero-oligomeric core complexes consisting of a variable number of ClpP and ClpR subunits.[52] Information about the specific functions of chloroplast Clp proteases is scarce. However, except for the plastid-encoded variety ClpP1, all other family members seem to be essential for the survival of the cell. Downregulation of individual components causes severe developmental problems.[53] Only recently, the first candidate substrate proteins, mainly housekeeping enzymes and chaperones, have been identified.[54]

ENDOPLASMIC RETICULUM

In contrast to the two other organelles mentioned earlier, mitochondria and chloroplasts, the ER is not of bacterial ancestry. However, it is enclosed by a phospholipid bilayer that separates the lumenal compartment from the cytosol and restricts diffusion of macromolecules. In particular, proteins can cross the ER membrane only at specific pore sites that interact with ribosomes and form a protein-conducting channel.[55] Owing to its specific biochemical function as a transit compartment in vesicular traffic, very stringent PQC mechanisms are required. In fact, a typical and well-studied example of a defect in ER PQC is the human respiratory disease cystic fibrosis. Here, a mutated chloride channel is removed by the PQC system before it can reach its normal localization in the plasma membrane.[56] Surprisingly, no internal proteolytic systems have been identified in the ER lumen. Instead, the ER employs a complicated machinery to identify nonfunctional polypeptides and export them to the cytosol where they are degraded by the proteasome, a process summarized as ER-associated protein degradation (ERAD).[4,5] However, the basic biochemical problems of PQC, like the specific recognition of substrates and the decision between repair or removal, remain the same as in other compartments. Owing to the complexity of the process, this review describes the principal reactions and indicates a few novel results. Three separate functional processes can be distinguished: (a) the discrimination between functional and nonfunctional polypeptides, (b) the retrotranslocation to the cytosol, and (c) the targeting of substrates for proteasomal degradation.

Substrate Discrimination and the Role of Carbohydrates

ER proteins are synthesized at cytosolic ribosomes and translocated into the organelle in a cotranslational mechanism.[55] The insertion of transmembrane domains into the membrane and the folding of hydrophilic domains at the lumenal and cytosolic sides also occur already during the translocation process. ER proteins destined for the plasma membrane or the extracellular medium contain specific oligosaccharide modifications (N-glycosylation) that are added at a very early step during translocation. The oligosaccharide side chains are extensively modified during the further passage through the ER and the Golgi compartments. The folding state of the polypeptide chain seems to be the main property that is monitored by the PQC system of the ER and determines the eventual fate of the substrate protein. The ER contains several types of proteins with chaperone activities that are capable of specifically recognizing nonnative protein structures. The major chaperone involved in PQC is the lumenal Hsp70 BiP. It is the first chaperone that interacts with the incoming polypeptide chain, stabilizes the imported proteins against aggregation, and assists its folding to the native conformation.[57] Owing to its affinity to nonnative protein segments, BiP acts also as an important component of the recognition mechanism of soluble ERAD substrates that are damaged or misfolded.[58]

Interestingly, the first modifications of added carbohydrate chains are coupled to the folding state of the respective substrate proteins. The N-glycosylated proteins interact with a dedicated chaperone machinery consisting of the proteins calnexin, a membrane-integrated lectin, and its binding partner calreticulin. The calnexin–calreticulin system is capable of interacting with the carbohydrate side chains of ER proteins as long as they have not been fully matured. Removal of a terminal glucose by glucosidase II, which has been proposed to be the final folding monitor, allows fully folded substrates to exit the calnexin–calreticulin system. Not fully folded proteins can be reglucosylated and bind to calnexin again, closing the so-called calnexin cycle.[59] In case of a terminally misfolded substrate, the removal of a different mannose residue leads to the entry of the polypeptide into the ERAD system. The de-mannosylated structures are recognized by a mannosidase-related lectin called ER degradagion-enhancing α-mannosidase-like protein (EDEM) that assists in the translocation of ERAD substrates back to the cytosol.[60]

Retrotranslocation through the ER Membrane

The decisive step of PQC in ER proteins is the specific translocation process that makes proteins with misfolded or damaged lumenal domains accessible to cytosolic proteolysis. Hence, a channel has to exist that makes lumenal proteins or domains of membrane proteins accessible to the cytosolic proteasome. Surprisingly, the initial experimental evidence indicated that the Sec61 translocation channel, responsible for the import of proteins into the ER, plays also a major role in the ERAD process.[58,61] Essentially, the same channel complex seems to facilitate both the forward movement of nascent polypeptides and the retrograde movement of damaged proteins through the ER membrane. However, the situation seems to be more complex since other membrane proteins like derlin-1 have also been proposed to act as retrotranslocation channels.[62] It is likely that the existence of different potential channel complexes reflects specificities for different types of ERAD substrate proteins.

Dislocation and Proteasomal Targeting of ERAD Substrates

Membrane transport of polypeptide chains usually requires an external energy source, predominantly in form of ATP-hydrolysis. The proteasome as the responsible protease is a large hetero-oligomeric complex that consists of the core protease complex and the lid structure. Among others, the lid contains ATP-dependent components required for substrate unfolding and transfer to the proteolytic core.[1] Although degradation of misfolded ER proteins is mediated by the proteasome, the ATP-dependent enzymes of the lid do not seem to be required. Instead, a different ATPase from the ubiquitous AAA protein family, CDC48/p97, has been identified as the energy-dependent component responsible for the extraction of ERAD substrates from the ER membranes.[63,64]

On the basis of the dependence for proteasomal degradation, polyubiquitination of ERAD substrate proteins is a prerequisite for their eventual proteolysis. Attachment of ubiquitin molecules by E3-type ubiquitin ligases is the crucial step that targets a substrate protein for final degradation. Several ubiquitin ligases have been identified to be involved in the ERAD processes. The ubiquitin ligases together with its cofactors form membrane-integrated protein systems that essentially link the substrate discrimination system of the ER with the CDC48/p97 extraction motor on the cytosolic face of the ER membrane.[65,66] Recent experiments analyzing the multiple protein interactions occurring at this step of the ERAD process in yeast cells have revealed distinct ERAD pathways that are specific for the topological orientations of their substrate proteins and contain defined protein components of the ubiquitin-labeling system.[67,68] The so-called ERAD-L pathway is responsible for the degradation of substrates with lumenal lesions and contains the E3 ligases Hrd1 and Hrd3 together with the E2 enzymes Ubc1 and Ubc7. Substrate discrimination is performed by several lumenal proteins: Kar2 (BiP), ER mannosidase I and the protein Yos9, a quality-control lectin that interacts also with Kar2. The details of the ERAD substrate recognition mechanism still have to be established; however, it seems that the Hrd3 ligase is the main recognition factor for nonnative polypeptides while the lectin Yos9 ensures that only terminally misfolded proteins are degraded.[69] The ERAD-C pathway for degradation of membrane proteins with problems in cytosolic domains seems to be more simple, comprising only the E3-ligase Doa10 and the E2s Ubc6 and Ubc7. Both pathways require the function of CDC48 and its cofactors Npl4 and Ufd1 that are responsible for the membrane association of the soluble AAA machine. In addition, a separate pathway for proteins with misfolded TM domains seems to exist, but the involved components have not been clarified.

SUMMARY

Despite the many differences in organellar PQC, several common principles can be observed: (a) functional PQC is always achieved in a close collaboration between molecular chaperones and proteases, resulting in a complex equilibrium between refolding, aggregation, and proteolysis of substrate polypeptides; (b) the final fate of affected polypeptides is mainly dependent on the intrinsic folding properties that determine their affinities to chaperone proteins; (c) proteolytic activities are usually provided by specific proteases that build large oligomeric protein complexes, forming a chamber where the proteolytic sites are shielded from the environment. Access of substrates to this proteolytic chamber is tightly regulated, in most cases in an ATP-dependent manner.

REFERENCES

1. PICKART, C.M. & R.E. COHEN. 2004. Proteasomes and their kin: proteases in the machine age. Nat. Rev. Mol. Cell Biol. **5:** 177–187.
2. ADAM, Z. & A.K. CLARKE. 2002. Cutting edge of chloroplast proteolysis. Trends Plant Sci. **7:** 451–456.
3. LANGER, T. & W. NEUPERT. 1996. Regulated protein degradation in mitochondria. Experientia **52:** 1069–1076.
4. MEUSSER, B., C. HIRSCH, E. JAROSCH & T. SOMMER. 2005. ERAD: the long road to destruction. Nat. Cell Biol. **7:** 766–772.
5. RÖMISCH, K. 2005. Endoplasmic reticulum-associated degradation. Annu. Rev. Cell Dev. Biol. **21:** 435–456.
6. KWONG, J.Q., M.F. BEAL & G. MANFREDI. 2006. The role of mitochondria in inherited neurodegenerative diseases. J. Neurochem. **97:** 1659–1675.
7. WIEDEMANN, N., A.E. FRAZIER & N. PFANNER. 2004. The protein import machinery of mitochondria. J. Biol. Chem. **279:** 14473–14476.
8. GAKH, O., P. CAVADINI & G. ISAYA. 2002. Mitochondrial processing peptidases. Biochim. Biophys. Acta **1592:** 63–77.
9. VOOS, W. & K. RÖTTGERS. 2002. Molecular chaperones as essential mediators of mitochondrial biogenesis. Biochim. Biophys. Acta **1592:** 51–62.
10. WAGNER, I., H. ARLT, VAN DYCK, L., *et al.* 1994. Molecular chaperones cooperate with PIM1 protease in the degradation of misfolded proteins in mitochondria. EMBO J. **13:** 5135–5145.
11. MOGK, A., T. TOMOYASU, P. GOLOUBINOFF, *et al.* 1999. Identification of thermolabile *Escherichia coli* proteins: prevention and reversion of aggregation by DnaK and ClpB. EMBO J. **18:** 6934–6949.
12. LEIDHOLD, C., B.V. JANOWSKY, D. BECKER, *et al.* 2006. Structure and function of Hsp78, the mitochondrial ClpB homolog. J. Struct. Biol. **156:** 149–164.
13. SCHMITT, M., W. NEUPERT & T. LANGER. 1996. The molecular chaperone Hsp78 confers compartment-specific thermotolerance to mitochondria. J. Cell Biol. **134:** 1375–1386.
14. VON JANOWSKY, B., T. MAJOR, K. KNAPP & W. VOOS. 2006. The disaggregation activity of the mitochondrial ClpB homolog Hsp78 maintains Hsp70 function during heat stress. J. Mol. Biol. **357:** 793–807.
15. VAN DYCK, L., M. DEMBOWSKI, W. NEUPERT & T. LANGER. 1998. Mcx1p, a ClpX homologue in mitochondria of *Saccharomyces cerevisiae*. FEBS Lett. **438:** 250–254.
16. RÖTTGERS, K., N. ZUFALL, B. GUIARD & W. VOOS. 2002. The ClpB homolog Hsp78 is required for the efficient degradation of proteins in the mitochondrial matrix. J. Biol. Chem. **277:** 45829–45837.
17. VAN DYCK, L., D.A. PEARCE & F. SHERMAN. 1994. PIM1 encodes a mitochondrial ATP-dependent protease that is required for mitochondrial function in the yeast *Saccharomyces cerevisiae*. J. Biol. Chem. **269:** 238–242.
18. WANG, N., S. GOTTESMAN, M.C. WILLINGHAM, *et al.* 1993. A human mitochondrial ATP-dependent protease that is highly homologous to bacterial Lon protease. Proc. Natl. Acad. Sci. USA **90:** 11247–11251.
19. REP, M., J.M. VAN DIJL, K. SUDA, *et al.* 1996. Promotion of mitochondrial membrane complex assembly by a proteolytically inactive yeast Lon. Science **274:** 103–106.

20. LIU, T., B. LU, I. LEE, et al. 2004. DNA and RNA binding by the mitochondrial Lon protease is regulated by nucleotide and protein substrate. J. Biol. Chem. **279:** 13902–13910.
21. BOTA, D.A. & K.J. DAVIES. 2002. Lon protease preferentially degrades oxidized mitochondrial aconitase by an ATP-stimulated mechanism. Nat. Cell Biol. **4:** 674–680.
22. VON JANOWSKY, B., K. KNAPP, T. MAJOR, et al. 2005. Structural properties of substrate proteins determine their proteolysis by the mitochondrial AAA+ protease Pim1. Biol. Chem. **386:** 1307–1317.
23. LU, B., T. LIU, J.A. CROSBY, et al. 2003. The ATP-dependent Lon protease of Mus musculus is a DNA-binding protein that is functionally conserved between yeast and mammals. Gene **306:** 45–55.
24. BOTA, D.A., J.K. NGO & K.J. DAVIES. 2005. Downregulation of the human Lon protease impairs mitochondrial structure and function and causes cell death. Free Radic. Biol. Med. **38:** 665–677.
25. MAJOR, T., B. VON JANOWSKY, T. RUPPERT, et al. 2006. Proteomic analysis of mitochondrial protein turnover: identification of novel substrate proteins of the matrix protease Pim1. Mol. Cell Biol. **26:** 762–776.
26. SAUER, R.T., D.N. BOLON, B.M. BURTON, et al. 2004. Sculpting the proteome with AAA(+) proteases and disassembly machines. Cell **119:** 9–18.
27. ZHAO, Q., J. WANG, I.V. LEVICHKIN, et al. 2002. A mitochondrial specific stress response in mammalian cells. EMBO J. **21:** 4411–4419.
28. LEONHARD, K., J.M. HERRMANN, R.A. STUART, et al. 1996. AAA proteases with catalytic sites on opposite membrane surfaces comprise a proteolytic system for the ATP-dependent degradation of inner membrane proteins in mitochondria. EMBO J. **15:** 4218–4229.
29. GRAEF, M. & T. LANGER. 2006. Substrate specific consequences of central pore mutations in the i-AAA protease Yme1 on substrate engagement. J. Struct. Biol. **156:** 101–108.
30. BIENIOSSEK, C., T. SCHALCH, M. BUMANN, et al. 2006. The molecular architecture of the metalloprotease FtsH. Proc. Natl. Acad. Sci. USA **103:** 3066–3071.
31. ITO, K. & Y. AKIYAMA. 2005. Cellular functions, mechanism of action, and regulation of FtsH protease. Annu. Rev. Microbiol. **59:** 211–231.
32. NOLDEN, M., S. EHSES, M. KOPPEN, et al. 2005. The m-AAA protease defective in hereditary spastic paraplegia controls ribosome assembly in mitochondria. Cell **123:** 277–289.
33. TATSUTA, T., S. AUGUSTIN, M. NOLDEN, et al. 2007. m-AAA protease-driven membrane dislocation allows intramembrane cleavage by rhomboid in mitochondria. EMBO J. **26:** 325–335.
34. JEYARAJU, D.V., L. XU, M.C. LETELLIER, et al. 2006. Phosphorylation and cleavage of presenilin-associated rhomboid-like protein (PARL) promotes changes in mitochondrial morphology. Proc. Natl. Acad. Sci. USA **103:** 18562–18567.
35. KÄSER, M., M. KAMBACHELD, B. KISTERS-WOIKE & T. LANGER. 2003. Oma1, a novel membrane-bound metallopeptidase in mitochondria with activities overlapping with the m-AAA protease. J. Biol. Chem. **278:** 46414–46423.
36. FALKEVALL, A., N. ALIKHANI, S. BHUSHAN, et al. 2006. Degradation of the amyloid beta-protein by the novel mitochondrial peptidasome, PreP. J. Biol. Chem. **281:** 29096–29104.

37. KAMBACHELD, M., S. AUGUSTIN, T. TATSUTA, et al. 2005. Role of the novel metallopeptidase Mop112 and saccharolysin for the complete degradation of proteins residing in different subcompartments of mitochondria. J. Biol. Chem. **280**: 20132–20139.
38. IGAKI, T., Y. SUZUKI, N. TOKUSHIGE, et al. 2007. Evolution of mitochondrial cell death pathway: proapoptotic role of HtrA2/Omi in Drosophila. Biochem. Biophys. Res. Commun. **356**: 993–997.
39. BOSTON, R.S., P.V. VIITANEN & E. VIERLING. 1996. Molecular chaperones and protein folding in plants. Plant Mol. Biol. **32**: 191–222.
40. SOLL, J. & E. SCHLEIFF. 2004. Protein import into chloroplasts. Nat. Rev. Mol. Cell Biol. **5**: 198–208.
41. SCHRODA, M., J. KROPAT, U. OSTER, et al. 2001. Possible role for molecular chaperones in assembly and repair of photosystem II. Biochem. Soc. Trans. **29**: 413–418.
42. LEE, C., S. PRAKASH & A. MATOUSCHEK. 2002. Concurrent translocation of multiple polypeptide chains through the proteasomal degradation channel. J. Biol. Chem. **277**: 34760–34765.
43. ADAM, Z., A. RUDELLA & K.J. VAN WIJK. 2006. Recent advances in the study of Clp, FtsH and other proteases located in chloroplasts. Curr. Opin. Plant Biol. **9**: 234–240.
44. ADAM, Z., I. ADAMSKA, K. NAKABAYASHI, et al. 2001. Chloroplast and mitochondrial proteases in Arabidopsis. A proposed nomenclature. Plant Physiol. **125**: 1912–1918.
45. YU, F., S. PARK & S.R. RODERMEL. 2004. The Arabidopsis FtsH metalloprotease gene family: interchangeability of subunits in chloroplast oligomeric complexes. Plant J. **37**: 864–876.
46. ZALTSMAN, A., N. ORI & Z. ADAM. 2005. Two types of FtsH protease subunits are required for chloroplast biogenesis and Photosystem II repair in Arabidopsis. Plant Cell **17**: 2782–2790.
47. CLAUSEN, T., C. SOUTHAN & M. EHRMANN. 2002. The HtrA family of proteases: implications for protein composition and cell fate. Mol. Cell **10**: 443–455.
48. LINDAHL, M., C. SPETEA, T. HUNDAL, et al. 2000. The thylakoid FtsH protease plays a role in the light-induced turnover of the photosystem II D1 protein. Plant Cell **12**: 419–431.
49. HAUSSUHL, K., B. ANDERSSON & I. ADAMSKA. 2001. A chloroplast DegP2 protease performs the primary cleavage of the photodamaged D1 protein in plant photosystem II. EMBO J. **20**: 713–722.
50. CHASSIN, Y., E. KAPRI-PARDES, G. SINVANY, et al. 2002. Expression and characterization of the thylakoid lumen protease DegP1 from Arabidopsis. Plant Physiol. **130**: 857–864.
51. ZHENG, B., T. HALPERIN, O. HRUSKOVA-HEIDINGSFELDOVA, et al. 2002. Characterization of chloroplast Clp proteins in Arabidopsis: localization, tissue specificity and stress responses. Physiol. Plant **114**: 92–101.
52. PELTIER, J.B., D.R. RIPOLL, G. FRISO, et al. 2004. Clp protease complexes from photosynthetic and non-photosynthetic plastids and mitochondria of plants, their predicted three-dimensional structures, and functional implications. J. Biol. Chem. **279**: 4768–4781.
53. ZHENG, B., T.M. MACDONALD, S. SUTINEN, et al. 2006. A nuclear-encoded ClpP subunit of the chloroplast ATP-dependent Clp protease is essential for early development in *Arabidopsis thaliana*. Planta **224**: 1103–1115.

54. SJÖGREN, L.L., T.M. STANNE, B. ZHENG, et al. 2006. Structural and functional insights into the chloroplast ATP-dependent Clp protease in Arabidopsis. Plant Cell **18:** 2635–2649.
55. OSBORNE, A.R., T.A. RAPOPORT & B. VAN DEN BERG. 2005. Protein translocation by the Sec61/SecY channel. Annu. Rev. Cell Dev. Biol. **21:** 529–550.
56. ZHANG, Y., G. NIJBROEK, M.L. SULLIVAN, et al. 2001. Hsp70 molecular chaperone facilitates endoplasmic reticulum-associated protein degradation of cystic fibrosis transmembrane conductance regulator in yeast. Mol. Biol. Cell **12:** 1303–1314.
57. KLEIZEN, B. & I. BRAAKMAN. 2004. Protein folding and quality control in the endoplasmic reticulum. Curr. Opin. Cell Biol. **16:** 343–349.
58. PLEMPER, R.K., S. BÖHMLER, J. BORDALLO, et al. 1997. Mutant analysis links the translocon and BiP to retrograde protein transport for ER degradation. Nature **388:** 891–895.
59. MOLINARI, M., C. GALLI, O. VANONI, et al. 2005. Persistent glycoprotein misfolding activates the glucosidase II/UGT1-driven calnexin cycle to delay aggregation and loss of folding competence. Mol. Cell **20:** 503–512.
60. MOLINARI, M., V. CALANCA, C. GALLI, et al. 2003. Role of EDEM in the release of misfolded glycoproteins from the calnexin cycle. Science **299:** 1397–1400.
61. KALIES, K.U., S. ALLAN, T. SERGEYENKO, et al. 2005. The protein translocation channel binds proteasomes to the endoplasmic reticulum membrane. EMBO J. **24:** 2284–2293.
62. YE, Y., Y. SHIBATA, C. YUN, et al. 2004. A membrane protein complex mediates retro-translocation from the ER lumen into the cytosol. Nature **429:** 841–847.
63. YE, Y., H.H. MEYER & T.A. RAPOPORT. 2001. The AAA ATPase Cdc48/p97 and its partners transport proteins from the ER into the cytosol. Nature **414:** 652–656.
64. JAROSCH, E., C. TAXIS, C. VOLKWEIN, et al. 2002. Protein dislocation from the ER requires polyubiquitination and the AAA-ATPase Cdc48. Nat. Cell Biol. **4:** 134–139.
65. NEUBER, O., E. JAROSCH, C. VOLKWEIN, et al. 2005. Ubx2 links the Cdc48 complex to ER-associated protein degradation. Nat. Cell Biol. **7:** 993–998.
66. GAUSS, R., T. SOMMER & E. JAROSCH. 2006. The Hrd1p ligase complex forms a linchpin between ER-lumenal substrate selection and Cdc48p recruitment. EMBO J. **25:** 1827–1835.
67. CARVALHO, P., V. GODER & T.A. RAPOPORT. 2006. Distinct ubiquitin-ligase complexes define convergent pathways for the degradation of ER proteins. Cell **126:** 3613–3673.
68. DENIC, V., E.M. QUAN & J.S. WEISSMAN. 2006. A luminal surveillance complex that selects misfolded glycoproteins for ER-associated degradation. Cell **126:** 349–359.
69. GAUSS, R., E. JAROSCH, T. SOMMER & C. HIRSCH. 2006. A complex of Yos9p and the HRD ligase integrates endoplasmic reticulum quality control into the degradation machinery. Nat. Cell Biol. **8:** 849–854.

Oxygen, Hypoxia, and Stress

CORMAC T. TAYLOR[a] AND JACQUES POUYSSEGUR[b]

[a]*UCD Conway Institute, University College Dublin, Ireland*

[b]*Institute of Signaling, Developmental Biology and Cancer Research, CNRS, University of Nice, Centre Antoine Lacassagne, Nice, France*

> ABSTRACT: Since cyanobacteria began to photosynthesize and introduce the colorless and odorless gas oxygen into the earth's atmosphere some 2.5 billion years ago, human evolution has been intrinsically linked to this critical molecule. Initially, the electrophilic chemical properties of oxygen rendered it a formidable toxic challenge to organisms; however, eukaryotic cells, following the incorporation of bacterial-derived mitochondria, evolved to make beneficial use of the chemical properties of molecular oxygen as the final electron acceptor in the highly efficient production of cellular energy supplies in the form of adenosine triphosphate. Because of both its necessity for eukaryotic life and its reactive chemical nature, however, a delicate balance exists between the supply of oxygen to a cell/tissue/organism and the beneficial or harmful outcome. In this minireview, we shall discuss the role of oxygen in metabolism with a particular emphasis on outcomes when oxygen supply is significantly altered. Furthermore, we will describe endogenous mechanisms that have evolved to protect cells and tissues during such adverse conditions and may prove useful as novel therapeutic targets in a range of disease states where oxygen-related stress occurs.
>
> KEYWORDS: oxidative stress; hypoxia; adaptation

INTRODUCTION

Oxygen in Metabolism

Eukaryotic cells generate adenosine triphosphate (ATP) primarily through the oxidative metabolism of glucose obtained from carbohydrates in food. Glucose is converted to pyruvate by the glycolytic pathway in the cellular cytoplasm. These pyruvate molecules are converted into CO_2 and a two-carbon acetyl group that combines with coenzyme A (CoA) to form acetyl CoA, which feeds into the mitochondrial Krebs cycle in the generation of the electron carrier

---

Address for correspondence: Cormac Taylor, Ph.D., UCD Conway Institute, UCD, Belfield, D4, Ireland. Voice: +353-1-716-6732; fax: 353-1-716-6701.

Cormac.taylor@ucd.ie

NADH, which donates its electrons to the process of oxidative phosphorylation. Oxidative phosphorylation occurs in the mitochondrial membrane and involves the activity of four protein complexes (complex I–IV). Electrons are passed down through a series of cytochromes, resulting in the generation of a proton gradient across the inner mitochondrial membrane, which serves as the driving force for the synthesis of ATP. This process involves the consumption of oxygen by cytochrome c oxidase (complex IV) and the generation of ATP and hydrogen peroxide (H_2O_2). Collectively, these three processes (glycolysis, the Krebs cycle, and oxidative phosphorylation) are extremely efficient and yield a total of 38 molecules of ATP for each molecule of glucose consumed. Notably, this process is not 100% efficient and involves the generation of reactive oxygen species (ROS), which, if excessive, may act as physiologic signals or mediators of cell damage and death.

In the steady state, approximately 90% of available oxygen is consumed by mitochondria during the production of ATP through oxidative phosphorylation.[1] The ATP produced in this process goes on to fuel the vast majority of active cellular processes. Thus, because the chemical reduction of molecular oxygen is the primary source of metabolic energy for virtually all eukaryotic cells, a constant oxygen supply is critical for continued cell function and survival.[2] Hypoxic stress, which occurs when oxygen demand exceeds supply, represents a severe threat to continued cell, tissue, and organism survival. Conversely, as outlined earlier, a significant consequence of oxidative phosphorylation is the generation of ROS. For this reason, excessive or discontinuous oxygen consumption can also have significant pathologic implications through the generation of ROS. Thus, fluctuating oxygen levels in tissues can represent a significant threat to continued cellular function. Later we will outline the various conditions where adverse reactions to oxygen-related stress may occur.

OXYGEN AND DISEASE

Hypoxia

In the steady state (normoxia), the vast majority of oxygen consumed by a cell is used by the mitochondria in the generation of ATP by oxidative phosphorylation. This leads to the generation of sufficient cellular ATP levels to facilitate normal physiologic function. The remainder "spare" oxygen is used for non-mitochondrial processes and is consumed by nonmitochondrial dioxygenases. When the oxygen supply to a tissue is compromised (or demand is sufficiently increased), a situation occurs where insufficient ATP levels prevent a cell from carrying out its normal function leading to a state of metabolic crisis.[2]

Hypoxic stress occurs in a wide range of conditions and may be acute or chronic in nature. Perhaps the purest form of whole-body hypoxia occurs in altitude sickness, wherein rapid exposure to decreased atmospheric oxygen leads to whole-body hypoxia. A similar condition occurs in carbon monoxide poisoning, where the oxygen-carrying capacity of hemoglobin is significantly altered. Surgical ischemia due to the temporary ligation of major blood vessels can also result in tissue hypoxia. Tissue hypoxia is also present in a diverse range of more clinically relevant disorders. For example, vascular diseases, including stroke and atherosclerosis, cause decreased blood supply to tissues, resulting in hypoxia. Chronic hypoxia can also occur in chronic inflammatory disease, where extensive and prolonged inflammation and fibrosis can lead to a breakdown in the microvascular architecture with decrease in perfusion efficiency and subsequent hypoxia. The relationship between hypoxia and inflammation will be further discussed at this symposium by Cormac Taylor. A number of encephalopathies are a result of prolonged ischemia leading to hypoxic stress in brain tissue. Finally, as a solid tumor develops, it outgrows the local blood supply, resulting in significant tumor hypoxia, an event that may be critical in the development of the tumor.[13] This topic will be further discussed during the symposium by Jacques Pouyssegur. Thus, hypoxic stress is a pathologically relevant stimulus that occurs in a diverse range of disease states.

Hyperoxia

Hyperoxia arises when excessive levels of oxygen are delivered to a cell or tissue and is primarily associated with the administration of exogenous oxygen in a clinical setting. This is primarily the case when supraphysiologic oxygen supplementation is used in the prevention or treatment of hypoxemia and acute respiratory failure. It is clear, however, that prolonged exposure to hyperoxia can cause significant pathology, particularly in the pulmonary epithelium, where cells are exposed to the highest PO_2 levels.[3] Hyperoxic stress results in the induction of inflammatory pathways in the lung characterized by elevated proinflammatory cytokine levels, increased leukocyte infiltration, impaired gas exchange, and pulmonary edema, resulting in injury and death in pulmonary epithelial cells. Critically, it is thought that in spite of the superior pulmonary antioxidant capacity, the generation of ROS, including superoxide (O_2^-), H_2O_2, and hydroxyl radical (HO) as a result of increased mitochondrial oxygen consumption and/or NADPH oxidase activity is an initiating signal in this pathophysiologic response. ROS produced during hyperoxia may directly chemically damage cellular macromolecules leading to cellular damage and death. Alternatively, ROS-dependent transcriptional responses can lead to the induction of inflammatory pathways that exacerbate this toxic response. For example, the activation of inflammatory pathways by hyperoxia-induced ROS can lead to the signals necessary to increase leukocyte infiltration, which may serve as a further source of ROS generation. The net effect of these events is the induction of multiple cell death pathways, including necrosis and apoptosis in pulmonary epithelial cells, leading to the pathophysiologic outcome of pulmonary disease. The intracellular signaling pathways involved in the hyperoxia-dependent induction of cell death have been expertly reviewed elsewhere.

Hypoxia/Reoxygenation

Reoxygenation occurs when the delivery of oxygen to a tissue that has been interrupted is rapidly reintroduced.[4,5] Reoxygenation-dependent injury is typically associated with conditions where blood flow is diminished or removed and then rapidly reintroduced, including thrombolytic therapy in myocardial infarction, acute renal failure, and post-transplantation injury. Collectively, this response is referred to as reperfusion injury and has been associated with the reoxygenation-dependent induction of Fas-ligand–dependent cell death pathways, increased production of ROS, and irreversible mitochondrial dysfunction—all leading to enhanced cell death through apoptosis and necrosis. The source of ROS production in reoxygenation injury may be from either resident cells or inflammatory cells that have infiltrated a tissue during the ischemic period. Interestingly, nonlethal reoxygenation/reperfusion can lead to a state of protection for a tissue against subsequent bouts of reperfusion injury termed ischemic preconditioning.

Intermittent Hypoxia

Intermittent hypoxia occurs when a cell or tissue is exposed to repetitive cycles of hypoxia and reoxygenation and is primarily associated with the repetitive nocturnal oxygen desaturations experienced by patients suffering from obstructive sleep apnea syndrome (OSAS).[6] It has recently become accepted that OSAS is a major risk factor for cardiovascular disease through increased expression of systemic inflammatory mediators such as tumor necrosis factor-α and, as such, the molecular mechanisms underlying this effect has become an area of intense investigation. Recent studies have demonstrated that the intermittent hypoxia associated with OSAS is associated with the production of ROS during repetitive bouts of reoxygenation.

Thus, although the primary pathological insult associated with hypoxia is associated with the decreased availability of oxygen (and subsequently ATP) to fuel physiologic processes leading to metabolic crisis and loss of function, hyperoxia, reoxygenation, and intermittent hypoxia are more likely associated with the generation of chemically potent ROS. Because oxygen deprivation and ROS production are encountered often in both physiologic and pathophysiologic states, it is not surprising that over the course of evolution, we have developed the ability to respond to such environmental stresses with the induction of specific protective pathways. Some of these pathways will be outlined later.

ENDOGENOUS PROTECTIVE MECHANISMS

Adaptation to Hypoxia

Hydroxylases and the Hypoxia Inducible Factor-1 Pathway

Over the course of evolution, we have developed the ability to adapt to hypoxia through the induction of a specific transcriptional pathway governed by the hypoxia inducible factor-1 (HIF-1), a transcription factor that regulates the expression of genes promoting angiogenesis, vasodilatation, oxidative phosphorylation, glycolysis, and erythropoesis.[7] The induction of such genes leads to increased tissue perfusion and anaerobic metabolism, thus maintaining ATP levels and forming a critical adaptive pathway in dealing with a hypoxic threat.

The mechanism by which HIF-1 is activated in hypoxia is relatively well understood.[2] HIF-1α is constitutively synthesized at a high level in normoxia, but its level is repressed by members of the 2-oxoglutarate–dependent dioxygenase superfamily, namely, the prolyl hydroxylases (PHDs). Three PHD isoforms that regulate HIF-dependent transcriptional activity have been described to date (PHD1, PHD2, and PHD3). Oxygen-dependent modification of specific proline residues within consensus LxxLAP motifs (Pro402 and Pro564) in HIF-1α by these enzymes, primarily the PHD2 isoform, results in targeting of HIF-1α for ubiquitination via an E3 ligase complex initiated by the binding of the Von Hipple Lindau protein (pVHL) and subsequent proteasomal degradation. A further hydroxylation of Asn803 in the transactivation domain of HIF-1α by factor-inhibiting HIF (FIH), an asparagine hydroxylase, represents a second mechanism of oxygen-dependent repression through inhibition of transactivation uncovered by Murray Whitelaw, who will present at this symposium.[8] Similar mechanisms exist for HIF-2α. The hypoxic sensitivity of the HIF pathway is achieved by the absolute requirement of hydroxylases for molecular oxygen as a cosubstrate (with iron and the Krebs cycle intermediate 2-oxoglutarate). Therefore, these enzymes act as true oxygen sensors. Inhibition of this pathway in hypoxia with the resultant stabilization and transactivation of HIF-α subunits represents a paradigm for oxygen sensing and hypoxia-responsive alterations in gene expression. Thus, the rapid hypoxia-dependent activation of the HIF-1 pathway leads to adaptive changes that promote tissue survival during hypoxia.

ATP-Dependent Signaling

A second protective mechanism that has evolved to enhance survival during hypoxia involves the initiation of signaling pathways that sense an alteration in cellular energy status as reflected by ATP levels. Hypoxia decreases mitochondrial ATP generation through decreased respiratory activity. ATP depletion is accompanied by a rise in its precursor AMP, thus increasing the AMP:ATP ratio, which leads to the activation of AMP-activated kinase (AMPK) through phosphorylation at Thr172.[9] The kinase upstream of AMPK primarily responsible for its phosphorylation/activation is the tumor suppressor LKB1. AMPK is a critical regulator of cellular energy homeostasis, the activation of which promotes catabolic pathways, including glucose transport, gluconeogenesis, respiration, and the use of alternative energy sources to oxygen, as well as downregulating anabolic pathways. These events are critical in maintaining cellular ATP levels. It has been hypothesized that the AMPK pathway may be involved in the beneficial effects of exercise and may represent a new therapeutic target in metabolic diseases such as diabetes.

Adaptation to Oxidative Stress

Antioxidant Systems

An important mechanism by which cells adapt to oxidant stress is to transcriptionally upregulate a distinct array of cytoprotective genes responsible for buffering the cells' antioxidant capacity. These genes act to maintain glutathione content and conjugational activity and are also responsible for the detoxification of damaging electrophilic by-products of oxidant stress and include glutathione S-transferases (GSTs), aldehyde dehydrogenases (ALDH) and NAD(P)H:quinone oxidoreductase 1 (NQO1). A master regulator of this specific antioxidant phenotype is the transcription factor Nrf2.[10] This transcription factor is held in the cytoplasm by a cytoskeletal associated specific inhibitory protein KEAP1 under conditions of normal cellular redox state, where Nrf2 is continuously targeted to proteasomal degradation. Under conditions of oxidative stress, cysteine residues within the hinge region of KEAP1 become modified through mechanisms that involve thiol oxidation, resulting in a conformational change in KEAP1 with the loss of Nrf2 binding and proteasomal targeting. Nrf2 then accumulates and localizes to the nucleus, where it heterodimerizes with specific cofactors, including members of the maf protein family, and coordinates the upregulation of cytoprotective genes through the initiation of transactivation at antioxidant response elements (AREs) within the regulatory regions of these genes.

A second important regulator of the cells' antioxidant armory is the AP-1 transcription factor family member junD, which regulates the basal expression of a number of antioxidant genes and reduces angiogenesis. Indeed junD$^{-/-}$ cells accumulate H_2O_2, reducing the availability of FeII to the HIF–PHDs that target HIF-1α for degradation. This mechanism accounts for the enhanced VEGF-A expression and increased angiogenesis in junD-deficient cells.[14]

OXYGEN AND REDOX SENSING

A key question that remains incompletely answered is how cells sense a change in oxygen levels or oxidative stress and respond by initiating the adaptive/protective responses outlined earlier. While this remains an area of intensive investigation, some recent studies have shed light on cellular oxygen- and redox-sensing mechanisms. Although the direct dioxygen dependence of the hydroxylase enzymes represents a clear and direct link between the hypoxic environment and the activation of the HIF system, there are complicating factors that likely modify this sensing mechanism, including the codependence of the hydroxylases on Krebs cycle intermediates, Fe^{2+} and potentially ROS. Furthermore, intracellular oxygen gradients are determined by the rate of mitochondrial oxygen consumption by cells, a factor that can be manipulated by inhibitors of respiration such as nitric oxide.[11] Thus, although this oxygen-sensing system represents a direct link between the local oxygen levels and activation of the HIF pathway, these coregulating factors likely allow the level of oxygen at which cells respond to hypoxia to differ between tissues.

The oxygen-sensing pathways outlined earlier relate primarily to conditions where there exists a sustained period of hypoxia, allowing transcriptionally mediated adaptive and protective responses to occur. A separate critical response to arterial hypoxemia involves the induction of an acute physiological response, resulting in increased rate and depth of pulmonary ventilation. A central event in this process involves excitation of the carotid body that signals to the respiratory centers located in the brainstem. Although it is clear that a hypoxia-dependent depression of potassium channel activity in cells of the carotid body is central to this response, the nature of the oxygen sensor in this system remains an area of hot debate.[12] In fact, it is unlikely in this system that a single oxygen-sensing pathway exists. Candidate sensors that will be discussed during this symposium (by Paul Kemp) include mitochondria, AMPK, and haemoxygenase-2.

THERAPEUTIC POTENTIAL OF OXYGEN-SENSING PATHWAYS

As outlined earlier, recent advances have shed significant light on our understanding of how cells respond to stress associated with insufficient oxygen supply or oxidative stress. This information is allowing the development of novel pharmacologic intervention strategies in a range of disease states. For example, the HIF-1 pathway is an attractive therapeutic target in ischemic disease, where the promotion of HIF-1-dependent adaptation with pharmacologic hydroxylase inhibitors would be predicted to enhance tissue survival. Conversely, a developing tumor that becomes hypoxic as it outgrows the local blood supply utilizes the HIF pathway to facilitate tumor development. Clearly, inhibiting the HIF pathway in cancer is of therapeutic potential. Finally, the Nrf-2 pathway can be activated by organic compounds such as sulforaphane, which may be used to enhance the expression of endogenous antioxidant mechanisms that may be protective in disease states associated with extensive oxidative stress.

REFERENCES

1. ROLFE, D.F. & G.C. BROWN. 1997. Cellular energy utilization and molecular origin of standard metabolic rate in mammals. Physiol. Rev. **77**: 731–758.
2. BUNN, H.F. & R.O. POYTON. 1996. Oxygen sensing and molecular adaptation to hypoxia. Physiol. Rev. **76**: 839–885.
3. ZAHER, T.E., E.J. MILLER, D.M. MORROW, et al. 2007. Hyperoxia-induced signal transduction pathways in pulmonary epithelial cells. Free Radic. Biol. Med. **42**: 897–908.
4. LEONARD, M.O., N.E. KIERAN, K. HOWELL, et al. 2006. Reoxygenation-specific activation of the antioxidant transcription factor Nrf2 mediates cytoprotective gene expression in ischemia-reperfusion injury. FASEB J. **20**: 2624–2626.
5. SAIKUMAR, P., Z. DONG, J.M. WEINBERG & M.A. VENKATACHALAM. 1998. Mechanisms of cell death in hypoxia/reoxygenation injury. Oncogene **17**: 3341–3349.
6. PRABHAKAR, N.R., R.D. FIELDS, T. BAKER & E.C. FLETCHER. 2001. Intermittent hypoxia: cell to system. Am. J. Physiol. Lung Cell Mol. Physiol. **281**: L524–L528.
7. SCHOFIELD, C.J. & P.J. RATCLIFFE. 2005. Signalling hypoxia by HIF hydroxylases. Biochem. Biophys. Res. Commun. **338**: 617–626.
8. LANDO, D., D.J. PEET, D.A. WHELAN, et al. 2002. Asparagine hydroxylation of the HIF transactivation domain a hypoxic switch. Science **295**: 858–861.
9. TOWLER, M.C. & D.G. HARDIE. 2007. AMP-activated protein kinase in metabolic control and insulin signaling. Circ. Res. **100**: 328–341.
10. MOTOHASHI, H. & M. YAMAMOTO. 2004. Nrf2-Keap1 defines a physiologically important stress response mechanism. Trends Mol. Med. **10**: 549–557.
11. HAGEN, T., C.T. TAYLOR, F. LAM & S. MONCADA. 2003. Redistribution of intracellular oxygen in hypoxia by nitric oxide: effect on HIF1alpha. Science **302**: 1975–1978.
12. KEMP, P.J. 2006. Detecting acute changes in oxygen: will the real sensor please stand up? Exp. Physiol. **91**: 829–834.
13. BRAHIMI-HORN, M.C. & J. POUYSSEGUR. 2007. Harnessing the hypoxia-inducible factor in cancer and ischemic disease. Biochem. Pharmacol. **73**: 450–457.
14. GERALD, D., E. BERRA, Y.M. FRAPART, et al. 2004. JunD reduces tumor angiogenesis by protecting cells from oxidative stress. Cell **118**: 781–794.

Review on Bacterial Stress Topics

ANNA MARIA GIULIODORI,[a] CLAUDIO O. GUALERZI,[a] SARA SOTO,[b] JORDI VILA,[b] AND MARÍA M. TAVÍO[c]

[a]*Laboratory of Genetics, Department of Biology MCA, University of Camerino, Camerino (MC), 62032 Italy*

[b]*Microbiology Department, Hospital Clinic of Barcelona, 08036 Spain*

[c]*Faculty of Health Sciences, University of Las Palmas de Gran Canaria, 35016 Spain*

ABSTRACT: A complex network of regulatory systems ensures a coordinated and effective response to different types of stress that can act on a bacterium. Bacterial stress response generates changes that influence efflux system and virulence factor expression. Thus, partial or total loss of pathogenicity islands in uropathogenic *Escherichia coli* can be induced by SOS-dependent or SOS-independent pathways related to selection of quinolone-resistant mutants. Likewise, hyperosmolarity and some chemicals, including fluoroquinolones, salicylate, nonantimicrobial medicaments like diazepam and anti-inflammatory drugs are all able to induce an increased active efflux, cyclohexane tolerance, loss of porins, and decreased susceptibility to multiple antimicrobials in enterobacterial strains, suggesting that bacterial response to the stress caused by an increase in osmolarity might be linked to the development of the multidrug-resistant phenotypes. Finally, a sudden downshift of the growth temperature (cold-shock) triggers a drastic reprogramming of bacterial gene expression to allow cell survival under the new unfavorable conditions. The strategy developed by *E. coli* to reach this goal consists in the induction of a set of (cold-shock) genes whose expression is regulated at both transcriptional and posttranscriptional levels.

KEYWORDS: pathogenicity islands; quinolones; hyperosmotic stress; multidrug resistance; cold shock; transcriptional and posttranscriptional regulations

INTRODUCTION

Bacterial stress response can be defined as a cascade of alterations in gene expression and protein activity for the purpose of surviving extreme and rapidly

Address for correspondence: María M. Tavío, Microbiology, Clinical Science Department, Faculty of Health Sciences, University of Las Palmas de G.C., Dr. Pasteur, 35016 Las Palmas de Gran Canaria, Spain. Voice: 34928453405; fax: 34928451416.
mtavio@dcc.ulpgc.es

changing and potentially damaging conditions sensed by bacteria, and which results in the cells becoming broadly stress resistant or eliminating the stress agent and/or mediating repair of cell injury.[1]

σ^s, or RpoS, is a sigma subunit of RNA polymerase in *Escherichia coli* and the master regulator of the general stress response, which is triggered by various stressful conditions and, in turn, activates more than 70 σ^S-dependent genes, resulting in resistance to hyperosmolarity, among other stresses.[1]

RpoS and other bacterial stress response regulators influence virulence and biofilm formation, and most of them may contribute to the development of multidrug resistance, an issue of considerable clinical relevance.[2]

Furthermore, several studies have demonstrated that quinolone-resistant uropathogenic *E. coli* strains have fewer virulence factors than their susceptible counterparts,[3] and quinolones may induce some bacterial stress responses, suggesting pathways evolved by bacteria to economize some molecular resources in order to accomplish a more effective survival after stress response.

Multidrug resistance is frequently associated with transcriptional regulators involved in bacterial stress response, such as *marA*, *soxS*, *sdiA*, etc.[4] Such regulators can activate the overexpression of efflux systems like AcrAB, whose natural function might be the transport of quorum-sensing signals that would trigger *rpoS*.[5] The *in vitro* induction of multidrug resistance in enterobacteria strains has been associated with inducer levels that are kept close to their own minimal inhibitory concentrations, suggesting the involvement of mechanisms for bacterial response to increasing medium osmolarity in the development of multidrug-resistant phenotypes.[6] In fact, several sigma and quorum-sensing regulators, which seem to cooperate in the management of hyperosmotic stress, can promote the increased expression of multidrug transporters in *E. coli*.[1,5]

The most studied adaptive response to a temperature change is the heat-shock response while relatively less is known about the cellular response to the cold, a phenomenon that has been observed in many prokaryotes and eukaryotes. Unlike heat shock, which involves the activity of a sigma factor (σ^{32}) other than the vegetative one, the osmotic stress, and the control of virulence gene expression, *E. coli* cells have evolved a σ^s-independent mechanism to cope with the effects caused by sudden lowering of the environmental temperature. The main features of the cold-shock (CS) response in *E. coli*, which is the better-characterized bacterium from this point of view, will be briefly described here.

The Cold-Shock Response in E. coli
Anna M. Giuliodori & Claudio O. Gualerzi

When an exponential culture of the mesophilic bacterium *E. coli* is transferred from 37°C to a temperature below 20°C, the cells transiently stop growing and enter an acclimation phase in which the synthesis of a small set of CS proteins is induced, whereas that of most of the other gene products is repressed. At the end of the acclimation phase, which can last one to several hours depending on the temperature, the synthesis of CS proteins declines and cellular growth resumes at a rather slow rate.[7] The list of the *E. coli* CS-induced proteins so far identified, mainly by proteomic approaches, is essentially constituted by nucleic acid–binding proteins involved in different cellular processes like RNA degradation, transcription, DNA replication and supercoiling, translation, and ribosome maturation, as well as by five members of the Csp family of *E. coli*.[7] CspA and its homologs are small proteins (65–70 aa) identified in all types of bacteria except for archaea and cyanobacteria. *E. coli* contains nine paralogues belonging to the *csp* family, probably generated as a result of a number of gene duplication events. Among these proteins CspA, CspB, CspE, CspG, and CspI are cold-inducible, while CspC, CspD, CspF, and CspH are not. Notably, the nucleic acid–binding domain of the CS proteins, called cold-shock domain (CSD), is the most evolutionary conserved nucleic acid–binding domain within prokaryotes and eukaryotes.[8] The list of CS proteins also includes two protein chaperones, namely, TF and Hsc66. A recent global transcript profiling of *E. coli* during cold shock has revealed that the level of other transcripts encoding molecular chaperones such as *htpG*, *mopA*, *mopB*, and *ppiA* (encoding HtpG, GroEL, GroES, and peptidyl-prolyl-*cis-trans* isomerase, respectively) transiently increases after cold shock.[9] In addition, the same study has underlined the induction of many genes involved in sugar transport and metabolism, as well as in membrane synthesis and function. Therefore, the overall data seem to suggest that the CS response is intended for (a) dealing with unfavorable secondary structures of nucleic acids induced/stabilized by the cold, which are expected to hinder basic functions such as transcription, ribosome assembly, and translation; (b) opposing the CS-induced decrease in membrane fluidity; (c) accumulating sugars displaying a protective effect against the low temperature, such as trehalose; and (d) helping protein folding at low temperatures.

The mechanisms that regulate the CS gene induction can be either transcriptional or posttranscriptional but, although most frequently both levels are involved, they do not seem to have the same relevance for all CS genes. Two main mechanisms of posttranscriptional regulation have been identified: selective cold-shock–induced stabilization of the transcripts of CS genes and "translational bias" whereby at low temperature translation of the CS mRNAs is preferentially favored while that of the non-CS mRNAs is disfavored. Unlike the heat-shock response, transcriptional induction was found to play a relevant role only for some of the genes under study.[7] Examples of these types of regulatory circuits can be found in the mechanisms underlying the CS induction of *infA* and *cspA*. InfA, encoding the translation initiation factor IF1, is transcribed in *E. coli* from two promoters (P1 and P2) yielding two mRNAs that differ for the length of their 5′-untranslated region (5′UTR). While under normal growth conditions *infA* is predominantly transcribed from P2 promoter, after cold-shock *infA* is *de novo* transcribed preferentially from the otherwise less used P1 promoter. A possible reason for this "promoter preference shift" is that transcription from P1 is stimulated by increasing concentrations of the CS protein CspA, which has essentially no effect on the P2 activity.[10]

As to the expression of *cspA*, its CS induction is mainly controlled at a posttranscriptional level since its mRNA, which is extremely unstable at 37°C, is drastically stabilized immediately after cold shock. Furthermore, the stability of *cspA* mRNA during cold shock is modulated in such a way that the rate of decay of *cspA* mRNA increases again toward the end of the acclimation phase, when the accumulation of the CS proteins is no longer necessary. This mechanism of regulation is not restricted to *cspA* mRNA but is active also for other CS transcripts.[7]

Finally, the CS expression of both *cspA* and *infA* is favored by the above-mentioned translational bias. Recent data indicate that this bias is partly due to intrinsic features of CS mRNAs, which make them prone to translation at low temperature. Preliminary results seem indeed to indicate that *cspA* mRNA undergoes conformational changes whereby the ribosome binding site becomes more accessible at low temperature compared to that at 37°C. Thus, *cspA* mRNA might behave as an RNA thermometer, whose conformation changes in response to temperature.[11] Furthermore, translation of CS templates is selectively stimulated by *trans*-acting factors whose level increases on cold shock. The most important *trans*-acting factor involved in the CS translational bias is IF3, whose level with respect to ribosomes increases during cold shock together with that of the other two initiation factors (IF1 and IF2).[12] IF3 is able to stimulate selectively, at low temperature and in a dose-dependent manner, the translation of CS mRNAs, while translation of non-CS mRNAs displays little or no IF3 dependence.[12] In particular, at low temperature, IF3 stimulates the rate of "30S initiation complexes" formation with CS mRNAs while inducing the formation of nonproductive 70S initiation complexes with non-CS mRNAs.[13] Furthermore, it has been shown that a larger-than-normal amount of IF1 and IF3 is needed at low temperature to overcome the increased tendency of the ribosomal subunits to associate.[10,13] These results suggest a mechanistic model that can explain the translational bias. On lowering the temperature, the formation of initiation complexes with non-CS mRNA becomes rate limiting, thus leading to a transient block of translation. This in turn triggers the CS induction of PY. This protein, whose likely function is that of "storing" the ribosomes as 70S monomers when the cell requires a reduced number of translating ribosomes would sequester a large proportion of 70S deriving from polysome dissociation in a functionally inactive state.[14] The transient increase of the IFs level would counteract the tendency of the few pY-free 70S monomers to remain stably associated, thus maintaining a pool of free subunits capable of initiating the translation of CS mRNAs that are accumulating in the cells owing to their increased stability. Furthermore, IF3 might reduce the translation of non CS mRNAs. Finally, the synthesis of RNA chaperones, helicases, and protein chaperones would help the cell adapt to the low temperature and to resume growth at the end of the acclimation phase.

Quinolones and Loss of Virulence Factors in E. coli
Sara Soto & Jordi Vila

It is well known that resistance to quinolones in *E. coli* is an increasing problem in Spain and other countries.[15] Quinolone-resistant uropathogenic *E. coli* strains express fewer virulence factors than quinolone-susceptible strains,[16] and this phenomenon might be particularly frequent among strains of the B2 phylotype.[17] Among the virulence factors, the hemolysin and cytotoxic necrotizing factor were more frequently found in quinolone-susceptible than in quinolone-resistant *E. coli* isolates.[16] These virulence factors are located in one or two pathogenicity islands, depending on the strain. Herein, we found that quinolones increase the frequency of loss of some pathogenicity island (PAI) in phylotype B2 *E. coli* strains. The loss of PAI-I was observed, whereas the loss of PAI-IV was not. Spontaneous loss of hemolytic capacity was not observed in any of the three tested strains in the absence of quinolone exposure.

Further studies are necessary to elucidate the mechanism involved in the induction of the loss of PAIs by quinolones. The role of the SOS system was analyzed because it is well known that quinolones induce SOS system response (DNA repair mechanism) and this response could favor the splitting of bacteriophages or related sequences from the chromosome.[18] In this regard, Shaikh and Tarr demonstrated that quinolones promote the excision of stx_2 bacteriophages as well as complete and truncated stx_1 bacteriophages.[19] Nevertheless, in our study when *recA* was truncated, the knock-out mutant did not lose the PAI in presence of quinolones.

These findings suggest that quinolones induce the *in vitro* loss of pathogenicity islands in uropathogenic *E. coli* strains when they are under the effect of subinhibitory concentrations of these antimicrobial agents. This loss can be partial or total depending on each strain, and, in some cases, the SOS system can be implicated. Nevertheless, the role of stress response regulators in response to quinolones, resulting in loss of virulence factors, must not be overlooked. In this sense, quinolones induce not only SOS system response but also heat-shock response.[20] Moreover, envelope stress regulators influence virulence factor expression.[21] Otherwise, the deletion processes may also play a role in the adaptation of uropathogenic *E. coli* strains during certain stages of infection. Therefore, genetic flexibility of pathogenic microbes may create selective advantages over other, less flexible organisms, and may finally result in proper replication in host organisms or other ecological niches. Although quinolone resistance is not necessary to lose PAIs, quinolone-resistant *E. coli* strains have likely been in previous contact with these antimicrobial agents, thus favoring the loss of the PAI.

Osmotic Stress and Other Stressors Inducing Multidrug Resistance
María M. Tavío

Intrinsic multiple-antibiotic resistance in bacteria is a clinically important problem, since it plays the role of a fundamental stepping stone to higher levels of resistance in clinical isolates. In this type of resistance, reduced penetration of antibiotics as well as an increase in active efflux results in low-level, broad resistance to distinct classes of antimicrobial agents.[22]

Multidrug resistance in *E. coli* has frequently been linked to the expression of transcriptional activators such as MarA, SoxS, or Rob, which belong to the XylS-AraC class of positive transcriptional regulators.[4] Other transcriptional regulators may also induce efflux pump expression, as in the case with SdiA, which is homologous to the LuxR family of quorum-sensing transcription factors, and whose amplification has a global impact on several functions of bacterial cell, such as cell septation (*ftsQAZ* cell division genes).[23] This suppressor of division inhibition (SdiA) regulates cell division in a cell density–dependent or quorum-sensing manner, in which AcrAB seems to have the natural function of pumping out quorum-sensing signals that do not diffuse easily on their own. AcrAB is an *E. coli* multidrug efflux transporter that allows passage of fluoroquinolones. Interestingly, a 4-quinolone (2-heptyl-3-hydroxy-4-quinolone) is produced by *Pseudomonas aeruginosa* and functions as a quorum-sensing signal.[23] Indeed, sublethal concentrations of fluoroquinolones and other antimicrobials may induce both multidrug-resistant phenotypes and quorum-sensing regulator *lux* genes.[5] Nevertheless, *in vitro* nonantimicrobial chemotherapeutic agents such as diazepam and anti-inflammatory drugs, among others, also induce the multidrug resistance phenotypes in enterobacteria. The inducing effect of such drugs is dose dependent in such a way that as they near the MIC, there is an increased cyclohexane tolerance, decreased susceptibility to distinct antimicrobials and loss of OmpF in the bacterial cell.[6,22]

Taking into account that osmotic stress due to increased osmotic pressure outside the lipid bilayer surrounding the bacterial cell depends on the concentration of the solute but not on its identity, a possible involvement of hyperosmotic stress response regulators in the induction of multidrug resistance phenotypes must not be overlooked. In this sense, high osmolarity of the medium triggers the expression of regulators like *rpoS* or *cpxR* that, in turn, activate efflux pump expression. Although, *tolC* upregulation (the outer-membrane channel that joins with the AcrAB or AcrEF efflux systems) is responsible for increased cyclohexane tolerance in *E. coli*, this still has not been associated with *rpoS* or *cpxR*. However, TolC as well as AcrAB and AcrEF efflux systems are overexpressed when *sdiA* upregulation is induced.[24] Recently, we have found *sdiA* overexpression in *E. coli* multidrug-resistant mutants *in vitro* selected with suprainhibitory concentrations of fluoroquinolones, ceftazidime, or diazepam.[4] In *E. coli,* the amplification of *sdiA*, a positive activator of *ftsQAZ* genes that are essential for septation, results in mitomycin C resistance as well as in a global impact on gene expression, including increased transcript levels of components of efflux systems such as *acrD*, *acrE*, *acrF*, *acrA*, *acrB*, and *tolC*.[24] In fact, multiple data suggest that regulation of multidrug transporters has a strong relationship to (i) stress responses, (ii) the growth phase, and (iii) quorum sensing.

E. coli DNA mycroarrays have not identified *micF* or *ompF* as genes under *sdiA* regulation.[24] On the other hand, both *ompC* and *ompF* genes are members of the Cpx regulon, and CpxR activation results in a strong decrease in *ompF* expression (not requiring *micF*) and an increase in *ompC* expression, and it also leads to an increased level of mutation owing at least partly to direct repression of the *ung* DNA repair gene.[25] Perhaps suprainhibitory concentrations of antimicrobials such as fluoroquinolones or ceftazidime, and chemicals like salicylate or diazepam, might activate the Cpx regulon, resulting in *ung* gene repression mediated by CpxR and consequently favoring *in vitro* selection of multidrug-resistant mutants.[4] Otherwise, the sigma factors σ^S and σ^E (a cooperator of σ^S in the management of hyperosmotic stress) are also involved in the regulation of porin expression in *E. coli*, although activation of σ^E results in decreased expression of both *ompF* and *ompC*.[26]

Therefore, a complex network of bacterial stress regulators would cooperate during bacterial response to an increase in medium osmolarity caused by antimicrobial or nonantimicrobial agents, resulting in induced multidrug-resistant phenotypes. Trials aimed to elucidate the role of the bacterial stress response regulatory systems in the changes associated with the development of intrinsic multiple-antimicrobial resistance might be a good starting point for a future better understanding of the concerning problem of multidrug resistance in clinical or environmental bacterial isolates.

CONCLUSIONS AND PERSPECTIVES

Transcriptional and Posttranscriptional Control of Cold-Shock Gene Expression

In spite of the failure in identifying the actual signal triggering the induction of the CS stimulon, and of the large number of questions that remain open, the study of the CS response, probably one of the most complex adaptive responses known in bacteria, has shed light on novel mechanisms of gene regulation and produced results of considerable interest for their potential applications, mainly allowing the expression of recombinant proteins under conditions limiting the danger of the formation of inclusion bodies.

Quinolones Induce the Loss of Virulence Factors in E. coli

Subinhibitory concentrations of quinolones might *in vitro* induce a bacterial stress response, resulting in the loss of virulence factors located in pathogenicity islands in uropathogenic *E. coli* strains.

Multidrug Resistance Induced by an Increasing Medium Osmolarity

The close relationship between high concentrations of inducers and selection or induction of multidrug-resistant strains suggests the involvement of hyperosmotic stress response regulators in such resistant phenotypes.

The role of such regulators in the appearance of multidrug-resistant strains in clinical isolates or in the environment, and the pathway by which they integrate into the network of stress regulator systems, may be helpful in thoroughly understanding how antimicrobial resistance occurs.

ACKNOWLEDGMENTS

The cold-shock response in *E. coli*: The costs of this investigation were partially supported by Italian MIUR grants (PRIN 2005 to Cynthia L. Pon and PRIN 2005 to COG). Quinolones and loss of virulence factors in uropathogenic *E. coli*: This research was supported by work contract from National Health System CP05/00140. Osmotic stress and other stressors as inducers of multidrug resistance: This work was partially supported by subsidy 2002/199 awarded by Canary Government.

REFERENCES

1. HENGGE-ARONIS, R. 2002. Signal transduction and regulatory mechanisms involved in control of the σ^S (RpoS) subunit of RNA polymerase. Microbiol. Mol. Biol. Rev. **66**: 373–395.
2. HENGGE-ARONIS, R. 2000. A role for the σ^S subunit of RNA polymerase in the regulation of bacterial virulence. Adv. Exp. Med. Biol. **485**: 85–93.
3. SOTO, S.M., M.T. JIMÉNEZ DEANTA & J. VILA. 2006. Quinolones induce partial or total loss of pathogenicity islands in uropathogenic *Escherichia coli* by SOS-dependent or -independent pathways, respectively. Antimicrob. Agents Chemother. **50**: 649–653.
4. TAVÍO, M.M., V.D. AQUILI, J. SÁNCHEZ, *et al.* Main role of *sdiA*, *tolC* and *ftsI* genes in increased resistance to broad-spectrum cephalosporins in two *Escherichia coli* mutants. Submitted for publication.
5. YANG, S., C.R. LÓPEZ & E.L. ZECHIEDRICH. 2006. Quorum sensing and multidrug transporters in *Escherichia coli*. Proc. Natl. Acad. Sci. USA **103**: 2386–2391.

6. TAVÍO, M.M., V.D. AQUILI, L. MACIÁ, et al. Four anti-inflammatory drugs as well as sodium chloride induce cyclohexane tolerance, enhanced active efflux, repression of OmpF synthesis and development of Mar phenotype in a Escherichia coli AG100 strain. Submitted for publication.
7. GUALERZI, C.O., A.M. GIULIODORI & C.L. PON. 2003. Transcriptional and post-transcriptional control of cold-shock genes. J. Mol. Biol. **331:** 527–539.
8. YAMANAKA, K., L. FANG & M. INOUYE. 1998. The CspA family in Escherichia coli: multiple gene duplication for the stress adaptation. Mol. Microbiol. **27:** 247–255.
9. PHADTARE, S. & M. INOUYE. 2004. Genome-wide transcriptional analysis of the cold shock response in wild-type and cold-sensitive, quadruple-*csp*-deletion strains of Escherichia coli. J. Bacteriol. **186:** 7007–7014.
10. GIANGROSSI, M., A. BRANDI, A.M. GIULIODORI, et al. 2007. Cold-shock-induced *de novo* transcription and translation of *infA* and role of IF1 during cold adaptation. Mol. Microbiol. **64:** 807–821.
11. NARBERHAUS, F., T. WALDMINGHAUS & S. CHOWDHURY. 2006. RNA thermometers. FEMS Microbiol. Rev. **30:** 3–16.
12. GIULIODORI, A.M., A. BRANDI, C.O. GUALERZI, et al. 2004. Preferential translation of cold-shock mRNAs during cold adaptation. RNA **10:** 265–276.
13. GIULIODORI, A.M., M. GIANGROSSI, A. BRANDI, et al. 2007. Cold-stress-induced *de novo* expression of *infC* and role of IF3 in cold-shock translational bias. RNA In press.
14. VILA-SANJURJO, A., B.S. SCHUWIRTH, C.W. HAU, et al. 2004. Structural basis for the control of translation initiation during stress. Nat. Struct. Mol. Biol. **11:** 1054–1059.
15. MARTÍNEZ-MARTÍNEZ, L., F. FERNÁNDEZ, & E.J. PEREA. 1999. Relationship between haemolysis production and resistance to fluoroquinolones among clinical isolates of Escherichia coli. J. Antimicrob. Chemother. **43:** 277–279.
16. VILA, J., K. SIMON, J. RUIZ, et al. 2002. Are quinolone-resistant uropathogenic Escherichia coli less virulent? J. Infect. Dis. **186:** 1039–1042.
17. HORCAJADA, J.P., S. SOTO, A. GAJEWSKI, et al. 2005. Quinolone resistant uropathogenic Escherichia coli from phylogenetic group B2 have fewer virulence factors than their susceptible counterparts. J. Clin. Microbiol. **43:** 2962–2964.
18. PHILLIPS, I., E. CULEBRAS, F. MORENO, et al. 1987. Induction of the SOS response by new 4-quinolones. J. Antimicrob. Chemother. **20:** 631–638.
19. SHAIKH, N. & P.I. TARR. 2003. Escherichia coli O157:H7 shiga toxin-encoding bacteriophages: integrations, excisions, truncations, and evolutionary implications. J. Bacteriol. **185:** 3596–3605.
20. VANBOGELEN, R.A., P.M. KELLEY & F.C. NEIDHARDT. 1987. Differential induction of heat shock, SOS, and oxidation stress regulons and accumulation of nucleotides in Escherichia coli. J. Bacteriol. **169:** 26–32.
21. RAIVIO, T.L. 2005. Envelope stress responses and Gram-negative bacterial pathogenesis. Mol. Microbiol. **56:** 1119–1128.
22. TAVÍO, M.M., J. VILA, M. PERRILLI, et al. 2004. Enhanced active efflux, repression of porin synthesis and development of Mar phenotype by diazepam in two enterobacteria strains. J. Med. Microbiol. **53:** 1119–1122.
23. RAHMATI, S., S. YANG, A.L. DAVIDSON, et al. 2002. Control of the AcrAB multidrug efflux pump by quorum-sensing regulator SdiA. Mol. Microbiol. **43:** 677–685.

24. WEI, Y., J.-M. LEE, D.R. SMULSKI, *et al.* 2001. Global impact of *sdiA* amplification revealed by comprehensive gene expression profiling of *Escherichia coli*. J. Bacteriol. **183:** 2265–2272.
25. DOREL, C., P. LEUJENE & A. RODRIGUE. 2006. The Cpx system of *Escherichia coli*, a strategic signaling pathway for confronting adverse conditions and for settling biofilm communities? Res. Microbiol. **157:** 306–314.
26. BATCHELOR, E., D. WALTHERS, L.J. KENNEY, *et al.* 2005. The *Escherichia coli* CpxA-CpxR envelope stress response system regulates the expression of the porins OmpF and OmpC. J. Bacteriol. **187:** 5723–5731.

Variation in Stress Responses within a Bacterial Species and the Indirect Costs of Stress Resistance

THOMAS FERENCI AND BENY SPIRA

School of Molecular and Microbial Biosciences, The University of Sydney, NSW 2006, Australia

ABSTRACT: Bacteria can exhibit high levels of resistance to one or more environmental stresses such as temperature, osmolarity, radiation, pH, starvation, as well as resistance to noxious chemicals and antibiotics. Yet evolution has not optimized stress resistance in all bacteria to all stresses. Even within a species like *Escherichia coli*, stress resistance is not constant between strains, suggesting that selection for stress resistance is under counterselection in some environments. The tradeoffs associated with stress resistance in *E. coli* are due to more than the direct cost of resistance mechanisms. A significant indirect cost is that high stress resistance is associated with a reduced ability to compete for poor growth substrates like acetate or even good substrates like glucose at suboptimal concentrations. High stress resistance also decreases the ability to use inorganic nutrients like phosphate. This tradeoff between self-preservation and nutritional competence, called the SPANC balance, is likely to be the major selective influence in natural populations. Another cost of high stress resistance in *E. coli* is an elevated mutation rate and the increased generation of deleterious mutations. Directional adaptations in SPANC balance and mutation rate are environment-dependent. The most common variations in SPANC are due to polymorphisms in the levels of global regulators RpoS and ppGpp between different strains. High levels favor stress resistance, and low levels allow better nutrition. The intimate association of RpoS/ppGpp with stress resistance and SPANC balancing influences numerous cellular processes and bacterial properties, including virulence.

KEYWORDS: stress responses; sigma factor RpoS; tradeoffs in bacterial regulation; adaptive mutations; stringent response; ppGpp

INTRODUCTION

Although it is tempting to think that bacteria are optimally adapted after billions of years of evolution, some features of bacteria are subject to constant

Address for correspondence: T. Ferenci, School of Molecular and Microbial Biosciences G08, The University of Sydney, NSW 2006, Australia. Voice: (61 2) 93514277; fax: 61 2 9351 4571.
tferenci@mail.usyd.edu.au

and conflicting challenges. The bacterial lifestyle is subject to pressures in environments that are not under bacterial control. Homeostasis is therefore important to free-living bacteria, and this discussion considers the selection pressures impacting on bacterial stress responses. Of particular interest is that different strains of bacteria like *Escherichia coli* exhibit a range of stress-resistance properties.[1,2] Why the variation?

Most environmental stresses such as nonoptimal temperature, osmolarity, oxygen radicals, radiation, pH, and starvation elicit a general stress response in *E. coli*, controlled by two central regulators that integrate multiple signals. One regulator is the sigma factor RpoS or σ^S,[3] and the other is the alarmone ppGpp.[4,5] The concentration of both these controllers depends on growth rate and other stress signals, as considered below. High cellular concentrations of the two regulators are seen at greatly suboptimal growth rates, at doubling times of approximately 7 h or more.[6,7]

Recently, it has become apparent that the stress resistance exhibited by bacteria is subject to a considerable cost beyond the synthesis of stress protectants. The downside of high stress resistance through RpoS is a reduction in the capacity of *E. coli* to compete vegetatively in environments containing nutrients at low concentrations.[8] In other words, there is a wide-ranging tradeoff between the processes of self-preservation and nutritional competence, called the SPANC balance. A consequence of this tradeoff is that different environments with particular stress levels and nutrient concentrations impose different selective pressures on RpoS and possibly ppGpp. High levels of σ^S can contribute to high stress resistance, as shown in FIGURE 1A, but with the reduced ability of the resistant strain to grow vegetatively.[9] Elevated σ^S not only hinders growth with carbon sources,[9] but also the acquisition of inorganic nutrients like phosphate (B. Spira and T. Ferenci, unpublished data). As shown in FIGURE 1B, different isolates of *E. coli* overcome this cost by mutational modifications in *rpoS* structure or expression, resulting in reduced stress resistance behavior. As discussed in detail for RpoS (σ^S) elsewhere,[8,10] bacteria switching environments with different SPANC balances accounts for much of the frequent polymorphisms in the *rpoS* gene in the species *E. coli* and *Salmonella enterica*. Here we focus on the second global stress regulator, ppGpp, and discuss recent findings indicating that ppGpp is also subject to diversifying selection in both laboratory and natural environments.

THE ROLE OF ppGpp IN *E. coli*

ppGpp is a cellular alarmone that accumulates in bacteria undergoing nutritional stress, such as amino acid, carbon, or phosphate limitation.[11–13] Accumulation of ppGpp triggers the "stringent response," in which a radical decrease in ribosome synthesis results in a general inhibition in protein synthesis and growth arrest.[4]

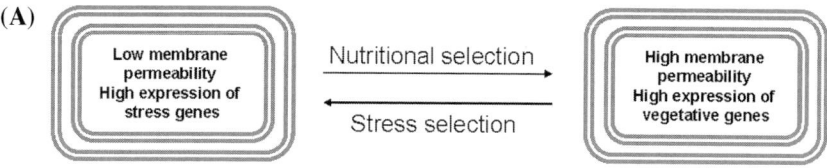

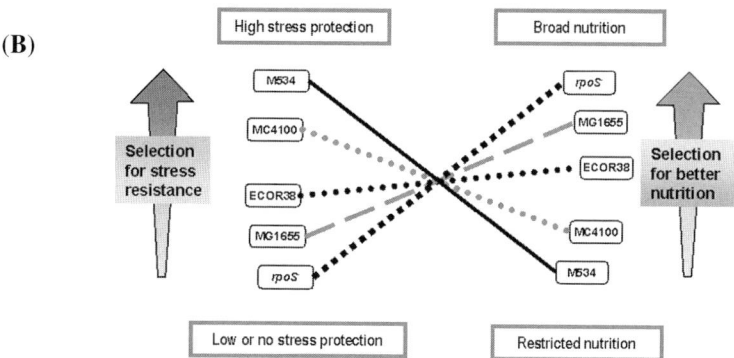

FIGURE 1. In **(A)**, the differences in properties of *E. coli* that contribute to either high resistance to stress or fitness in vegetative growth with low nutrient concentrations or poor carbon sources are shown. *E. coli* strains with different stress-resistance and nutritional properties are commonly found,[9] and the examples of strains in **(B)** show that M534, which is highly resistant to acid or starvation, can metabolize 20 less carbon sources than MG1655, which is less resistant to stresses. The differences are mainly due to distinct endogenous levels of RpoS in strains of *E. coli*.[8,23]

ppGpp is synthesised by two proteins, encoded by the genes *relA* and *spoT*. Synthesis of ppGpp under amino acid starvation is totally dependent on *relA*. The RelA protein is bound to the ribosomes, where it senses the presence of uncharged tRNAs that trigger the RelA-dependent synthesis of ppGpp. *relA* mutants display a relaxed phenotype in response to amino acid limitation, which is characterized mainly by a further decrease in ppGpp concentration and the continuous production of stable RNA.[4] SpoT is a cytosolic bifunctional enzyme, which is responsible for both ppGpp synthesis and degradation. Under nutritional stresses other than amino acid limitation, accumulation of ppGpp is largely *relA*-independent, as the augment in ppGpp concentration is mainly due to inhibition of the SpoT hydrolytic activity.[14]

The level of ppGpp is also important in the bacterial response to stress, because it helps to control the amount of σ^S in the cell. ppGpp is involved in a severalfold increase in the cellular concentration of σ^S during nutritional stress or upon entry in the stationary phase. The absence of ppGpp impairs or severely delays the accumulation of σ^S.[15] ppGpp positively affects the efficiency of *rpoS* translation in the stationary phase and under stress conditions as well as *rpoS* basal expression under conditions of optimal growth.[16,17] Because ppGpp is needed to obtain normal levels of *rpoS* expression, and because high levels of σ^S impair bacteria nutritional competence,[9] it is expected that ppGpp would also affect the SPANC equilibrium by modulating the levels of σ^S in the cell. ppGpp is also likely to have a direct effect on bacteria nutritional ability, as it binds directly to RNA polymerase and affects either positively or negatively the expression of many genes.[4,18] However, the possibility that ppGpp may affect the SPANC equilibrium has not been hitherto seriously contemplated.

POLYMORPHISM OF ppGpp AND SPANC IN LABORATORY STRAINS OF *E. coli*

The *E. coli* lineage used in this laboratory as a model for the study of SPANC equilibrium is the widely used MC4100 strain.[19] Its carries a *relA1* mutation and, compared to other *E. coli* strains, it was found to express a high level of σ^S under nutrient limitation.[9] Despite its *relA1* mutation, we found that MC4100 produces as much ppGpp as the *relA*$^+$ wild-type strain MG1655. The sequence of the MC4100 *relA* gene was as expected for a *relA1* mutant, but the sequence of its *spoT* gene revealed two separate mutations in its ORF. One mutation is an insertion of a Gln-Asp dipeptide at position 84 of the protein and another mutation is the substitution of a histidine by a tyrosine residue at position 255.

To test if the mutations in *spoT* are responsible for the relatively high ppGpp and RpoS content of MC4100, the *spoT* gene of MC4100 was substituted by a wild-type *spoT*$^+$. The resulting strain was more sensitive to amino-triazole (AT), a histidine analog[20] which causes growth inhibition that may be relieved by increasing the level of ppGpp, suggesting that the mutated SpoT did contribute to elevate ppGpp in MC4100. In line with the role of ppGpp in RpoS expression, the phenotypes of MC4100 dependent on the status of *rpoS*, such as glycogen production and the inability to grow on the poor carbon substrate acetate,[9] were also altered by the allele of *spoT* present in MC4100 (FIG. 2).

In steady-state growth under glucose limitation, strain MC4100 tends to accumulate mutations in *rpoS*, as a way of relieving the repression of genes related to growth.[21] Strains such as MG1655 that express low levels of *rpoS* do not accumulate mutations in this gene. MC4100 *spoT*$^+$ still accumulates mutations in *rpoS* under glucose limitation, but to a lesser degree than its parent. For instance, after 48 h of growth under glucose limitation, about 70% of MC4100 colonies were *rpoS* mutants, while in its *spoT*$^+$ derivative only 1% of the progeny were *rpoS*$^-$. Given that the rate of *rpoS* loss is proportional to the intrinsic level of σ^S in the cell,[9] this is also consistent with the conclusion that MC4100 *spoT*$^+$ also produces less σ^S than its parent. Altogether, the

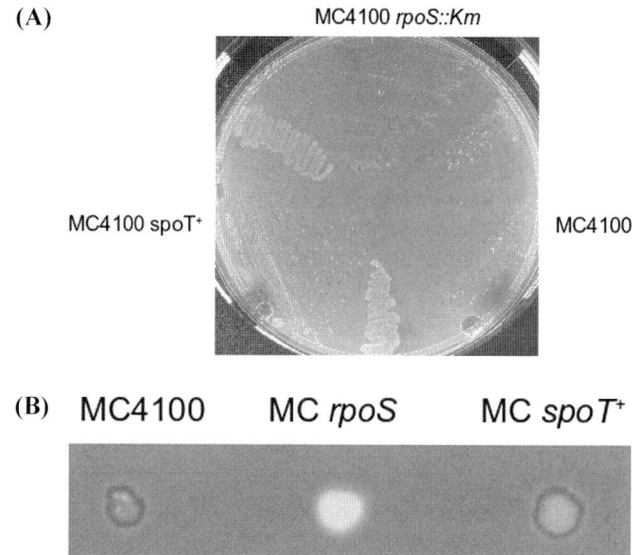

FIGURE 2. *rpoS*-related phenotypes of MC4100 and its *spoT*$^+$ derivative. **(A)** Growth on an acetate minimal medium plate is shown after 3 days at 37°C. MC4100 growth is slow in patches, indicative of growth differences due to RpoS levels in bacteria.[9] The individual colonies in the MC4100 sector (right) are *rpoS* mutants; the *rpoS* knockout (top) and the spoT replacement strains (left) are shown. **(B)** Iodine staining of bacterial patches, indicative of glycogen production; the darkness of staining is proportional to RpoS levels.[21]

evidence indicates that MC4100 *spoT*$^+$ accumulates less ppGpp than the MC4100 previously used in our laboratory. The decrease in ppGpp shifted the SPANC equilibrium by increasing the nutritional competence of the strain in utilizing an alternative carbon source such as acetate and in expressing lower amounts of σ^S, as indicated by the decrease in glycogen production and reduced level of *rpoS* loss in a chemostat.

Our unpublished data suggest that not all MC4100 lab lineages contain the double mutation in *spoT*. As was already documented for *rpoS* polymorphisms,[22] laboratory storage conditions can result in divergence between strains kept in different laboratories or over extended periods. Inadvertent selection for better growth or high levels of stress (as in thawing frozen stocks) can select changes in the SPANC balance. In general though, both RelA and SpoT have very conserved protein sequences. A BLAST search showed that 21 of the 29 SpoT proteins of *E. coli* and *Shigella* strains in the sequence database are 100% identical. The other eight strains have one or two amino acid substitutions and only one K-12 derivative in the database carries a Gln-Asp insertion at position 84, the same insertion that is present in MC4100. As for RelA, 24 of 28 strains are 100% identical, while the other four have only single amino acid substitution.

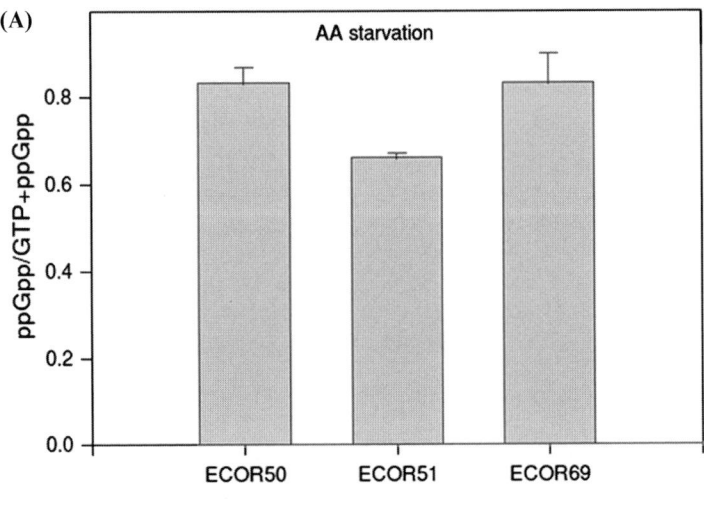
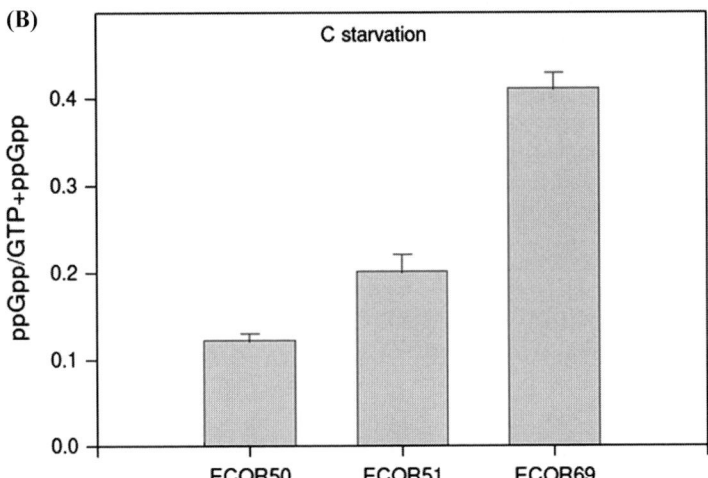

FIGURE 3. ppGpp level of three strains from the ECOR collection of diverse *E. coli* isolates.[24] Bacteria grown in minimal medium were labeled with ^{32}P for two generations and then treated for 30 min with either **(A)** 1% serine hydroxamate (amino acid starvation[25]) or **(B)** 2% α-methyl-glucoside (carbon starvation[26]). After separation by thin-layer chromatography, the ppGpp as a proportion of total guanosine phosphate was estimated by phosphorimaging in duplicate extracts.

POLYMORPHISM OF ppGpp AND SPANC IN NATURAL ISOLATES OF *E. coli*

So how much SPANC selection pressure is ppGpp under in nature and how variable is ppGpp in natural isolates? We found a significant variability in

SpoT-dependent ppGpp level under carbon limitation among *E. coli* natural isolates, but almost no differences in ppGpp accumulation due to RelA under amino acid starvation (FIG. 3). As shown for three examples of ECOR isolates from different sources, there was a nearly fourfold range of ppGpp levels in response to carbon starvation under identical conditions. However, gene sequence analysis of some of these strains did not reveal significant polymorphism in the *spoT* gene. This suggests that differences in ppGpp levels between natural isolates are probably due to polymorphism in extragenic regulatory genes or stress signal processing, rather than in *spoT* itself. Whatever the mechanism behind ppGpp variation, it seems that in natural environments, intrinsic cellular concentrations of ppGpp are under diversifying selective pressures, probably as a result of changes to the SPANC equilibrium.

CONCLUSIONS

A possible source of variation in stress resistance within the species *E. coli* is the intrinsic heterogeneity in ppGpp as well as RpoS levels. We find that ppGpp concentrations are nonconstant under identical conditions in different strains of *E. coli*. Both laboratory strains and natural isolates are inherently changeable, probably because of the constant SPANC selection pressure toward resistance or better nutrition. Differences in ppGpp influence growth phenotypes on poor carbon sources, so the SPANC effects initially ascribed to RpoS are at least partially ppGpp-dependent. A more accurate assessment is that both ppGpp and RpoS contribute to the particular SPANC setting in a particular strain or isolate. This finding has major consequences, because RpoS and ppGpp influence numerous cellular processes and bacterial properties, including virulence and stress resistance in pathogenic strains.

ACKNOWLEDGMENTS

This work was supported by Fundação de Amparo a Pesquisa do Estado de São Paulo (FAPESP- Brazil) and an Endeavour Research Fellowship (Australia).

REFERENCES

1. BENITO, A. *et al.* 1999. Variation in resistance of natural isolates of *Escherichia coli* O157 to high hydrostatic pressure, mild heat, and other stresses. Appl. Environ. Microbiol. **65:** 1564–1569.
2. WHITING, R.C. & M.H. GOLDEN. 2002. Variation among *Escherichia coli* O157:H7 strains relative to their growth, survival, thermal inactivation, and toxin production in broth. Int. J. Food Microbiol. **75:** 127–133.

3. LOEWEN, P.C. & R. HENGGE-ARONIS. 1994. The role of the sigma factor sigma(S) (KatF) in bacterial global regulation. Annu. Rev. Microbiol. **48:** 53–80.
4. CASHEL, M. *et al.* 1996. The stringent response. *In Escherichia coli* and Salmonella: Cellular and Molecular Biology. F.C. Neidhardt, Ed.: 1458–1496. ASM. Washington, DC.
5. MAGNUSSON, L.U., A. FAREWELL & T. NYSTROM. 2005. ppGpp: a global regulator in *Escherichia coli*. Trends Microbiol. **13:** 236–242.
6. NOTLEY, L. & T. FERENCI. 1996. Induction of RpoS-dependent functions in glucose-limited continuous culture: what level of nutrient limitation induces the stationary phase of *Escherichia coli*? J. Bacteriol. **178:** 1465–1468.
7. TEICH, A. *et al.* 1999. Growth rate related concentration changes of the starvation response regulators sigma(S) and ppGpp in glucose-limited fed-batch and continuous cultures of *Escherichia coli*. Biotechnol. Prog. **15:** 123–129.
8. FERENCI, T. 2005. Maintaining a healthy SPANC balance through regulatory and mutational adaptation. Mol. Microbiol. **57:** 1–8.
9. KING, T. *et al.* 2004. A regulatory trade-off as a source of strain variation in the species *Escherichia coli*. J. Bacteriol. **186:** 5614–5620.
10. FERENCI, T. 2003. What is driving the acquisition of *mutS* and *rpoS* polymorphisms in *Escherichia coli* ? Trends Microbiol. **11:** 457–461.
11. CASHEL, M. & J. GALLANT. 1969. Two compounds implicated in the function of the RC gene of *Escherichia coli*. Nature **221:** 838–841.
12. LAZZARINI, R.A., M. CASHEL & J. GALLANT. 1971. On the regulation of guanosine tetraphosphate levels in stringent and relaxed strains of *Escherichia coli*. J. Biol. Chem. **246:** 4381–4385.
13. SPIRA, B., N. SILBERSTEIN & E. YAGIL. 1995. Guanosine 3′,5′-bispyrophosphate (ppGpp) synthesis in cells of *Escherichia coli* starved for Pi. J. Bacteriol. **177:** 4053–4058.
14. MURRAY, K.D. & H. BREMER. 1996. Control of SpoT-dependent ppGpp synthesis and degradation in *Escherichia coli*. J. Mol. Biol. **259:** 41–57.
15. GENTRY, D.R. *et al.* 1993. Synthesis of the stationary-phase sigma factor sigma s is positively regulated by ppGpp. J. Bacteriol. **175:** 7982–7989.
16. BROWN, L. *et al.* 2002. DksA affects ppGpp induction of RpoS at a translational level. J. Bacteriol. **184:** 4455–4465.
17. HIRSCH, M. & T. ELLIOTT. 2002. Role of ppGpp in *rpoS* stationary-phase regulation in *Escherichia coli*. J. Bacteriol. **184:** 5077–5087.
18. REDDY, P.S., A. RAGHAVAN & D. CHATTERJI. 1995. Evidence for a ppGpp-binding site on *Escherichia coli* RNA polymerase: proximity relationship with the rifampicin-binding domain. Mol. Microbiol. **15:** 255–265.
19. CASABADAN, M.J. 1976. Transposition and fusion of the *lac* genes to selected promoters in *Escherichia coli* using bacteriophage Lambda and Mu. J. Mol. Biol. **104:** 541–555.
20. RUDD, K.E. *et al.* 1985. Mutations in the *spoT* gene of *Salmonella typhimurium*: effects on his operon expression. J. Bacteriol. **163:** 534–542.
21. NOTLEY-MCROBB, L., T. KING & T. FERENCI. 2002. *rpoS* mutations and loss of general stress resistance in *Escherichia coli* populations as a consequence of conflict between competing stress responses. J. Bacteriol. **184:** 806–811.
22. ATLUNG, T., H.V. NIELSEN & F.G. HANSEN. 2002. Characterisation of the allelic variation in the *rpoS* gene in thirteen K12 and six other non-pathogenic *Escherichia coli* strains. Mol. Genet. Genom. **266:** 873–881.

23. BHAGWAT, A.A. *et al.* 2006. Functional heterogeneity of RpoS in stress tolerance of enterohemorrhagic *Escherichia coli* strains. Appl. Environ. Microbiol. **72:** 4978–4986.
24. OCHMAN, H. & R.K. SELANDER. 1984. Standard reference strains of *Escherichia coli* from natural populations. J. Bacteriol. **157:** 690–693.
25. METZGER, S. *et al.* 1989. Characterization of the *relA1* mutation and a comparison of *relA1* with new *relA* null alleles in *Escherichia coli*. J. Biol. Chem. **264:** 21146–21152.
26. GENTRY, D.R. & M. CASHEL. 1996. Mutational analysis of the *Escherichia coli spoT* gene identifies distinct but overlapping regions involved in ppGpp synthesis and degradation. Mol. Microbiol. **19:** 1373–1384.

Molecular Mechanisms of Light Stress of Photosynthesis

IMRE VASS, KRISZTIÁN CSER, AND OTILIA CHEREGI

Institute of Plant Biology, Biological Research Center, Hungarian Academy of Sciences, Szeged, Temesvári krt., Hungary

ABSTRACT: Photosynthesis is the basic energy conversion process on Earth, which makes possible the utilization of the energy of sunlight for living organisms. However, light is not only the basic driving force of photosynthesis, but also an important stress factor at the same time. Light-induced decline of photosynthetic activity, generally denoted as photoinhibition, is a general phenomenon in all oxygenic photosynthetic organism under conditions when the metabolic processes cannot keep up with the electron flow produced by the primary photoreactions. Although light-induced damage occurs in all pigmented photosynthetic complexes the primary site of photoinhibition is the photosystem II (PSII) complex, which performs light-driven oxidation of water to protons and oxygen. The main factors, which are responsible for the light sensitivity of photosystem II, are excited pigment molecules, oxygen, manganese, as well as electron donors with high-oxidizing potential. Photosystem II can be efficiently protected from photodamage by the combination of harmless dissipation of absorbed light energy, nonradiative charge recombination, and repair of damaged reaction center complexes, making possible the safe utilization of light, the highly energetic substrate of photosynthesis.

KEYWORDS: light stress; photosynthesis; reactive oxygen species

INTRODUCTION

Photosynthesis is driven by light, a highly energetic and potentially dangerous substrate, which can affect many components of the photosynthetic apparatus and induce secondary destructive processes. Therefore, light represents an important abiotic stress factor for plants, algae, and cyanobacteria whose survival depends on their ability to adapt to the ever-changing light conditions. Light-induced decline of photosynthetic activity, termed as photoinhibition affects mainly the photosystem II (PSII) complex, which performs light-driven oxidation of water, although photosystem I (PSI), which performs

Address for correspondence: Imre Vass, Institute of Plant Biology, Biological Research Center, 6726 Szeged, Temesvári krt. 62, Hungary. Voice: +36-62-599-700; fax: +36-62-433-434.
imre@brc.hu

the reduction of NADP, can also be damaged by light under low temperature conditions.[1–4] One of the important products of photosynthetic electron transport is oxygen produced by PSII, which can form highly toxic species when interacts with excited pigments or redox components. A further problem is created by the very high oxidation potentials of PSII donor components, reaching 1.25 V,[5] which can induce oxidative damage in their protein environment. Different aspects of photoinhibition have been covered by extensive reviews.[6–10] Here we summarize the progress that took place during the last few years in our understanding of the molecular mechanisms of light-induced damage of the photosynthetic apparatus.

THE PS II COMPLEX

PSII catalyses the light-driven oxidation of water, evolving one molecule of O_2 and four protons per two H_2O molecules oxidized. The PSII complex, which is embedded in the thylakoid membrane, is consisted of more than 20 protein subunits, about 40 chlorophyll molecules, as well as protein-bound plastoquinone, Mn, Ca, Cl, and Fe, which act as redox cofactors of light-driven electron transfer.[11,12] The reaction center of PSII is made up by the heterodimer of the D1 and D2 proteins, that bind or contain the redox cofactors involved in light-induced electron transport (FIG. 1). The site of water oxidation is situated on the lumenal side of the thylakoid membrane, containing four Mn ions and the redox active tyrosine, Tyr-Z.[13–17] One Ca^{2+} and one Cl^- ion are also required for catalytic activity and appear to be located in the vicinity of the Mn cluster.[14] The structure of PSII has recently been determined by X-ray crystallography at 3.5–3.8 Å resolution,[11,12,18,19] which facilitates structure-based understanding of the PSII function.

During light-driven PSII electron transfer the primary event is charge separation between the excited reaction center chlorophyll (P_{680}^*) and pheophytin (Phe) molecule producing the primary charge separated state ($P_{680}^+Phe^-$), which is followed by rapid charge stabilization processes. On the so-called acceptor side of PSII the electron from reduced Phe is transferred to the first, Q_A, and second, Q_B, plastoquinone (PQ) electron acceptor, and then to the mobile pool PQ in the lipid phase of the thylakoid membrane. On the so-called donor side, P_{680}^+ is reduced by Tyr-Z, a redox-active tyrozine of the D1 protein, which then extracts an electron from the water-oxidizing complex[13] (FIG. 1.). Water oxidation is catalyzed by a cluster of four Mn ions, which undergo light-induced changes in their oxidation states, called S-states. The complex cycles through five S-states denoted as $S_0,\ldots,S_4$ and oxygen is released during the $S_3 \to S_4 \to S_0$ transition, in which S_4 is a short-lived intermediate (for reviews see Refs. 14, 20, 21).

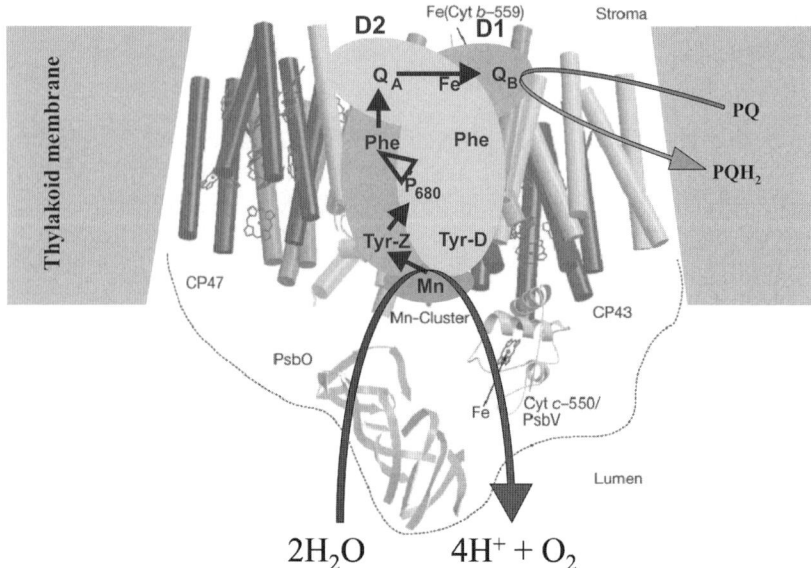

FIGURE 1. Schematic representation of the protein structure and main redox components of the PSII complex. D1 and D2 show the protein subunits of the reaction center, which bind the redox cofactors of light-induced electron transfer: P680, the primary Chl electron donor, the primary pheophytin (Phe) electron acceptor, the quinone electron acceptors Q_A, Q_B, and PQ, the redox-active tyrosine molecules, Tyr-Z and Tyr-D, and the catalytic Mn cluster of water oxidation.

Photodamage by Visible Light

A large body of evidence demonstrates that under strong illumination by visible light a series of light-induced modifications occur at the acceptor side PSII, which eventually lead to the inhibition of electron transport. In an initial step a light-induced conformational change at the PSII acceptor side slows down the Q_A^- to Q_B electron transfer step and leads to an irreversible change of the D1 protein.[22] This effect is probably initiated by the light-induced increase in the reduction level of the plastoquinone (PQ) molecules in the lipid phase of the membrane when the main electron sinks, such as the Calvin cycle and photorespiration, are unable to keep up with the electron flow from PSII. Under anaerobic conditions, when reduction of the PSII acceptor side is enhanced overreduction of the PQ pool creates a situation when the binding site of the secondary quinone electron acceptor (Q_B) is left unoccupied.[23–25] This leads to the stabilization of the reduced primary quinone electron acceptor (Q_A^-),[25] and finally Q_AH_2 that may leave its binding site.[25]

A very important consequence of light-induced modifications at the acceptor side of PSII is the formation of $^3P_{680}$ promoting states of the PSII reaction center.[24,25] In the presence of molecular oxygen the interaction of $^3P_{680}$ with $^3O_2$ results in the formation of highly reactive singlet oxygen,[26] which can irreversibly damage its environment. The validity of the acceptor-side induced photoinhibitory mechanism has been unambiguously demonstrated by the detection of singlet oxygen formation in various PSII preparations[27,28] and intact leaves.[29]

Light sensitivity of PSII increases dramatically under conditions when electron donation from the Mn cluster of water oxidation is unable to keep up with the rate of withdrawal of electrons from $P_{680}{}^+$. Such conditions are created by the complete removal of the Mn cluster by Tris washing or NH_2OH treatment[30,31] or by the isolation of PSII reaction center particles.[32] The main mechanism of photoinactivation of PSII with impaired donor-side electron transport is the generation of long-lived highly oxidizing $P_{680}{}^+$ and Tyr-Z^+ radicals that could not be reduced in the absence of sufficient electron flow from the Mn cluster.[30,33,34]

Under illumination by a series of light flashes photodamage and D1 protein degradation,[35] as well as singlet oxygen production[36] is correlated with the amount of $S_2Q_B{}^-$ and $S_3Q_B{}^-$ charge pairs, which are formed in the initial charge separation events, as well as with the dark interval allowed for charge recombination. These stable charge separated states recombine in darkness and form the primary radical pair $P_{680}{}^+Phe^-$ in the singlet or triplet spin configuration. The singlet radical pair recombines to $^1P_{680}{}^*(Phe)$, whereas the triplet radical pair recombines to $^3P_{680}{}^*(Phe)$ whose interaction with ground state $^3O_2$ leads then to the formation of $^1O_2$. Triplet forming charge recombination takes place both at low- and high-intensity illumination, which can explain photodamage induced in a wide intensity range.

PHOTODAMAGE BY ULTRAVIOLET LIGHT

Although physiologically relevant sunlight consists mainly of the photosynthetically active spectral range PAR (400–700 nm) the shorter wavelength ultraviolet radiation is highly damaging for the photosynthetic apparatus. Comparison of the characteristics of PSII damage induced by UV-A (315–400 nm) and UV-B (290–315 nm) radiation, shows that these UV spectral ranges inhibit PSII by very similar or identical mechanisms, which target primarily the water-oxidizing complex.[37–40] The damaging effects of the UV-C (200–290 nm) region are expected to be similar to that of UV-B and UV-A see, Ref.41 but this notion is not fully proven.

Light-sensitized damage of the Mn cluster has been suggested to occur not only in the UV, but also in the whole visible spectral range.[42] However, the model that suggests Mn as the sole sensitizer of photodamage[42] cannot explain the detailed action spectra of photoinhibition, which show a peak in the red region that matches Chl absorption.[43,44]

Accumulation of PSII centers with impaired electron donation from the Mn cluster to Tyr-Z^+ has been demonstrated under UV-A and UV-B illumination,[39,40,45] and also after exposure of *Synechocystis* 6803 cells to unfiltered sunlight.[46] From these results it was proposed that photoinhibition by sunlight can be initiated by a UV-induced inhibition of PSII donor side followed by the destruction of the PSII reaction center via the donor-side-induced mechanism of photoinhibition.[46] This idea was further developed by suggesting that not only UV light, but to a smaller extent blue and green light can also inactivate the water-oxidizing complex followed by red light-driven destruction of the PSII reaction center.[47]

ACTIVE OXYGEN PRODUCTION UNDER LIGHT STRESS

Production of reactive oxygen species (ROS) in the photosynthetic apparatus is a well-documented phenomenon. Singlet oxygen production has been shown by EPR spin trapping[27,48] or chemiluminescence[28] in isolated PSII preparations under the conditions of photoinhibitory illumination by continuous light or by single turnover flashes.[36] Singlet oxygen can also be detected by fluorescence spin traps both in isolated PSII preparations and in intact leaves.[29] Under conditions of donor-side photoinhibition the dominating species are hydroxyl radicals,[27] which represent the main ROS also under UV photoinhibition.[49] OH· radical production at the acceptor side of PSII has also been proposed.[50] In this process the interaction of O_2^- and the nonheme iron is proposed to lead to the formation of bound peroxide intermediates, which then results in OH· formation. In addition, O_2^- formed at the acceptor side of PSII is proposed to lead to the H_2O_2 production, whose interaction with free metal irons results in OH· formation via the Fenton reaction.

MECHANISMS OF PHOTOPROTECTION

Photodamage to PSII in chloroplasts can be efficiently prevented by excitation quenching in the light-harvesting antenna, which consequently dissipates excess excitation energy harmlessly as heat. This phenomenon includes processes termed nonphotochemical quenching (NPQ) of chlorophyll fluorescence that can be induced within a time scale of seconds to minutes as reviewed recently.[51,52] Crucial for induction of nonphotochemical quenching is a decrease of pH in the thylakoid lumen upon illumination, thereby activating the thermal dissipation as a feedback mechanism, which, in turn, downregulates the photosynthetic electron transport. Two proteins besides the LHCII antenna, the violaxanthin deepoxidase in the thylakoid lumen[53] and the PsbS protein in the thylakoid membrane, have an essential role in the development of maximal of NPQ.[52] As outlined above charge recombination that leads to the formation of triplet radical pair is an important mechanism of singlet oxygen production. Protection against this recombination-type photoinhibition can be achieved by regulating the balance of triplet-producing and nontriplet-producing pathways.[54]

Rapid turnover of the PSII D1 reaction center protein after photodamage can also be regarded as one of the most important photoprotective mechanisms. No detectable damage to PSII occurs in photosynthesizing cells at light intensities below those saturating the photosynthetic carbon fixation. Indeed, under these conditions the damage and repair are in balance and only at higher light intensities the rate of damage exceeds the rate of repair resulting in measurable photoinhibition of PSII.[8] A critical step in the D1 protein repair cycle is the degradation of the light-damaged D1 copies, which involves FtsH type proteases both in visible[55] and UV light.[56]

CONCLUDING REMARKS

Photoinhibition of oxygenic photosynthesis is a general phenomenon, which occurs as an unavoidable consequence of dealing with light, and represents the damaging consequence of light acting as stress factor. Due to the complexity of the underlying events we do not yet understand each aspect of the process. However, the recent developments in exploring the structure of the PSII complex[11,12] are expected to lead us to a general understanding of how plants manage safe utilization of light.

ACKNOWLEDGMENTS

This work was partly supported by grants from EU (MRTN-CT-2003–505069).

REFERENCES

1. TERASHIMA, I., S. FUNAYAMA & K. SONOIKE. 1994. The site of photoinhibition in leaves of *Cucumis sativus* L. at low temperatures in Photosystem I, not Photosystem II. Planta **193:** 300–306.
2. SONOIKE, K., I. TERASHIMA, M. IWAKI & S. ITOH. 1995. Destruction of Photosystem I iron-sulfur centers in leaves of *Cucumis sativus* L. by weak illumination at chilling temperatures. FEBS Lett. **362:** 235–238.
3. SONOIKE, K., M. KAMO, Y. HIHARA, T. HIYAMA & I. ENAMI. 1997. The mechanism of the degradation of *psaB* gene product, one of the photosynthetic reaction center subunits of Photosystem I, upon photoinhibition. Photosynth. Res. **53:** 55–63.
4. SCHELLER, H.V. & A. HALDRUP. 2005. Photoinhibition of Photosystem I. Planta **221:** 5–8.

5. RAPPAPORT, F., M. GUERGOVA-KURAS, P.J. NIXON, et al. 2002. Kinetics and pathways of charge recombination in photosystem II. Biochemistry **41:** 8508–8517.
6. POWLES, S.B. 1984. Photoinhibition of photosynthesis induced by visible light. Annu. Rev. Plant Physiol. **35:** 15–44.
7. ADIR, N., H. ZER, S. SHOCHAT & I. OHAD. 2003. Photoinhibition – a historical perspective. Photosynth. Res. **76:** 343–370.
8. ARO, E.-M., I. VIRGIN & B. ANDERSSON. 1993. Photoinhibition of photosystem II. Inactivation, protein damage and turnover. Biochimica et Biophysica Acta **1143:** 113–134.
9. CRITCHLEY, C. & A.W. RUSSEL. 1994. Photoinhibition of photosynthesis *in vivo*: the role of protein turnover in Photosystem II. Physiol. Plantarum. **92:** 188–196.
10. ÖQUIST, G., W.S. CHOW & J.M. ANDERSON. 1992. Photoinhibition of photosynthesis represents a mechanism for the long-term regulation of Photosystem II. Planta. **186:** 450–460.
11. FERREIRA, K.N., T.M. IVERSON, K. MAGHLAOUI, et al. 2004. Architecture of the photosynthetic oxygen-evolving center. Science **303:** 1831–1838.
12. LOLL, B., J. KERN, W. SAENGER, et al. 2005. Towards complete cofactor arrangement in the 3.0 Å resolution structure of Photosystem II. Nature **438:** 1040–1044.
13. ANDERSSON, B. & S. STYRING. 1991. Photosystem II: Molecular organization, function, and acclimation. Curr. Top. Bioenerg. **16:** 1–81.
14. DEBUS, R.J. 1992. The manganese and calcium ions of photosynthetic oxygen evolution. Biochim. Biophys. Acta **1102:** 269–352.
15. TOMMOS, C. & G.T. BABCOCK. 2000. Proton and hydrogen currents in photosynthetic water oxidation. Biochim. Biophys. Acta **1458:** 199–219.
16. RENGER, G. 2001. Photosynthetic water oxidation to molecular oxygen: apparatus and mechanism. Biochim. Biophys. Acta **1503:** 210–228.
17. NUGENT, J.H.A., A.M. RICH & C.W. EVANS. 2001. Photosynthetic water oxidation: towards a mechanism. Biochim. Biophys. Acta **1503:** 138–146.
18. ZOUNI, A., H.T. WITT, J. KERN, et al. 2001. Crystal structure of photosystem II from *Synechococcus elongatus* at 3.8 A resolution. Nature **409:** 739–743.
19. KAMIYA, N. & J.-R. SHEN. 2003. Crystal structure of oxygen-evolving photosystem II from *Thermosynechococcus vulcanus* at 3.7-A resolution. Proc. Natl. Acad. Sci. USA. **100:** 98–103.
20. RENGER, G. 1997. Mechanistic and structural aspects of photosynthetic water oxidation. Physiol. Plantarum. **100:** 828–841.
21. HOGANSON, C.W. & G.T. BABCOCK. 1997. A metalloradical mechanism for the generation of oxygen from water in photosynthesis. Science **277:** 1953–1956.
22. OHAD, I., N. ADIR, H. KOIKE, et al. 1990. Mechanism of photoinhibition *in vivo*. A reversible light-induced conformational change of reaction center II is related to an irreversible modification of the D1 protein. J. Biol. Chem. **265:** 1972–1979.
23. SETLIK, I., S.I. ALLAKHVERDIEV, L. NEDBAL, et al. 1990. Three types of photosystem II photoinactivation. I. Damaging processes on the acceptor side. Photosynth. Res. **23:** 39–48.
24. VAN MIEGHEM, F.J.E., W. NITSCHKE, P. MATHIS & A.W. RUTHERFORD. 1989. The influence of the quinone-iron electron acceptor complex on the reaction centre photochemistry of photosystem II. Biochimica et Biophysica Acta **977:** 207–214.
25. VASS, I., S. STYRING, T. HUNDAL, et al. 1992. Reversible and irreversible intermediates during photoinhibition of photosystem II: stable reduced Q_A species promote chlorophyll triplet formation. Proc. Natl. Acad. Sci. USA **89:** 1408–1412.

26. DURRANT, J.R., L.B. GIORGI, J. BARBER, et al. 1990. Characterization of triplet states in isolated photosystem II reaction centres: oxygen quenching as a mechanism for photodamage. Biochimica et Biophysica Acta **1017**: 167–175.
27. HIDEG, É., C. SPETEA & I. VASS. 1994. Singlet oxygen and free radical production during acceptor- and donor-side-induced photoinhibition. Studies with spin trapping EPR spectroscopy. Biochimica et Biophysica Acta **1186**: 143–152.
28. MACPHERSON, A.N., A. TELFER, J. BARBER & T.G. TRUSCOTT. 1993. Direct detection of singlet oxygen from photosystem II reaction centres. Biochimica et Biophysica Acta **1143**: 301–309.
29. HIDEG, É., T. KÁLAI, K. HIDEG & I. VASS. 1998. Photoinhibition of photosynthesis *in vivo* results in singlet oxygen production. Detection via nitroxide-induced fluorescence quenching in broad bean leaves. Biochemistry **37**: 11405–11411.
30. BLUBAUGH, D.J. & G.M. CHENIAE. 1990. Kinetics of photoinhibition in hydroxylamine-extracted photosystem II membranes: relevance to photoactivation and sites of electron donation. Biochemistry **29**: 5109–5118.
31. CHEN, G.-X. & G.M. CHENIAE. 1992. Photoinhibition of hydroxylamine-extracted photosystem II membranes: studies of the mechanism. Biochemistry **31**: 11072–11083.
32. SEARLE, G.F.W., A. TELFER, J. BARBER & T.J. SCHAAFSMA. 1990. Millisecond time-resolved EPR of the spin-polarised triplet in the isolated Photosystem II reaction centre. Biochimica et Biophysica Acta. **1016**: 235–243.
33. ECKERT, H.-J., B. GEIKEN, J. BERNADING, et al. 1991. Two sites of photoinhibition of the electron transfer in oxygen evolving and Tris-trated PSII membrane fragments from spinach. Photosynth. Res. **27**: 97–108.
34. JEGERSCHOLD, C., I. VIRGIN & S. STYRING. 1990. Light-dependent degradation of the D1 protein in photosystem II is accelerated after inhibition of the water splitting reaction. Biochemistry **29**: 6179–6186.
35. KEREN, N., A. BERG, P.J.M. VAN KAN, et al. 1997. Mechanism of photosystem II photoinactivation and D1 protein degradation at low light: the role of back electron flow. Proc. Natl. Acad. Sci. USA **94**: 1579–1584.
36. SZILÁRD, A., L. SASS, É. HIDEG & I. VASS. 2005. Photoinactivation of Photosystem II by flashing light. Photosynth. Res. **84**: 15–20.
37. RENGER, G., M. VÖLKER, H.J. ECKERT, et al. 1989. On the mechanism of Photosystem II deterioration by UV-B irradiation. Photochem. Photobiol. **49**: 97–105.
38. TURCSÁNYI, E. & I. VASS. 2000. Inhibition of photosynthetic electron transport by UV-A radiation targets the Photosystem II complex. Photochem. Photobiol. **72**: 513–520.
39. VASS, I., E. TURCSÁNYI, E. TOULOUPAKIS, et al. 2002. The mechanism of UV-A radiation-induced inhibition of photosystem II electron transport studied by EPR and chlorophyll fluorescence. Biochemistry **41**: 10200–10208.
40. LARKUM, A.W.D., M. KARGE, F. REIFARTH, et al. 2001. Effect of monochromatic UV-B radiation on electron transfer reactions of Photosystem II. Photosynth. Res. **68**: 49–60.
41. TREBST, A. & E. PISTORIUS. 1965. Photosynthetische reaktionene in UV-bestrahlten Chloroplasten. Zeitschrift für Naturforschung. **20**: 885–889.
42. HAKALA, M., I. TUOMINEN, M. KERANEN, et al. 2005. Evidence for the role of the oxygen-evolving manganese complex in photoinhibition of Photosystem II. Biochim. Biophys. Acta **1706**: 68–80.
43. JONES, L.W. & B. KOK. 1966. Photoinhibition of chloroplast reactions. I. Kinetics and action spectra. Plant Physiol. **41**: 1037–1043.

44. SANTABARBARA, S., I. CAZZALINI, A. RIVADOSSI, et al. 2002. Photoinhibition *in vitro* involves weakly coupled chlorophyll–protein complexes. Photochem. Photobiol. **75:** 613–618.
45. IWANZIK, W., M. TEVINI, G. DOHNT, et al. 1983. Action of UV-B radiation on photosynthetic primary reaction in spinach chloroplasts. Physiol. Plantarum. **58:** 401–407.
46. SICORA, C., Z. MÁTÉ & I. VASS. 2003. The interaction of visible and UV-B light during photodamage and repair of Photosystem II. Photosynth. Res. **75:** 127–137.
47. OHNISHI, N., S.I. ALLAKHVERDIEV, S. TAKAHASHI, et al. 2005. Two-step mechanism of photodamage to Photosystem II: Step 1 occurs at the oxygen-evolving complex and step 2 occurs at the photochemical reaction center. Biochemistry **44:** 8494–8499.
48. HIDEG, É., I. VASS, T. KÁLAI & K. HIDEG. 2000. Singlet oxygen detection with sterically hindered amine derivatives in plants under light stress. Methods Enzymol. **319:** 77–85.
49. HIDEG, E. & I. VASS. 1996. UV-B induced free radical production in plant leaves and isolated thylakoid membranes. Plant Sci. **115:** 251–260.
50. POSPÍSIL, P., A. ARATO, A. KRIEGER-LISZKAY & A.W. RUTHERFORD. 2004. Hydroxyl radical generation by Photosystem II. Biochemistry **43:** 6783–6792.
51. HORTON, P. & A. RUBAN. 2005. Molecular design of the Photosystem II light-harvesting antenna: photosynthesis and photoprotection. J. Exp. Bot. **56:** 365–373.
52. NIYOGI, K.K., X.-P. LI, V. ROSENBERG & H.-S. JUNG. 2005. Is PsbS the site of non-photochemical quenching in photosynthesis? J. Exp. Bot. **56:** 375–382.
53. DEMMIG-ADAMS, B., W.W. ADAMS III, D.H. BAKER, et al. 1996. Using chlorophyll fluorescence to assess the fraction of absorbed light allocated to thermal dissipation of excess excitation. Physiol. Plant. **98:** 253–264.
54. KRIEGER-LISZKAY, A. & A.W. RUTHERFORD. 1998. Influence of herbicide binding on the redox potential of the quinone acceptor in photosystem II.: Relevance to photodamage and phytotoxicity. Biochemistry **37:** 17339–17344.
55. SILVA, P., E. THOMPSON, S. BAILEY, et al. 2003. FtsH is involved in the early stages of repair of Photosystem II in *Synechocystis* sp PCC 6803. Plant Cell **15:** 2152–2164.
56. CHEREGI, O., C. SICORA, P.B. KÓS, et al. 2007. The role of the FtsH and Deg proteases in the repair of UV-B radiation-damaged Photosystem II in the cyanobacterium *Synechocystis* PCC 6803. Biochim. Biophys. Acta **1767:** 820–828.

The Plant Host–Pathogen Interface

Cell Wall and Membrane Dynamics of Pathogen-Induced Responses

BRAD DAY[a] AND TERRY GRAHAM[b]

[a]*Department of Plant Pathology, Michigan State University, East Lansing, Michigan, USA*

[b]*Department of Plant Pathology, Ohio State University, Columbus, Ohio, USA*

ABSTRACT: Perception of pathogens by their hosts is the outcome of a highly coordinated and sophisticated surveillance network, tightly regulated by both host and pathogen elicitors, effectors, and signaling processes. In this article, we focus on two relatively well-studied host–pathogens systems, one involving a bacterial–plant interaction (*Pseudomonas syringae*–Arabidopsis) and the other involving an oomycete–plant interaction (*Phytophthora sojae*–soybean). We discuss the status of current research related to events occurring at the host–pathogen interface in these two systems, and how these events influence the organization and activation of resistance responses in the respective hosts. This recent research has revealed that in addition to the previously identified resistance machinery (R-proteins, molecular chaperones, etc.), the dynamics of the cell wall, membrane trafficking, and the actin cytoskeleton are intimately associated with the activation of resistance in plants. Specifically, in Arabidopsis, a possible connection between the actin machinery and R-protein- mediated induction of disease resistance is described. In the case of the *P. sojae*–soybean interaction, we describe the fact that a classical basal resistance elicitor, the cell wall glucan elicitor from the pathogen, can directly activate host hypersensitive cell death, which is apparently modulated in a race-specific manner by the presence of R genes in the host.

KEYWORDS: disease resistance; defense; innate immunity; effector; elicitor; hypersensitive cell death; *Arabidopsis thaliana*; Glycine max, soybean; *Pseudomonas syringae*; *Phytophthora sojae*; isoflavone; phytoalexins; cytoskeleton; actin

---

Address for correspondence: Terry Graham, Ph.D., Department of Plant Pathology, Ohio State University, 201 Kottman Hall, 2021 Coffey Road, Columbus, OH, USA. Voice: +1-614-292-1375; fax: +1-614-292-4455.
graham.1@osu.edu

INTRODUCTION

Plants have evolved two primary defense systems to combat pathogen attack.[1] One of the plant's first responses to invading microbes is *basal resistance*, also sometimes called general resistance or innate immunity. Basal resistance is triggered by the recognition of pathogen-associated molecular patterns (PAMPs). PAMPs are defense elicitors often associated with the cell surface of pathogens. While often polymeric, active fragments of PAMPs are released upon contact with the host. While the molecular mechanisms underlying basal resistance are currently not well understood, its induction is believed to be associated with MAP kinase signaling, transcriptional induction of pathogen responsive protein and secondary product defense genes, deposition of polymeric wall reinforcements (e.g., callose, lignin, and other phenolic polymers) at sites of infection and, ultimately, abrogation of pathogen growth. In addition to basal defenses induced by PAMPs, plants also defend themselves by *effector-triggered plant resistance*. As the primary tenants of gene-for-gene resistance, these effectors play dual functions as both virulence and avirulence factors.[2,3] In the absence of the cognate resistance (R) proteins, effectors can function to disable host basal defenses and to release nutrients from host cells, rendering the host susceptible to pathogen proliferation. However, if these effectors are recognized by plant surveillance systems, usually the R gene and/or associated host proteins, they activate defense responses, usually manifested as a form of programmed cell death called the hypersensitive response (HR).

In this article we discuss these two broad forms of resistance and their regulation in two very different, but relatively well-delineated host–pathogen systems. The first is that between the bacterial pathogen *Pseudomonas syringae* and its host, *Arabidopsis thaliana*. The second involves the interactions between the oomycetic pathogen, *Phytophthora sojae* and its host, *Glycine max*. In both cases we present evidence for interesting new findings on the mechanisms underlying these classical forms of resistance. In the case of the former system, a possible connection between the actin machinery and R-protein-mediated induction of disease resistance in *Arabidopsis* is described. In the case of *P. sojae*–soybean interactions, we describe the fact that a classical PAMP, the cell wall glucan elicitor from the pathogen, can directly activate cell death, which is apparently modulated in a race-specific manner by the presence of R genes in the host.

PSEUDOMONAS SYRINGAE-ARABIDOPSIS THALIANA: RESISTANCE, EFFECTORS, AND ACTIN

Gram-negative pathogens of both plants and animals share a number of common features, which has aided not only in the elucidation of their respective activities, but also in determining host targets and mechanisms of resistance signaling.[2-4] For example, it has recently been demonstrated that *Pseudomonas syringae* effector proteins AvrPphB and AvrRpt2 are members of a class of type III secreted cysteine proteases, functionally related to the YopT effector from *Yersinia pestis*, a bacterial pathogen of humans. YopT is a cysteine protease that cleaves the Rho family GTPase, which causes the disruption of the actin cytoskeleton and contributes to the inhibition of phagocytosis of the pathogen.[5] As we will discuss below, research in our lab has identified a possible connection between the actin machinery and R-protein-mediated induction of disease resistance in *Arabidopsis*.

Effector-Triggered Plant Resistance: R-Proteins

Numerous R genes have been cloned from a wide range of plant species.[6] The largest class cloned to date is the family encoding proteins that contain a nucleotide-binding (NB) site and leucine-rich repeat (LRR) domain. Interestingly, the nucleotide-binding motifs in plants share sequence similarities with regions of apoptosis regulators, such as *CED4* from *Caenorhabditis elegans* and *Apaf-1* from humans.[6] This suggests that R protein function may require, at least in part, the activity associated with ATP binding and/or hydrolysis,[7] and by analogy, may serve as a signal transducer of cell death-related responses. The carboxy-terminal domain of this class of resistance proteins is represented by the LRR, which is typically 20–30 amino acids in length, and appears to be involved in the formation of protein–protein interactions. The NB-LRR class of *R* genes can be further divided into coiled-coil (CC)-NB-LRR and toll-interleukin-1 receptor (TIR)-NB-LRR according to their N-terminal domain. Evidence suggests that the N terminus influences the requirement for downstream defense response components. In the model plant system *Arabidopsis thaliana*, over 150 proteins are predicted to be NB-LRR proteins. Collectively, this class of R-proteins determines resistance to bacterial, viral, fungal, and oomycete pathogens. The best-characterized members of the NB-LRR class include members of the CC-NB-LRR subclass: RPS2, RPM1, and RPS5, *Arabidopsis* R-proteins specifying resistance to *P. syringae* carrying the bacterial effectors AvrRpt2, AvrRpm1/AvrB, and AvrPphB, respectively.

Indirect Pathogen Recognition: Surveillance

Although many R genes and their corresponding pathogen effectors have been cloned, the biochemical and genetic relationship(s) between each pair is largely unknown. Previously, plant resistance proteins have been hypothesized to serve as receptors directly interacting with pathogen effectors acting as ligands. However, research over the last 5 years has uncovered a complex surveillance mechanism that coordinates resistance responses in *Arabidopsis* to a multitude of pathogens.[1,6,8] Contrary to the ligand–receptor model, it is now evident that bacterial effector recognition and signaling have likely evolved as an indirect mechanism whereby a resistance protein monitors the perturbation of a third plant protein by the enzyme activity of the effector. This process of indirect recognition leads to the activation of plant defense responses.

The best-characterized example of the activation of resistance by way of monitoring bacterial effector activity is that of the *Arabidopsis* protein RIN4. RIN4 is monitored by at least two R-proteins, RPM1 and RPS2. RPM1 and RPS2 have each been shown to physically associate with RIN4 *in planta*.[9–13] The *Arabidopsis* protein RPM1 recognizes two unrelated *P. syringae* effector proteins, AvrRpm1 and AvrB.[14–15] When AvrRpm1 or AvrB are delivered to the plant cell, RIN4 is hyperphosphorylated by a yet to be identified kinase. This phosphorylation in turn leads to the activation of RPM1-mediated resistance. Thus, although RPM1 resistance is activated in the presence of either AvrB or AvrRpm1, it is activated through an indirect mechanism (i.e., detection of the modified state of RIN4). It has recently been shown that AvrRpm1 inhibits basal defense responses, presumably through its modification of RIN4 and other host targets.[1] Activation of this signaling mechanism requires the activity of the resistance-associated protein NDR1.

Loci Required for Defense Signaling

Genetic screens to identify suppressors of plant resistance genes have identified several important loci including: *NDR1, EDS1, PBS1, PAD1–4, RAR1,* and *Hsp90*.[1,6] The *NDR1* locus is required for RPS2, RPM1, and RPS5 function (members of the coiled-coil + nucleotide-binding site + leucine-rich repeat class of proteins), while EDS1 is required for RPS4 function. The *RAR1* gene is required by all four resistance loci, while the *PBS1* gene, which encodes a protein kinase, is only required for *RPS5* function.

Of the aforementioned loci, *EDS1* and *NDR1* are among the most identifiable for their respective (and somewhat divergent) roles in resistance mediated by members of the largest class of resistance proteins, the NB-LRR class. EDS1 has been extensively characterized for its contribution to the TIR-NB-LRR class of resistance proteins, among which include RPS4 and the loci conferring resistance to *Hyaloperonospora parasitica*, RPP2/4/5/21. Conversely, NDR1 has been shown to be required for the CC-NB-LRR class of resistance proteins.[16] Among these are RPS2, RPS5, and RPM1. Although EDS1 and NDR1 appear to function in divergent resistance pathways mediated by members of the NB-LRR family of resistance proteins, there are likely shared points of convergence; both in the initial perception of pathogens, as well as resistance signaling and cell death. While the role EDS1 plays in disease resistance has been better characterized, the function of NDR1 remains enigmatic.

Recent work has demonstrated that NDR1 is a plasma membrane-localized, glycosylphosphatidyl-inisotol (GPI)-anchored protein whose expression is required for resistance to *P. syringae* DC3000 expressing the bacterial effector proteins AvrRpt2, AvrB/AvrRpm1, and AvrPphB. However, the detailed biochemical mechanism whereby NDR1 transduces defense signaling is largely unknown.

Actin Dynamics and Resistance

A primary focus of research in our laboratory is the understanding of host processes both required for pathogen recognition, as well as those processes that are directly targeted by the pathogens themselves. Research in this area has led to the identification of dozens of host processes that are presumably targeted by pathogens during infection. Preliminary data in our lab suggest that one such process targeted by plant pathogens is the actin cytoskeleton.

Given the central role the actin machinery plays in the innate immune response in mammals, we hypothesized that the actin cytoskeleton in plants likely plays a central role in host defense responses, and too, may in fact represent a virulence target, much as is the case in mammals. As a general process, the actin cytoskeleton of plants has been demonstrated to play a role in a variety of processes, among which include membrane trafficking, flowering, development, and disease resistance. Taking the latter into consideration, the role of actin depolymerization in mammals in response to biotic stress has been well documented, and moreover, the specific targeting of the actin machinery by pathogenic bacteria is a well-characterized mechanism of pathogenicity.

Following the gene-for-gene model for defense activation, resistance in the host plant following perception of the *P. syringae* follows one of two courses. First, in the absence of recognition, the pathogen multiplies freely, leading to increased pathogen growth, disease, and ultimately death of the host plant. Conversely, when pathogen perception occurs, generally the result of a triggering of monitored host defenses (see above), then resistance is initiated, and pathogen growth is abrogated. To determine if the actin machinery is one such host process that has evolved as a pathogen virulence target, and too, may in fact be a general process that the host monitors for perturbations, we asked the question: Would a mutation resulting in the loss of critical machinery of the host actin cytoskeleton result in the plant's ability to detect pathogen infection? Moreover, does *P. syringae* express (and deliver via the T3SS) an effector candidate that does in fact target the actin machinery?

Using a forward genetics approach, we have identified the actin depolymerization machinery as a key regulator in the initiation of defense responses in *Arabidopsis*. Mutational analyses of several members of the actin depolymerization factor (i.e., ADF) family of proteins revealed a breakdown in the activation of defense responses following infection with the bacterial pathogen

P. syringae. Interestingly, the susceptibility phenotype associated with the ADF mutation(s) correlates with the activity of a specific bacterial effector protein: AvrPphB. As noted above, AvrPphB is related to the YopT family of effector proteins, a class of cysteine proteases, which specifically targets the actin machinery in mammalian cells, effectively shutting down the phagocytotic process. In the case of AvrPphB, while the exact mechanism is unknown, activity of the cysteine protease appears to target (either directly or indirectly) the actin cytoskeleton, effectively shutting down the cellular processes associated with intracellular trafficking. In short, our data suggest that in the case of *P. syringae-Arabidopsis*, depolymerization of the actin cytoskeleton is likely required for resistance.

A BASAL RESISTANCE ELICITOR FROM *PHYTOPHTHORA SOJAE* ACTIVATES HR CELL DEATH IN SOYBEAN

Early Foundations of the Soybean–P. sojae Interaction

The *P. sojae*–soybean association is one of the earliest host–pathogen systems closely examined at a physiological, cellular, and biochemical level. Very early work led to a thorough characterization of the secondary product pathways leading to the soybean phytoalexins,[17] the pathogen elicitors of these pathways[18] and the physiological events regulating responses to infection and elicitor treatment.[19,20] The cell wall glucan elicitor (WGE) from *P. sojae* was also one of the first PAMPs identified. Originally identified as an elicitor of the soybean phytoalexins, the glyceollins,[18] as we will see here it is a remarkably global defense elicitor.

The *P. sojae*–soybean association is also a very rich one genetically, with a series of Rps resistance genes providing race-specific resistance to many defined races of the pathogen. Although WGE was found to induce the soybean phytoalexins, the glyceollins, in a race-cultivar nonspecific manner,[18] making it a typical PAMP or elicitor of basal resistance responses, the glyceollins were nevertheless induced in a race-specific manner in infected tissues. Thus, the *P. sojae*–soybean interaction was from the very beginning an intriguing system in which to study the connections between race-specific and basal resistance.

Recent Advances in the Cellular Biochemistry of Soybean –P. sojae Interactions

Due to their simple cellular architecture and ease of manipulation, cotyledons have been an organ of choice for cellular, biochemical, and molecular work on soybean–*P. sojae* interactions. Their use was instrumental in defining the multiplicity of responses to infection and WGE treatment and their regulation (for reviews see Refs. 21, 22). Work with cotyledon tissues demonstrated that WGE-induced defense responses in soybeans are very global and include multiple secondary product (phenylpropanoid/isoflavone) and pathogenesis-related (PR) protein defense responses that are orchestrated in a sophisticated manner in different cell populations proximal and distal to the point of inoculation or elicitor treatment.[21–23] It was clear from these studies that the WGE from *P. sojae* was a central player in these various defense responses and a large number of studies focused on this elicitor. Secondary product responses to WGE include the formation of conjugates of the isoflavones daidzein and genistein.[24] Genistein is directly toxic to *P. sojae*,[25] and daidzein is a precursor for the phytoalexin, glyceollin, which also subsequently accumulates in WGE-treated tissues. Phenolic polymers (lignin and suberin) derived rapidly from early phenylpropanoid precursors can also accumulate to massive levels in elicitor-treated cells.[26] Thus, the secondary product responses include the formation of two antibiotics and the reinforcement of a potential cell wall barrier. WGE also leads to the activation of expression of genes for various PR proteins,[23] including PR-1a, PR-2, PR-4, PR-6, and PR-10.

In addition to the very global effects of WGE on the activation of defense responses, a possible connection of its activity to hypersensitive cell death was suggested in studies on a phenomenon called elicitation competency.[27] It was discovered that the activation of accumulation of the glyceollins by WGE required proximity of treated cells to either wounded or HR dying cells. In the absence of wounding or HR dying cells, WGE induced the accumulation of the isoflavone daidzein, the precursor of glyceollin. It was hypothesized that entry into the cell death program was required for the activation of elicitation competence. Other connections of WGE activity to race-specific resistance and HR cell death were suggested from genetic studies in which it was discovered that elicitation competence was strongly conditioned by the presence of several Rps resistance genes.[28]

Gene Silencing Reveals Unexpected Connections among Isoflavones, WGE, and the Activation of Race-Specific Hypersensitive Cell Death

In the past few years, we have employed *Agrobacterium rhizogenes*-based RNAi gene silencing[29,30] to extend the various findings in the model cotyledon system to roots, the primary and economically most important target for *P. sojae* infection. Gene silencing has been highly effective in allowing us to determine the importance of various candidate genes in both race-specific and basal resistance in roots and has confirmed all of the major aspects of defense deployment and regulation first described in cotyledons. However, it has also uncovered some unexpected connections between PAMP-induced basal resistance and race- specific resistance pathways.

Recently, we described that the silencing of isoflavone synthase causes a 95–98% reduction in root isoflavones[29] and enhanced susceptibility to *P. sojae*, including an apparent breakdown of race-specific resistance in silenced roots. While it has been established for many years that the isoflavone defenses are earlier and more strongly expressed in incompatible infected soybean tissues,[31] these results indicated that they may also participate in the *establishment* of race-specific resistance. We thus explored this phenomenon further. We first demonstrated that silencing of chalcone reductase, which led to an equally effective (ca. 95%) decrease in daidzein (but not genistein) pools in roots, also led to a complete breakdown in race-specific resistance,[30] suggesting that *daidzein is the critical isoflavone for expression of race-specific resistance.* To determine if the breakdown of race-specific resistance was accompanied by a loss of HR cell death, we examined the effects of chalcone reductase and isoflavone synthase silencing on cell death as measured by Evan's Blue vital staining or yellow autofluorescence and histochemical staining for hydrogen peroxide/peroxidase induction. All three of these often used protocols for following HR cell death demonstrated that isoflavone synthase or chalcone reductase silencing both led to complete suppression of cell death and the associated hydrogen peroxide/peroxidase activation in lines carrying resistance genes at the Rps 1 locus.[30]

Together, these results suggested a very tight association of cell death and race-specific resistance to isoflavone accumulation in these Rps lines. Moreover, they demonstrated a connection between isoflavone accumulation, AOS production, and/or increased peroxidase activity. The connection of isoflavones to cell death is also consistent with the fact that lactofen, which induces massive accumulations of the isoflavones[32] also upregulates isoflavone synthase and chalcone reductase mRNA prior to the induction of a form of programmed cell death in soybean.[33] The implications of these observations were very important in that they suggest that the *isoflavones are playing far more than a simple antibiotic and lesion-limiting defense role and in fact appear to play a more complex role in hypersensitive cell death and associated reactions.*

WGE Directly Induces Isoflavone-Mediated Cell Death in Soybean Roots and Silencing of Release of Elicitor Fragments from WGE Blocks the HR and WGE-Induced Cell Death

WGE is the major pathogen elicitor of daidzein and glyceollin.[24] We thus reasoned that it might participate in some way in isoflavone-mediated cell death. Roots of all soybean lines so far examined, regardless of the presence of an Rps resistance gene, showed a cell death response to WGE that was characterized by yellow autofluorescence (beginning within 24–36 h) and complete collapse of the tissue by 48 h.[30] As with yellow autofluorescent cell death in incompatible infections, silencing of CHR or IFS led to a complete suppression of WGE-induced cell death in roots, confirming that the isoflavone daidzein was required.

The PR-2 class of pathogenesis-related proteins encode endoglucanases that putatively release active elicitor fragments from fungal and oomycetic pathogen cell walls.[34] An ethylene-induced and elicitor-releasing endoglucanse has been purified and cloned from soybean.[35] While this PR-2 is thought to release active elicitor fragments from the intact cell wall glucan of *P. sojae*. Previously, we have examined the expression of this elicitor-releasing PR-2 in soybean in response to wounding and WGE treatment.[23] In cotyledons, it is constitutively expressed at relatively low levels and strongly upregulated by wounding and WGE. RNAi silencing of the expression of this PR-2 in soybean roots led to strong suppression of mRNA for the gene as measured by qRT-PCR. As did silencing of IFS and CHR, silencing this PR-2 also led to a complete breakdown of race-specific resistance and HR cell death in near isogenic lines carrying the Rps1c and Rps1k genes. Silencing also led to the complete suppression of elicitation of glyceollin by WGE in all soybean lines and to cell death responses to WGE. To test whether the PR-2 silenced phenotype could be complemented, we used a preparation of elicitor fragments prereleased from intact WGE by treatment with a cell-free extract of the PR-2 endoglucanase.[36] Enzymatically prereleased elicitor preparations elicited a very strong glyceollin response and cell death in PR-2 silenced roots, biochemically complementing the loss of PR-2 expression. These results suggest that PR-2 actually functions *in planta* in elicitor release, and that PR-2-mediated elicitor release is required for the expression of cell death and race-specific resistance in the soybean–*P. sojae* association. Since WGE is required for the *de novo* accumulation of daidzein and glyceollin, this is also highly consistent with the role of these isoflavonoids in the regulation of cell death and race-specific resistance.

Role of WGE in Induction of the Hypersensitive Response in Incompatible Infected Tissues

The activation of cell death by WGE and the effects of silencing PR-2 on the hypersensitive response in infected tissues, strongly suggest a role of WGE in initiation of hypersensitive cell death in infected tissues. However, given the fact that WGE induces cell death even in lines carrying no known R genes, suggests that either undiscovered R genes exist in the so-called universally susceptible cultivar (Williams) used in our studies, or that they function upstream of the R gene interactions with the corresponding Avr gene effectors released by *P. sojae*. The latter possibility is supported by the fact that *P. sojae* produces an additional potential effector protein, which functions as a virulence factor in suppression of the release of active elicitor fragments by the endoglucanase PR-2.[37] If this were the case, then in compatible interactions, *P. sojae* may suppress the activation of cell death by inhibiting release of elicitor fragments. In incompatible infections interactions of the R gene and Avr gene products may nullify this suppression of HR cell death.

ACKNOWLEDGMENTS

The authors would like to acknowledge funding from the following sources. BD: NSF CAREER Award (IOB-0641319). TG: The Ohio Soybean Council and the Illinois-Missouri Biotechnology Alliance.

REFERENCES

1. CHISHOLM, S.T., G. COAKER, B. DAY & B.J. STASKAWICZ. 2006. Host-microbe interactions: shaping the evolution of the plant immune response. Cell **124**: 803–814.
2. ALFANO, J.R. & A. COLLMER. 2004. Type III secretion system effector proteins: double agents in bacterial disease and plant defense. Annu. Rev. Phytopathol. **42**: 385–414.
3. NOMURA, K., M. MELOTTO & S.Y. HE. 2005. Suppression of host defense in compatible plant-Pseudomonas syringae interactions. Curr. Opin. Plant Biol. **8**: 361–368.
4. ZIPFEL, C. & G. FELIX. 2005. Plants and animals: a different taste for microbes? Curr. Opin. Plant Biol. **8**: 353–360.
5. SHAO, F., C. GOLSTEIN, J. ADE, *et al*. 2003. Cleavage of Arabidopsis PBS1 by a bacterial type III effector. Science **301**: 1230–1233.
6. DANGL, J.L. & J.D. JONES. 2001. Plant pathogens and integrated defence responses to infection. Nature **411**: 826–833.
7. TAMELING, W.I., S.D. ELZINGA, P.S. DARMIN, *et al*. 2002. The tomato R gene products I-2 and MI-1 are functional ATP binding proteins with ATPase activity. Plant Cell **14**: 2929–2939.
8. VAN DER BIEZEN, E.A. & J.D. JONES. 1998. Plant disease-resistance proteins and the gene-for-gene concept. Trends Biochem. Sci. **23**: 454–456.
9. MACKEY, D., B.F. HOLT III, A. WIIG & J.L. DANGL. 2002. RIN4 interacts with Pseudomonas syringae type III effector molecules and is required for RPM1-mediated resistance in Arabidopsis. Cell **108**: 743–754.
10. MACKEY, D., Y. BELKHADIR, J.M. ALONSO, *et al*. 2003. Arabidopsis RIN4 is a target of the type III virulence effector AvrRpt2 and modulates RPS2-mediated resistance. Cell **112**: 379–389.
11. AXTELL, M.J. & B.J. STASKAWICZ. 2003. Initiation of RPS2-specified disease resistance in Arabidopsis is coupled to the AvrRpt2-directed elimination of RIN4. Cell **112**: 369–377.
12. DAY, B, D. DAHLBECK, J. HUANG, *et al*. 2005. Molecular basis for the RIN4 negative regulation of RPS2 disease resistance. Plant Cell **17**: 1292–1305.
13. KIM, H.S., D. DESVEAUX, A.U. SINGER, *et al*. 2005. The *Pseudomonas syringae* effector AvrRpt2 cleaves its C-terminally acylated target, RIN4, from Arabidopsis membranes to block RPM1 activation. Proc. Natl. Acad. Sci. USA **102**: 6496–6501.
14. BISGROVE, S.R., M.T. SIMONICH, N.M SMITH, *et al*. 1994. A disease resistance gene in Arabidopsis with specificity for two different pathogen avirulence genes. Plant Cell **6**: 927–933.

15. KIM, M.G., L. DA CUNHA, A.J. MCFALL, *et al*. 2005. Two *Pseudomonas syringae* type III effectors inhibit RIN4-regulated basal defense in Arabidopsis. Cell **121**: 749–759.
16. CENTURY, K.S., E.B. HOLUB & B.J. STASKAWICZ. 1995. NDR1, a locus of *Arabidopsis thaliana* that is required for disease resistance to both a bacterial and a fungal pathogen. Proc. Natl. Acad. Sci. USA **92**: 6597–6601.
17. EBEL, J. 1986. Phytoalexin synthesis: the biochemical analysis of the induction process. Annu. Rev. Phytopathol. **24**: 235–264.
18. HAHN, M.G. 1996. Microbial elicitors and their receptors in plants. Annu. Rev. Phytopathology **34**: 387–412.
19. GRAHAM, T.L. & M.Y. GRAHAM. 1991. Cellular coordination of molecular responses in plant defense. Mol. Plant-Microbe Interact. **4**: 415–422.
20. GRAHAM, T.L. 1995. Cellular biochemistry of phenylpropanoid responses of soybean to infection by Phytophthora sojae. *In* Handbook of Phytoalexin Metabolism and Action. M. Daniel & R.P. Purkayastha, Eds.: 85–116. Marcel Dekker. New York, NY.
21. GRAHAM, T.L. & M.Y. GRAHAM. 1999. Role of hypersensitive cell death in conditioning elicitation competency and defense potentiation. Physiol. Molec. Plant Pathology **55**: 13–20.
22. GRAHAM, T.L. & M.Y. GRAHAM. 2000. Defense potentiation and elicitation competency: redox conditioning effects of salicylic acid and genistein. *In* Plant Microbe Interactions, Vol. 5. G. Stacey & N.T. Keen, Eds.: 181–220. APS Press. St. Paul, MN.
23. GRAHAM, M.Y., J. WEIDNER, K. WHEELER, *et al*. 2003. Induced expression of pathogenesis-related protein genes in soybean by wounding and the *Phytophthora sojae* cell wall glucan elicitor. Physiol. Molec. Plant Pathol. **63**: 141–149.
24. GRAHAM, T.L. & M.Y. GRAHAM. 1991. Glyceollin elicitors induce major but distinctly different shifts in isoflavonoid metabolism in proximal and distal soybean cell populations. Mol. Plant Microbe Interact. **4**: 60–68.
25. RIVERA-VARGAS, L.I., A.F. SCHMITTHENNER & T.L. GRAHAM. 1993. Soybean flavonoid effects on and metabolism by *Phytophthora sojae*. Phytochemistry **32**: 851–857.
26. GRAHAM, M.Y. & T.L. GRAHAM. 1991. Rapid accumulation of anionic peroxidases and phenolic polymers in soybean cotyledon tissues following treatment with Phytophthora megasperma f. sp. glycinea wall glucan. Plant Physiol. **97**: 1445–1455.
27. GRAHAM, M.Y. & T.L. GRAHAM. 1994. Wound-associated competency factors are required for the proximal cell responses of soybean to the Phytophthora sojae wall glucan elicitor. Plant Physiol. **105**: 571–578.
28. ABBASI, P.A., M.Y. GRAHAM & T.L. GRAHAM. 2001. Effects of soybean genotype on the glyceollin elicitation competency of cotyledon tissues to Phytophthora sojae glucan elicitors. Physiol. Molec. Plant Pathol. **59**: 95–105.
29. SUBRAMANIAN, S., M.Y. GRAHAM, O. YU & T.L. GRAHAM. 2005. RNA interference of soybean isoflavone synthase genes leads to silencing in tissues distal to the transformation site and to enhanced susceptibility to *Phytophthora sojae*. Plant Physiol. **137**: 1345–1353.
30. GRAHAM, T.L., M.Y. GRAHAM, S. SUBRAMANIAN & O. YU. 2007. RNAi silencing of genes for elicitation or biosynthesis of 5-deoxyisoflavonoids suppresses race-specific resistance and HR cell death in *Phytophthora sojae* infected tissues. Plant Physiol. **144**: 728–740.

31. GRAHAM, T.L., J.E. KIM & M.Y. GRAHAM. 1990. Role of constitutive isoflavone conjugates in the accumulation of glyceollin in soybean infected with Phytophthora megasperma. Molec. Plant Microbe Interact. **3:** 157–166.
32. LANDINI, S., M.Y. GRAHAM & T.L. GRAHAM. 2002. Lactofen induces isoflavone accumulation and glyceollin elicitation competency in soybean. Phytochemistry **62:** 865–874.
33. GRAHAM, M.Y. 2005. The diphenylether herbicide lactofen induces cell death and expression of defense-related genes in soybean. Plant Physiol. **139:** 1784–1794.
34. VAN LOON, L.C. & E.A. VAN STRIEN. 1999. The families of pathogenesis-related proteins, their activities, and comparative analysis of PR-1 type proteins. Physiol. Mol. Plant Pathol. **55:** 85–97.
35. TAKEUCHI, Y., M. YOSHIKAWA, G. TAKEBA, et al. 1990. Molecular cloning and ethylene induction of messenger RNA encoding a phytoalexin elicitor-releasing factor beta-1 3 endoglucanase in soybean. Plant Physiol. **93:** 673–682.
36. YOSHIKAWA, M., M. MATAMA & H. MASAGO. 1981. Release of a soluble phytoalexin elicitor from mycelial walls of *Phytophthora megasperma* var. sojae by soybean tissues. Plant Physiol. **67:** 1032–1035.
37. HAM, K.-S., S.-C. WU, A.G. DARVILL & P. ALBERSHEIM. 1997. Fungal pathogens secrete an inhibitor protein that distinguishes isoforms of plant pathogenesis-related endo-beta-1,3-glucanases. Plant J. **11:** 169–179.

Long-Term Acclimation of Plants to Elevated CO_2 and Its Interaction with Stresses

ZOLTÁN TUBA[a,b] AND HARTMUT K. LICHTENTHALER[c]

[a]*Plant Ecology Research Group of Hungarian Academy of Sciences, Faculty of Agricultural and Environmental Sciences, Szent István University, Gödöllö, Hungary*

[b]*Department of Botany and Plant Physiology, Faculty of Agricultural and Environmental Sciences, Szent István University, Gödöllö, Hungary*

[c]*Botanical Institute, Molecular Biology and Biochemistry of Plants, University of Karlsruhe, Karlsruhe, Germany*

ABSTRACT: Rising atmospheric carbon dioxide (CO_2) concentration and air temperature are of major concern when considering the possible effects of global climate change on vegetations. Although production has been found to increase in many cases, other experiments have also indicated increased hazards for plant growth because of the increased frequency of weather extremes, such as droughts, floods, and extreme temperatures. Thus at the same time elevated CO_2 and the extreme climatic events, intra- and interannual climatic variability alone can be foreseen as an indirect constraint, which separately influences significantly the carbon cycling in ecosystems, too. In the shorter term the effect of CO_2 is direct and is mediated by photosynthesis. In the longer term the effects of elevated CO_2 became more and more indirect and its effects are mediated by source-sink interactions within plants, resources (nutrients, water), temperature, microbes, herbivores, and land-use management practice. In fact, the plants can make use of their general stress coping mechanisms to avoid or compensate possible negative effects of elevated CO_2. One has to consider that all the classical abiotic, biotic, and anthropogenic stressors are threatening plant growth and development also under elevated CO_2, although at possibly different doses compared to ambient CO_2 concentrations. Therefore, the knowledge of the general stress coping, stress avoiding, and tolerance mechanisms is needed to understand the regulation of the plants' metabolism under normal and elevated CO_2 levels.

KEYWORDS: acclimation; environmental constraints; forests; grasslands; isoprenoids; photosynthesis; stress alleviator; stress detection

Address for correspondence: Zoltán Tuba, Department of Botany and Plant Physiology, Faculty of Agricultural and Environmental Sciences, Szent István University, Gödöllö, H-2103, Gödöllö, Páter K. u.1.Hungary. Voice: +36 28 522 075; fax: +36 28 410 804.
tuba.zoltan@mkk.szie.hu

INTRODUCTION

Global climate change appears to be the greatest ecological problem of the future. When considering prospects of mankind, sound, and detailed information about the unavoidable ecological effects of the global climatic change are inevitably needed. Unfortunately, this statement would be correct even if all the factors causing this change could be brought to a halt at once. This is because these factors will continue to exert their effect for a long time. For example, the carbon dioxide emitted into the atmosphere will have a long-term effect and this is one of the most influential ecological factors at global scale. The concentration of CO_2 in the atmosphere has been increasing in the two last centuries from about 280 ppm to a present value of 360 ppm, and is expected to reach more than twice the preindustrial concentration by the end of this century.[1] The increasing use of fossil fuels and deforestation causes severe impacts on climatic components, such as temperature and rainfall. There is general agreement among different climate models that the accumulation of CO_2 and other greenhouse gases (CH_4, N_2O, and chlorofluorocarbons) will cause an increase in mean global temperature at the surface of earth. Although debate has continued about this topic since the end of the 19th century, it is only in the recent years that it has become possible to model the process in a more realistic way.[2]

Rising atmospheric carbon dioxide concentration and air temperature are of major concern when considering the possible effects of global climate change on crops and vegetation and their magnitudes. The predicted rise in global surface temperature together with all its ecological consequences is mainly attributed to the increased level of CO_2 in the atmosphere.

The increasing concentration of CO_2 affects the plants directly, causing changes in their chemical composition, physiological processes, production, and fitness.[3,4] Consequently, the tolerance, reproduction, distribution, and abundance of plants will be altered.[5] Species composition and diversity of plant communities, the vegetation dynamics, and the structure of vegetation itself are all expected to change, too. All these changes in individual plants and in the vegetation will affect the subsistence of every living creature, so that of mankind.

As a consequence of increased photosynthesis at elevated CO_2, dry matter production of terrestrial C_3 plants is expected to increase. Reviews by Kimball[6] (1983) and Cure and Acock[7] (1986) of experiments done under a wide range of conditions have shown that a doubling of atmospheric CO_2 from 330 to 660 ppm may increase the productivity of C_3 species by an average of 33%. If these observations are extrapolated to the field, this means that the CO_2 enrichment since the start of the industrial revolution could be expected to have increased dry matter production of C_3 herbaceous plants by 7.5% to 9.0%. Other studies[8] reported that the average growth stimulation of C_3 plants is higher (40–44%) than 33% and the average increase of production in C_4

plants in response to a doubled CO_2 concentration is about 22–33%, which is lower than the average for C_3 plants.

Considering terrestrial ecosystems on a global scale forests and grasslands are the two largest vegetation formations. Forests cover the largest area of terrestrial vegetation on the earth. Grasslands cover 24% of the earth's vegetated area[9] and their area is further increasing with the land use changes (clearance of forests, urbanization, extensive agricultural practices). They occur over a broad range of climatic and soil conditions that vary from intensively managed sown pastures to natural grasslands. Grasslands similarly to other natural vegetation can largely affect the global cycling of carbon.[10] Despite their importance, the potential effects of climate change on grasslands have received much less attention than effects on other ecosystems, such as forests.[11,12]

A prediction of the effects of global change on plants and vegetation, like forests and grasslands may be based on studies of the effects of temperature, radiation, humidity on single species, or small plot vegetation.[13–15] However, scaling from these results,[16] usually obtained on plants or small plots growing in artificial substrates and/or controlled environments, to natural conditions in which competition[17] or symbiosis[18] may play a major role, is often misleading. An attempt to predict the consequences of increased air CO_2 concentration on crops and vegetation must therefore take account of both the direct effects of elevated CO_2 and the indirect effects mediated by climatic change, as well as their interaction, bearing in mind the complexity of the natural ecosystems in which plants are growing.

Although production has been found to increase in many cases, other experiments have also indicated increased hazards for plant growth because of the increased frequency of weather extremes, such as droughts, floods, and extreme temperatures.

ELEVATED AIR CO_2 CONCENTRATION AND STRESSES

Elevated CO_2 is a new challenge for plants and the ecosystem. Higher CO_2 levels may not be a real or strong stressor for plants, but they are a continuous strain that requires particular short-term acclimation and long-term adaptation responses. In fact, the plants can make use of their general stress coping mechanisms to avoid or compensate possible negative effects of elevated CO_2. One has to consider that all the classical abiotic, biotic, and anthropogenic stressors[19,20] are threatening plant growth and development also under elevated CO_2, although at possibly different doses compared to ambient CO_2 concentrations. Therefore, the knowledge of the general stress coping, stress avoiding, and tolerance mechanisms is needed to understand the regulation of the plants' metabolism under normal and elevated CO_2 levels.

The original stress concept of Hans (János) Selye in 1936[21] has also been extended to plants, and is today well defined to describe the action of unfavorable environmental constraints and stressors on plants as well as the plants' response via stress-avoiding and stress-coping mechanisms including special short-term acclimations and long-term adaptations (see the reviews Lichtenthaler 1996 and 1998, where more references are given).[19,20] According to this unifying stress concept, the plants respond to the action of single or combined stressors with a competent crisis management that consists of four phases: (1) the response phase (alarm reaction, beginning of stress), (2) the restitution phase (stage of resistance, hardening, repair processes at continuing stress), (3) end phase (stage of exhaustion at long-term high stress intensity when the adaptation capacity is overcharged), and (4) the regeneration phase (i.e., partial or full regeneration of physiological function upon removal of the stressors). Acute damage, chronic disease, or finally death will occur particularly in those plants that possess only low or no stress tolerance. There exist many stress-coping mechanisms that show up depending on the type, length, and dosis of the stressors, such as proline accumulation during drought and salinity, polyol accumulation (mannitol, sorbitol) at water stress conditions, formation of radical scavenging substances (ascorbate, glutathion, α-tocopherol), increase of the levels of superoxide dismutase, production of heat shock proteins, as well as accumulation of UV-absorbing phenols and flavonoids in the epidermis layer to protect the photosynthetic pigment apparatus in the green mesophyll cells against damaging UV radiation and excess light stress.

Sooner or later all stressors will affect either directly or indirectly the photosynthetic apparatus and its function as well as the many other central metabolic and biosynthetic activities of chloroplasts—other than photosynthetic carbon assimilation—such as *de novo* fatty acid biosynthesis, *de novo* porphyrin ring formation for biosynthesis of chlorophylls and all cellular cytochromes, biosynthesis of carotenoids and isoprenoids via the DOXP/MEP pathway.[22,23] This is the reason, why during evolution many stress-coping mechanisms were developed in chloroplasts, to protect their multiple metabolic activities. Such mechanisms comprise at high irradiance the partial photoinhibition (i.e., a destruction and rapid turnover of the D1-protein of PS2), an enhanced xanthophyll cycle activity, the *de novo* biosynthesis of β-carotene and zeaxanthin, as well as accumulation of α-tocopherol and reduced plastoquinone-9 in the osmiophilic plastoglobuli.[23] All these biosynthetic processes consume photosynthetically formed ATP and NADPH, thus helping to avoid overreduction of the photosynthetic apparatus. Elevated CO_2, however, interacts with such processes, since it causes a short-term enhanced photosynthetic CO_2 assimilation. Moreover, *de novo* biosynthesis and emission of volatile isoprenoids, such as isoprene or methylbutenol or the monoterpenes—all being formed via the chloroplastidic DOXP/MEP pathway of isoprenoid biosynthesis,[22,24] are now understood as a possibility to regulate the internal chloroplast and cell metabolism and to efficiently protect the plants' photosynthetic apparatus and thylakoid lipids at higher temperatures ($>20°C$) and excess irradiance from ozone and other reactive oxygen species.[25]

The available data show that under high incident photon fluxes and elevated temperatures rather high amounts of photosynthetically fixed carbon are channeled into the two hemiterpenes isoprene and methylbutenol, which causes a negative carbon balance. In fact, isoprene emissions from plants amounts to hundreds of millions of metric tons to the global atmosphere per year, the estimations range from 180 to ca. 450×10^{12} g carbon per year worldwide.[26–28] Since its emission essentially contributes to the photochemical formation of ozone in the atmosphere, it is by catalyzing ozone formation an additional stress factor for plant growth and development. Elevated CO_2 competes in the chloroplast with isoprene formation for photosynthetically formed ATP and NADPH and thus should be able to reduce isoprene emission. In fact, evidence has emerged that the amounts of isoprene emitted from a leaf significantly increase at below-ambient CO_2 levels and decrease when CO_2 levels rise to above-ambient.[29] Therefore, changes in isoprene and monoterpene emission of forests and grasslands due to climate variability and changes in atmospheric CO_2 are matter of intensive research.[30]

Plant stress detection: There exist various techniques to detect stress effects in plants. Measurement of chlorophyll fluorescence parameters and reflectance signatures from single leaf points are routine methods for early stress detection as reviewed before.[31] However, such single point measurements have the disadvantage that they provide only one information per each measurement. In the last decade the imaging techniques have been established as a diagnostic tool for detecting plant stress and screening chlorophyll fluorescence.[32–34] These imaging methods simultaneously provide the fluorescence information of whole leaves via several ten thousands or more than 100,000 leaf points (pixels) per measurement. Thus, they allow an excellent stress detection with a high statistical significance. A decline in photosynthetic activity due to water stress or the differences in photosynthetic activity between sun and shade leaves have been documented by fluorescence imaging,[35–37] and this imaging method should be able to also detect differences in photosynthetic activity between plants grown at ambient and elevated CO_2 levels. By including of the plants' blue and green fluorescence from the cell walls, together with the red and far-red chlorophyll fluorescence, in a multispectral fluorescence imaging is presently the best method for an early detection of stress symptoms of plants and the ecosystem.[33,34] Specifically, images of the fluorescence ratios blue/red and blue/far-red are very sensitive early indicators for strain and stress in plants.[34,38] Imaging multispectral fluorescence together with the plants' reflectance signals is the newest high-tech development in stress detection of plants.[39] These imaging techniques are further developed for the application in remote sensing of the ecosystem; they will be of great use in evaluating the effects of global warming and elevated CO_2 on plants and ecosystems.

It seems that production under elevated atmospheric CO_2 concentration is expected to increase in situations where growth limitation by CO_2 is significant when compared to the other main factors, such as nutrient availability or temperature. When one of the latter is seriously limiting plant growth, increased atmospheric CO_2 concentration is not expected to alleviate this limitation.

It is to expected that higher growth rate in response to elevated CO_2 will increase a plant's nutrient demand. Nitrogen and phosphorus are the nutrients most likely to be influenced by rising CO_2 since relatively high quantities of both are needed in the photoreductive C cycle and the photooxidative cycle.[40] When nitrogen is low or marginal, the acceleration of growth in elevated CO_2 may drive the plant into nitrogen deficiency. As a consequence, the nitrogen content of plant tissues is very commonly reduced.

The relatively low increase of the aboveground stand production under elevated CO_2 may have been the consequence of the enhanced C-allocation to the roots. This is a typical high CO_2 allocation response in order to increase both the C-sink capacity and nitrogen uptake in high CO_2 exposed plants and grassland stands.[41] Increased biomass allocation to the roots has been shown to be caused by the N-limitation. Under elevated CO_2, species of Leguminosae have been shown to alleviate the N-limitation on the growth of the grass species.[42] Therefore, the decline in the leaf nitrogen concentration under elevated CO_2 will be much milder in N_2-fixing species than in no N_2-fixing species.[43] In addition, there is a close link between the carbon and nitrogen cycles, which is supported by the findings that a decline in inorganic nitrogen availability in grass sward contributed to the increased below-ground allocation of the grass.[42]

For phosphorus, in contrast to nitrogen, higher tissue concentrations are required for an optimal productivity under elevated CO_2.[40,44] At the low P supply white clover plants grown at twice ambient $p(CO_2)$ lost their stimulation of photosynthesis, they accumulated nonstructural carbohydrates in the leaves and their growth rate was not stimulated as was in the case of high P grown plants.[44] This and other studies indicate a C-sink limitation of growth. This limitation could be overcome by mycorrhiza.[45]

Communities with decreased diversity can acquire less biomass and carbon than plant assemblages with greater diversity. This is in good agreement with earlier statement of Hunt *et al.*[46] (1997) that plant biomass productivity is severely constrained in plant communities limited by resources (water, nitrogen, phosphorus) and it seems likely that CO_2 effects may occur primarily through interactions with these limiting environmental factors and may take place without major increase in community productivity.

In cases where production has been found to increase, experiments have also indicated in grasslands increased hazards for plant growth because of the increased frequency of weather extremes, such as droughts, extreme temperatures, floods, etc. A greater response to CO_2 is often observed in dry years.[47] In water limited grasslands, elevated CO_2 results in greater water availability for longer periods during the growing season, therefore the hydrological and water economical consequences of elevated CO_2 in water limited habitats can be greater than the direct effect of high CO_2 on photosynthetic CO_2 assimilation and production.[48] In temperate grasslands, around 3 to 4°C increase in temperature may counterbalance the effect of elevated CO_2 on production.[49]

One of the significant impacts of elevated CO_2 on plants is to reduce stomatal conductance.[41] It is currently assumed that stomatal aperture decreases with increasing CO_2, with little or no effect on photosynthesis, but with a substantial increase in WUE, which is expected to range between 60 and 160% under doubled atmospheric CO_2 concentrations. This will be of most importance in the summer drought conditions experienced in the temperate continental and other seasonally drier climates. It is possible that the greatest impact of elevated CO_2 will be on plant water relations and drought survival rather than on photosynthetic productivity under these summer drought conditions.[14]

The effect of stomatal response on the total stand water use may be negligible if LAI is significantly increased by elevated CO_2[50] although there may be effects on the energy balance of the canopy. On the other hand, in forest trees there may be a substantial reduction in the leaf area per plant,[41] enhancing the effect of reduced stomatal conductance on canopy water use. It has been observed that reduction in transpiration may also reduce the water vapor mole fraction in the atmosphere, thus, in turn, stimulating transpiration.[51]

The protecting effect of elevated CO_2 against water stress under high temperatures[52] is related to the increased water use efficiency[53] and osmotic adjustment[54] in plants grown in high CO_2 environment. Under elevated CO_2 during water stress plants have greater ability to withstand water stress, which is usually related to the partial closure of stomata together with the occurrence of osmotic adjustment mechanisms. In particular, increased concentrations of solutes in leaves growing under elevated air CO_2 are credited with the maintenance of higher relative water content and turgor potential.[54] The altered leaf and individual plant water relations can modify the stand level water relation responses.[41]

Elevated CO_2 level proved to be beneficial also at the desiccation end of the dehydration–rehydration cycle (desiccation stress) in the moss *T. ruralis* and lichen *C. convoluta*. The high CO_2 exposure not only increased the overall carbon gain by about one-third but it also prolonged the net photosynthesis without influencing the rate of water loss.[15] Most likely, in desiccation-tolerant plants CO_2 is a limiting factor for carboxylation even at this low hydration level.[55]

According to Tuba *et al.*[56] (1999) the high CO_2 level might be beneficial for the CO_2 assimilation of the desiccation tolerant mosses with physiological level of heavy metals by improving their carbon balance. Elevated CO_2 concentration could ameliorate partly the deleterious effects of heavy metal stress as well, but response of the high CO_2 concentration to the heavy metals were very diverse.[57]

CONCLUDING REMARKS

In long-term elevated atmospheric CO_2 concentration cannot be considered as unambiguously beneficial for natural vegetation, like grasslands and field grown agronomic plants, but rather a profound change in carbon source to which the vegetation and outdoor crops have to acclimate to.[5,41] However, in an indirect way by interaction with other environmental and biological[58] factors elevated CO_2 can appear as a stress effect. Further, under current atmospheric CO_2 concentration vegetation and field grown agronomic plants continuously have to face multiple environmental constraints (limited resources, extreme temperatures, herbivores, pollution, fire, etc.), which occur as naturally induced stress effects.[19,59] On the other hand, there are several indications that plants may generally cope better with various environmental stresses at higher CO_2 concentration. It can help plants to cope better, for example, with drought,[60] heat and frost,[61] and salinity[62] when these stressors do not act too long and the stress-dosis is not too high that it fully overcharges the general adaptation and response capacity as well as the repair mechanisms of plants. Thus, at the same time elevated CO_2 can be foreseen as an indirect constraint and also as a stress alleviator.[5]

Many of experimental evidences support the prediction that the effects of elevated CO_2 will occur not as a direct beneficial effect of CO_2 on plants growing under natural environmental conditions. In the shorter term the effect of the CO_2 is direct and is mediated by photosynthesis. However, in the longer term the effects of elevated CO_2 become more and more indirect and its effects are mediated by source-sink interactions within plants, resources (nutrients, water), temperature, microbes, herbivores, and land use/management practice.[4,46] Although in agricultural plant production systems the situation is favorable, yet in these systems a few but not all of the environmental constraints can be controlled or their limiting effects reduced.

ACKNOWLEDGMENTS

This work was supported by VOP-AKF-3.1.1.-2004-05-0358/3.0, KlimaKKT (NKFP6-00079/2005, CARBOEUROPE - IP EU (GOCE-CT-2003-50557), NITROEUROPE - IP EU (017841 - GOCE) and the Czech-Hungarian Science and Technology Intergovernmental (2007–2008) projects.

REFERENCES

1. HOUGHTON, J.T., G.J. JENKINS & J.J. EPHRAUMS, Eds.: 1990. Climate change: the IPCC scientific assessment, Cambridge University Press, Cambridge.
2. AMTHOR, J.S. 1995. Terrestrial higher-plant response to increasing atmospheric [CO2] in relation to the global carbon cycle. Global Change Biol. **1**: 243–274.

3. DRAKE, B.G., M.A. GONZÁLEZ-MÁLER & S.P. LONG 1997. More efficient plants: a consequence of rising atmospheric CO2? Ann. Rev. Plant Physiol. Plant Mol. Biol. **48:** 609–639.
4. TUBA, Z. 2005. Ecological responses and adaptations of crops to rising atmospheric carbon dioxide. p 414. Haworth Press. New York, NY. (ISBN: 1–56022-120-8)
5. TUBA, Z., A. RASCHI, G.M. LANNINI, *et al*. 2003. Vegetations with various environmental constraints under elevated atmospheric CO2 concentrations. *In* Abiotic Stresses in Plants: Eds Luigi Sanità di Toppi and Barbara Pawlik-Skowronska, pp. 157–204. Kluwer Academic Publishers Dordrecht, Wand.
6. KIMBALL, B.A. 1983. Carbon dioxide and agricultural yield: an assemblage and analysis of 430 prior observations. Agronomy J. **75:** 779–788.
7. CURE, J.D. & B. ACOCK 1986. Crop responses to carbon dioxide doubling: a literature survey. Agriculture Forest Meteorol. **38:** 127–145.
8. POORTER, H. 1993. Interspecific variation in the growth response of plants to an elevated ambient CO2 concentration. Vegetation **104/105:** 77–97.
9. GOUDRIAAN, J. 1995. Global carbon cycle and carbon sequestration. In Carbon sequestration in the biosphere, A.M. Beran: Ed. pp. 3–18 Springer Verlag. NATO ASI Series, 133. Berlin, Heidelberg.
10. SALA, O.E., W.K. LAURENROTH & I.C. BURKE 1996. Carbon budgets of temperate grasslands and the effects of global change. *In* Global Change: Effects on Coniferous Forests and Grasslands, A.I. Breymeyer, D.O. Hall, J.M. Melillo & G.I. Agren. Eds.: pp. 101–120. John Wiley Chichester, New York, Brisbane, Toronto, Singapore.
11. EAMUS, D. & P.G. JARVIS. 1989. Direct effects of increase in the global atmospheric CO2 concentration on natural and commercial trees and forests. Adv. Ecol. Res. **19:** 1–55.
12. CEULEMANS, R. & M. MOUSSEAU. 1994. Effects of elevated atmospheric CO2 on woody plants. New Phytol. **127:** 425–446.
13. TUBA, Z., K. SZENTE, Z. NAGY, *et al*. 1996. Responses of CO2 assimilation, transpiration and water use efficiency to long-term elevated CO2 in perennial C3 xeric loess steppe species. J. Plant Physiol. **148:** 356–361.
14. TUBA, Z., Z.S. CSINTALAN, K. SZENTE, *et al*. 1998a. Carbon gains by desiccation tolerant plants at elevated CO2. Funct. Ecol. **12:** 39–44.
15. TUBA, Z., M.B. JONES, K. SZENTE, *et al*. 1998b. Some ecophysiological and production responses of grasslands to long-term elevated CO2 under continental and atlantic climates. Ann. N.Y. Acad. Sci. **851:** 241–250.
16. JARVIS, P.G. 1995. Scaling processes and problems. Plant Cell Environ. **18:** 1079–1089.
17. JONGEN, M. & M.B. JONES. 1997. Effects of elevated CO2 on species co-occurring in a neutral grassland. Abstracta Botanica **21:** 249–260.
18. SIMARD, S.W., D.A. PERRY, M.D. JONES, *et al*. 1997. Net transfer of carbon between ectomycorrhizal tree species in the field. Nature **388:** 579–582.
19. LICHTENTHALER, H.K. 1996. Vegetation stress: an introduction to the stress concept in plants. J. Plant Physiol. **148:** 4–14.
20. LICHTENTHALER, H.K. 1998. The stress concept in plants: an introduction. *In* Stress of Life: From Molecules to Man, P. Csermely Ed.: Ann. N. Y. Acad. Sci. **851:** 187–198.
21. SELYE, H. 1936. A syndrome produced by diverse nocuous agents. Nature **138:** 32.

22. LICHTENTHALER, H.K. 1999. The 1-deoxy-d-xylulose-5-phosphate pathway of isoprenoid biosynthesis in plants. Annu. Rev. Plant Physiol. Plant Mol. Biol. **50**: 47–65.
23. LICHTENTHALER, H.K. 2007. Biosynthesis, accumulation and emission of α-tocopherol, plastoquinone-9, carotenoids, isoprene and other isoprenoids in leaves under photosynthetic high irradiance conditions. Photosynth. Res. **93**. Doi: 10.1007/s11120-007-9204-y.
24. ZEIDLER, J. & H.K. LICHTENTHALER 2001. Biosynthesis of 2-methyl-3-buten-2-ol emitted from needles of Pinus ponderosa via the non-mevalonate DOXP/MEP pathway of isoprenoid formation. Planta **213**: 323–326.
25. AFFEK, H.P. & D. YAKIR. 2002. Protection by isoprene against singlet oxygen in leaves. Plant Physiol. **129**: 269–277.
26. RASMUSSEN, R.H. & M.A.K. KHALIL. 1998. Isoprene over the Amazon Basin. J. Geoph. Res. **93**: 1417–1421.
27. SHARKEY, T.D. 1996. Isoprene emission by plants and animals. Endeavour **20**: 74–78.
28. SHARKEY, T.D. & S. YEH 2001. Isoprene emission from plants. Annu. Rev. Plant Physiol. Plant Mol. Biol. **52**: 407–436.
29. ARNETH, A., Ü. NIIMETS, S. PRESSLEY, *et al*. 2007. Process-based estimates of terrestrial isoprene emissions: incorporating the effects of direct CO2-isoprene emission. Atmos. Chem. Phys. **7**: 31–53.
30. NAIK, V., C. DELIRE & D.J. WUEBBLES. 2004. Sensitivity of global biogenic isoprenoid emissions to climate variability and atmospheric CO2. J. Geophys. Res. **109**: DO6301,doi:10.1029/2003JD004236,2004.
31. LICHTENTHALER, H.K., O. WENZEL, C. BUSCHMANN & A. GITELSON. 1998. Plant stress detection by reflectance and fluorescence. *In* Stress of Life: From Molecules to Man, P. Csermely. Ed.: Ann. N. Y. Acad. Sci. **851**: 271–285.
32. LICHTENTHALER, H.K. & J.A. MIEHÉ 1997. Fluorescence imaging as a diagnostic tool for plant stress. Trends Plant Sci. **2**: 316–320.
33. BUSCHMANN, C. & H.K. LICHTENTHALER 1998. Principles and characteristics of multi-colour fluorescence imaging of plants. J. Plant Physiol. **152**: 297–314.
34. NEDBAL, L., J. SOUKUPOVA, D. KAFTAN, *et al*. 2000. Kinetic imaging of chlorophyll fluorescence using modulated light. Photosynth. Res. **66**: 3–12.
35. LICHTENTHALER, H.K. & F. BABANI. 2000. Detection of photosynthetic activity and water stress by imaging the red chlorophyll fluorescence. Plant Physiol. Biochem. **38**: 889–895.
36. LICHTENTHALER, H.K., F. BABANI, G. LANGSDORF & C. BUSCHMANN. 2000. Measurement of differences in red chlorophyll fluorescence and photosynthetic activity between sun and shade leaves by fluorescence imaging. Photosynthetica **38**: 521–529.
37. LICHTENTHALER, H.K., F. BABANI, G. LANGSDORF. 2007. Chlorophyll fluorescence imaging of photosynthetic activity in sun and shade leaves of trees. Photosynth. Res. **94**.
38. LICHTENTHALER, H.K., N. SUBHASH, O. WENZEL, & J.A. MIEHÉ, 1997. Laser-induced imaging of blue/red and blue/far-red fluorescence ratios, F440/F690 and F440/F740, as a means of early stress detection in plants. In: Proceedings of the International Geoscience and Remote Sensing Symposium IGARSS '97, Singapore, IEEE/USA, pp. 1799–1801.

39. LENK, S., L. CHAERLE, E.E. PFÜNDEL, *et al*. 2007. Multispectral fluorescence and reflectance imaging at the leaf level and its possible applications. J. Exp. Botany **58:** 807–814.
40. ROGERS, H.H., G.B. RUNION, S.A. PRIOR & H.A. TORBERT. 1999. Response of plants to elevated atmospheric CO2: Root growth, mineral nutrition and soil carbon. *In* Carbon Dioxide and Environmental Stress. Y. Luo, H.A. Mooney, Eds.: 215–245. Academic Press, San Diego.
41. JARVIS, P.G. 1993. Global change and plant water relations. *In* Water Transport in Plants under Climatic Stress, M. Borghetti, J. Grace & A. Raschi. Eds.: pp. 1–13 Cambridge University Press, Cambridge. Jarvis, P.G. (1995). Scaling processes and problems. Plant, Cell and Environ. 18: 1079–1089.
42. SOUSSANA, J.F., U.A. Hartwig & J.A. ARNONE. 1996a. The effects of elevated CO2 on symbiotic N-2 fixation: a link between the carbon and nitrogen cycles in grassland ecosystems. Plant Soil **187:** 321–332.
43. HARTWIG, U.A., A. LUSCHER, M. DAEPP, *et al*. 2000. Due to symbiotic N-2 fixation, five years of elevated atmospheric pCO(2) had no effect on the N concentration of plant litter in fertile, mixed grassland. Plant Soil **224:** 43–50.
44. ALMEIDA, J.P.F., A. LUSCHER, M. FREHNER, *et al*. 1999. Partitioning of P and the activity of root acid phosphatase in white clover (Trifolium repens L.) are modified by increased atmospheric CO2 and P fertilisation. Plant Soil **210:** 159–166.
45. FITTER, A.H., A. HEINEMEYER & P.L. STADDON 2000. The impact of elevated CO2 and global climate change on arbuscular mycorrhizas: mycocentric approach. New Phytologist **147:** 179–187.
46. HUNT, R., J.P. GRIME, S. DÍAZ, *et al*. 1997. Effects of elevated carbon dioxide on British native grassland species and communities. Abstracta Botanica **21:** 275–288.
47. OWENSBY, C.E., J.M. HAM, A.K. KNAPP & L.M. AUEN. 1999. Biomass production and species composition in a tallgrass prairie ecosystem after long-term exposure to elevated atmospheric CO2. Glob. Change Biol. **5:** 497–506.
48. CAMPBELL, B.D., D.M. STAFFORD SMITH & G.M. MCKEON. 1997. Elevated CO2 and water supply interactions in grasslands: a pastures and rangelands management perspective. Glob. Change Biol. **3:** 177–187.
49. JONES, M. & M. JONGEN. 1996. Sensitivity of temperate grasslands to elevated CO2 and the interaction with temperature and water stress. Agr. Food Sci. Finland **5:** 271–283.
50. DUGAS, W.A., M.L. HEUER, D. HUNSAKER, *et al*. 1994. Sap flow measurements of transpiration from cotton grown under ambient and enriched CO2 concentrations. Agricul. Forest Meteorol. **70:** 231–245.
51. MIGLIETTA, F. & J.S. AMTHOR 1996. Evapotranspiration and CO2. Weather **51:** 322.
52. THORNLEY, J.H.M., M.G.R. CANNELL & O.F.F. MET. 1996. Temperate forest responses to carbon dioxide, temperature and nitrogen: a model analysis. Plant Cell Environ. **19:** 1331–1348.
53. TUBA, Z., K. SZENTE, Z. NAGY, *et al*. 1996a. Responses of photosynthesis, transpiration and water use efficiency to long-term elevated CO2 in four perennial C3 dry loess grassland species. J. Plant Physiol. **148:** 356–361.
54. MENCONI, M., A. RASCHI, V. COSENTINO, *et al*. 1997. Water relations in alfalfa (Medicago sativa L.) plants grown under high carbon dioxide concentration. Abstracta Botanica **21:** 305–308.

55. TUBA, Z., Z.S. CSINTALAN, K. SZENTE, et al. 1998c. Carbon gains by desiccation tolerant plants at elevated CO2. Func. Ecol. **12:** 39–44.
56. TUBA, Z., M.C.F. PROCTOR & Z. TAKÁCS. 1999. Desiccation-tolerant plants under elevated air CO2: a review. Zeitschrift für Naturforschung **54:** 788–796.
57. TAKÁCS, Z., Z.S. CSINTALAN & Z. TUBA. 1999. Ecophysiological responses of the lichen, Cladonia convoluta to elevated CO2 level and simultaneous heavy metal treatment. Zeitschrift für Naturforschung **54:** 797–801.
58. KÖRNER, C.H., F.A. BAZZAZ & C.B. FIELD. 1996. The significance of biological variation, organism interactions and life histories in CO2 research. *In* Carbon dioxide, Populations and Communities, Eds.: C.H. Körner & F.A. Bazzaz, pp. 443–456. Academic Press. San Diego, CA.
59. GEIDER, R.J., E.H. DELUCIA, P.G. FALKOWSKI, et al. 2001. Primary productivity of planet earth: biological determinants and physical constraints in terrestrial and aquatic habitats. Global Change Biol. **7:** 849–882.
60. ARP, W.J., J.E.M. VAN MIERLO, F. BERENDSE & W. SNIJDERS 1998. Interactions between elevated CO2 concentration, nitrogen and water: effects on growth and water use of six perennial plant species. Plant Cell Environ. **21:** 1–11.
61. WAYNE, P.M., E.G. REEKIE & F.A. BAZZAZ 1998. Elevated CO2 ameliorates birch response to high temperature and frost stress: Implications for modelling climate-induced geographic range shifts. Oecologia **114:** 335–342.
62. REUVENI, J., J. GALE & M. ZERONI 1997. Differentiating day from night effects of high ambient CO2 on the gas exchange and growth of Xanthium strumarium L exposed to salinity stress. Ann. Bot. **79:** 191–196.

Heat Shock Proteins and Protection of the Nervous System

IAN R. BROWN

Center for the Neurobiology of Stress, University of Toronto at Scarborough, Toronto, Ontario, Canada

ABSTRACT: Manipulation of the cellular stress response offers strategies to protect brain cells from damage induced by ischemia and neurodegenerative diseases. Overexpression of Hsp70 reduced ischemic injury in the mammalian brain. Investigation of the domains within Hsp70 that confers ischemic neuroprotection revealed the importance of the carboxyl-terminal domain. Arimoclomol, a coinducer of heat shock proteins, delayed progression of amyotrophic lateral sclerosis (ALS) in a mouse model in which motor neurons in the spinal cord and motor cortex degenerate. Celastrol, a promising candidate as an agent to counter neurodegenerative diseases, induced expression of a set of Hsps in differentiated neurons grown in tissue culture. Heat shock "preconditioning" protected the nervous system at the functional level of the synapse and selective overexpression of Hsp70 enhanced the level of synaptic protection. Following hyperthermia, constitutively expressed Hsc70 increased in synapse-rich areas of the brain where it associates with Hsp40 to form a complex that can refold denatured proteins. Stress tolerance in neurons is not solely dependent on their own Hsps but can be supplemented by Hsps from adjacent glial cells. Hence, application of exogenous Hsps at neural injury sites is an effective strategy to maintain neuronal viability.

KEYWORDS: ischemia; neuroprotection; protein misfolding disorders; synapse: neurodegenerative diseases; extracellular Hsps

INTRODUCTION

In response to various forms of stress, cells activate a highly conserved heat shock response in which a set of heat shock proteins (Hsps) are induced, which play important roles in cellular repair and protective mechanisms. Evidence suggests that manipulation of the cellular stress response may offer strategies to protect brain cells from damage that is encountered following cerebral ischemia or during the progression of neurodegenerative diseases. This article highlights

Address for correspondence: Ian R. Brown, Ph.D., Center for the Neurobiology of Stress, University of Toronto at Scarborough, 1265 Military Trail, Toronto, Ontario, Canada M1C 1A4. Voice: 416-287-7413; fax: 416-287-7642.

ibrown@utsc.utoronto.ca

aspects of research in these areas that were presented at the symposium session on "Heat shock proteins and cerebral protection" at the 2007 World Conference on Stress held in Budapest in August 2007.

HSP70 AND PROTECTION FROM CEREBRAL ISCHEMIA

Overexpression of Hsp70

Stroke is the third leading cause of death in many countries. Experiments using both animal models of stroke and tissue culture systems have indicated that overexpression of Hsp70 reduced ischemic injury and protected both neurons and glial cells.[1–3] The protective actions of Hsp70 could include prevention of protein aggregation, refolding of partially denatured proteins, reduction of inflammatory responses, and inhibition of cell death pathways.[4–6]

Analysis of the Neuroprotective Domains of Hsp70

To investigate the domains within Hsp70 that are important for conferring ischemic protection, two mutants of wild-type Hsp70 have been employed-an ATPase-deficient point mutant and a deletion mutant lacking the ATPase domain and encoding the carboxyl-terminal domain.[7] Injection of constructs into the brain resulted in the transfection of both neurons and glial cells. The ATPase-deficient mutant and the deletion mutant containing the carboxyl-terminal domain were similar to wild-type Hsp70 in their ability to significantly reduce focal ischemic injury. Animals that overexpressed either of the two mutant proteins or wild-type Hsp70 had improved neurological scores, smaller infarcts, reduced protein aggregation, and decreased nuclear translocation of apoptosis-inducing factor compared to control animals. These results suggest that the carboxyl-terminal portion of Hsp70 is sufficient for neuroprotection against ischemia *in vivo*.[7] Similar results were obtained in a tissue culture model system of ischemic-like injury.[7] Further, experiments demonstrated that the protective effect of the carboxyl-terminal portion of Hsp70 is associated with maintenance of mitochondrial physiology during stress induced by glucose deprivation.[8]

Antiapoptotic and Anti-inflammatory Effects of Hsp70

Overexpression of Hsp70 has been reported to be associated with a decrease in apoptotic cell death, an increase in the expression of the antiapoptotic protein Bcl-2, a suppression of microglial/monocyte activation, and a reduction in matrix metalloproteinases.[9] These observations suggest that Hsp70 is a multifaceted protein capable of protecting brain cells from injury through a variety of mechanisms.

HSPS AND PROTEIN MISFOLDING DISORDERS

Upregulation of Hsps as a Strategy to Counter Neurodegenerative Diseases

Neurodegenerative disorders such as Alzheimer's disease, Parkinson's disease, and amyotrophic lateral sclerosis (ALS) have been termed "protein misfolding disorders" that are characterized by the neuronal accumulation of aggregation-prone misfolded proteins.[10–13] Overexpression studies on Hsp70, Hsp40, and Hsp27 have demonstrated protective effects of Hsps in several animal models of neurodegenerative diseases.[12] For example, overexpression of Hsp70 rescued neurons from β-amyloid-mediated toxicity in models of Alzheimer's disease.[14,15] In a *Drosophila* model of Parkinson's disease, expression of wild-type or mutant α-synuclein in dopaminergic neurons, resulted in the formation of inclusion bodies and a loss of neurons.[16] Coexpression of human Hsp70 prevented α-synuclein-mediated toxicity.[16] These studies involved the overexpression of a single Hsp. However, a set of Hsps is induced in the cellular stress response and induction of a set could provide an enhanced level of neuroprotection compared to overexpression of one type of Hsp. In mammalian cells induction of the set of stress-inducible Hsps is controlled at the transcriptional level by heat shock transcription factor 1 (HSF1), the master stress-inducible regulator.[17] This has stimulated interest in the regulation of Hsps by HSF1 and the investigation of pharmacologically active molecules that modulate HSF1 as a strategy to counter neurodegenerative diseases that generate misfolded aggregation-prone proteins[13,18]

A Coinducer of Heat Shock Proteins Delays Disease Progression in ALS Mice

Arimoclomol, a coinducer of heat shock proteins, is a potential therapeutic agent for the treatment of ALS, a rapidly progressive disease in which motor neurons in the spinal cord and motor cortex, degenerate, resulting in muscle weakness, paralysis, and death within 2–5 years of diagnosis. Using a mouse model of ALS, it has been demonstrated that application of arimoclomol delayed disease progression with marked improvement in hind limb muscle function, decreased cell death of motor neurons, and increased life span.[19–22] A significant delay in ALS disease progression was observed even when arimoclomol was administered after the visible onset of disease symptoms.[19] Arimoclomol prolonged the activation of HSF1, resulting in increased expression of Hsp70 and Hsp90 in motor neurons.[19] It has been reported that cultured motor neurons fail to induce Hsp70 after heat shock or exposure to excitotoxic glutamate, hence these neurons may be hampered in their ability to react to cellular stress.[23] The coinducer arimoclomol overcomes the inability of motor neurons to induce Hsp70 and rescues these neurons in the mouse model of ALS.[19] Hsps may protect against toxic proteins that arise during neurodegenerative diseases at levels that include prevention inappropriate protein interactions, promotion of protein degradation, sequestering of toxic protein forms, and blockage of signal events that lead to neuronal dysfunction and cell death.[12]

Neuroprotective Effects of Celastrol

Celastrol has been identified as a potential neuroprotective candidate in a collaborative drug screen that was designed to identify therapeutic agents against neurodegenerative diseases from a panel of 1,040 existing drugs (of which 75% are approved by the U.S. Food and Drug Authority).[24,25] Using tissue culture-based assays, candidate drugs were scored on their ability to suppress features associated with neurodegenerative diseases, such as protein aggregation. Celastrol is a quinine methide triterpene extracted from a plant of the Celastraceae family.[26] It has been used in traditional Chinese medicine for the treatment of ailments, such as inflammation[27] and rheumatoid arthritis[28] and also demonstrates immunosuppressive effects.[29] Celastrol has been reported to be exhibit protective properties in several animal models of neurodegenerative diseases, such as Parkinson's disease,[30] ALS,[31] and Hungtington's disease.[30,32] Celastrol induces Hsp70 by activation of heat shock transcription factor HSF1 in undifferentiated neuroblastoma cells.[33]

Neurodegenerative diseases have been termed "protein misfolding disorders" that are characterized by the neuronal accumulation of protein aggregates.[10–13] Manipulation of the cellular stress response involving induction of Hsps in differentiated neurons offers a therapeutic strategy to counter conformational changes in neuronal proteins that trigger pathogenic cascades resulting in neurodegenerative diseases. Hsps are protein repair agents that provide a line of defense against misfolded, aggregation-prone proteins.[12] However, neurons in the differentiated state, in both *in vivo* and *in vitro* systems, have been reported to be resistant to Hsp induction following conventional heat shock.[23,34–37]

Recent observations indicate that celastrol is capable of inducing a set of Hsps in differentiated neurons.[38] These studies were carried out on two cell lines of different species, namely human and rodent. The rationale for including these two differentiated neuronal lines was to explore species-specific differences in Hsp neuronal induction patterns that may influence the translation of observations on animal-based models of neurodegenerative diseases to the actual human condition.[38] Animal models of Alzheimer's disease, Parkinson's disease, and ALS have been developed in rodents. Our studies indicate that celastrol induced a wider set of potentially neuroprotective Hsps, including Hsp70B', in differentiated human neurons compared to differentiated rodent neurons.[38] Given this differential pattern of Hsp induction, celastrol may confer greater protective effects against neurodegenerative diseases in the human brain, exceeding its potential in the rodent-based neurodegenerative models mentioned above.[30–32]

Our study[38] highlights the fact that the spectrum of Hsp70 genes present in the human genome includes members, such as Hsp70B', which are not present in rodent genomes[39,40] and hence are not beneficial factors in animal models of neurodegenerative diseases to combat misfolded aggregration-prone proteins. The Hsp70B' gene arose after the divergence of rodents and humans.[39,40] Although Hsp70B' and stress-inducible Hsp70 share 84% sequence identity, differences in the substrate-binding pocket and activation profiles may confer Hsp70B' with a distinct cellular role.[40] Hsp70B' has not received attention in the field of human neurodegenerative diseases. Hsp70B' is a unique member of the human Hsp70 family of chaperones about which information is scarce.[40] Celastrol is a promising candidate as a therapeutic agent for neurodegenerative diseases with the attractive feature of upregulating a wider set of Hsps, including Hsp70B', in differentiated human neurons compared to differentiated rodent neurons.[38]

HSPS AND SYNAPTIC CONNECTIONS BETWEEN NEURONS

Protection of the Physiological Process of Synaptic Neurotransmission

Exposure to sublethal temperatures induces the expression of Hsps and protects tissues and organisms from cell death that would normally result from exposure to lethal temperatures and other forms of stress.[41] Heat shock "preconditioning" has been reported to protect neural cells against subsequent stressful stimuli in both tissue culture systems and in the intact nervous system.[42–47]

An intriguing question that has received comparatively little attention is whether heat shock "preconditioning" protects neural cells at the functional level. Synapses are critical points of information transfer between neurons and their functionality must be preserved during stressful conditions to prevent communication breakdown in the nervous system. Synaptic connections are recognized as being particularly vulnerable regions of neurons that are involved in the physiological process of neurotransmission that links neurons to functional networks.[48] Prior heat shock protected synaptic neurotransmission and synapses were able to function at temperatures that would normally be disruptive.[49] Selective overexpression of Hsp70 enhanced the level of synaptic protection at the functional level.[50] Use of a *Drosophila* mutant that fails to accumulate inducible Hsp70 revealed the compensatory upregulation of constitutively expressed Hsps and the preservation of synaptic thermoprotection.[51] Addition of exogenous Hsp70 to the medium of a slice preparation from the mammalian brain protected synaptic transmission from thermal stress as did "preconditioning" treatment.[52] Targeting Hsp70 to motor neurons caused structural plasticity of axonal terminals associated with increased transmitter release at neuromuscular junctions at high temperature.[53] This protected larval locomotor activity from hyperthermia in *Drosophila*. Biochemical isolation of synaptic fractions and immunocytochemistry at the electron microscope level have localized Hsp70, Hsc70, Hsp32, and Hsp27 to the synapse.[54,55] In summary, induction of the heat shock response protects the nervous system at the functional level and permits neurotransmission events to proceed at synaptic connections between neurons under stressful conditions.

Lipid rafts are specialized membrane microdomains that are enriched in neural tissue and serve as assembly and sorting platforms for signal transduction complexes including those involved with neurotransmission events at synapses. In addition to neurotransmitter receptors, a set of Hsps has been identified in lipid rafts isolated from neural tissue.[56] This set of stress proteins may play roles in maintaining the stability of raft-associated signal transduction complexes following neural stress.

Enhancement of Constitutively Expressed Hsc70 in Synapse-Rich Neural Areas after Hyperthermic Stress

Hsp70 is a multigene family composed of some members that are stress-inducible (Hsp70) and other members that are constitutively expressed (Hsc70). We have noted that Hsc70 protein is enriched in the mammalian nervous system compared to nonneural tissues and is present at high levels in neuronal cell bodies.[34,57] After thermal stress, overall levels of constitutive Hsc70 did not change in the brain as determined by Western blotting.[57,58] However, confocal immunocytochemistry detected an enhancement of Hsc70 in synapse-rich areas of the cerebral cortex where it associates with Hsp40 to form a complex that can refold denatured proteins.[58] These results could be interpreted to suggest that the heat shock response in the nervous system involves not only the synthesis of stress-inducible Hsps, but also translocation of constitutively expressed Hsc70 to synapse-rich areas where it could participate in neuroprotective mechanisms that preserve synaptic function during times of stress.[58] This protective scenario may be particularly relevant to differentiated neurons that characteristically exhibit high levels of Hsc70 and do not synthesize stress-inducible Hsp70 after thermal stress.[23,34,35,59,60]

As previously mentioned, heat shock "preconditioning" protects neural cells at the functional level of neurotransmission events at synapses.[49–51]

Investigations of neuroprotective mechanisms have tended to focus on stress-inducible Hsp70 that is induced by the "preconditioning" procedure. Our observations suggest that enhancement of constitutively expressed Hsc70 at synapses, and its co-localization at these sites with Hsp40, could play a role in synaptic protective mechanisms that preserve neurotransmission processes during times of stress.[58]

CONSTITUTIVE HEAT SHOCK PROTEINS AND NEURODEGENERATIVE DISEASES

We have reported that constitutively expressed Hsc70 protein is enriched in the mammalian nervous system compared to nonneural tissues and is present at high levels in neuronal cell bodies.[34,56,57] Recently, we have compared levels of Hsc70 in different classes of neurons that are affected in different neurodegenerative diseases.[61] Motor neurons of the spinal cord, that are impacted in a low frequency disease, such as ALS, exhibited very high levels of Hsc70 whereas neurons in the hippocampus and entorhinal cortex, which are affected in a high frequency disease, such as Alzheimer's, showed comparatively low levels of Hsc70.[61] Intermediate levels of Hsc70 were apparent in neurons of the substantia nigra what are impacted in an intermediate frequency disease, such as Parkinson's.

These three neurodegenerative diseases have been termed "protein misfolding disorders" that exhibit common molecular defects associated with the accumulation of protein aggregates and selective neuronal loss.[10–13] Hsps provide a line of defense against misfolded aggregation-prone proteins and among the most potent suppressors of neurodegeneration in animal models.[12,13,62] We suggest that variable levels of constitutively expressed Hsc70 in different classes of neurons may result in a variable buffering capacity against protein misfolding disorders, which correlates with the relative frequency of these diseases in the human population.[61] Neurons may rely on their constitutive levels of Hsc70 as a "preprotection" mechanism for defense against protein misfolding and aggregation that are induced by stressful stimuli or associated with neurodegenerative diseases.

NEUROPROTECTIVE EFFECTS OF EXTRACELLULAR HEAT SHOCK PROTEINS

Early work demonstrated that Hsps are transferred between cell types in the nervous system.[63] Thermal stress induced the synthesis of Hsp70 in glial cells located in the sheath that surrounds the squid giant axon and this glial Hsp70 was rapidly transported to the adjacent axonal process.[63] This "glial to neuron" transfer provides a mechanism for fast delivery of neuroprotective Hsp70 to cellular processes that are distant for the neuronal cell body. In the mammalian brain, thermal stress induced Hsp27 and Hsp32 in glial cells and the transport of these Hsps to "perisynaptic glial processes" that surround and nurture synaptic termini.[54] These glial stress proteins appear to be transferred to postsynaptic elements that provide a structural framework for signal transduction complexes involved in neurotransmission events that control communication between neurons.[54] The early work by Tytell[63] led to the hypothesis that stress tolerance in neurons is not solely dependent on their own Hsps, but can be supplemented by additional Hsps transferred from adjacent glial cells. Hence supplying exogenous Hsps at neural injury sites could be an effective strategy to maintain neuronal viability. This idea has been tested in a number of model systems.[64] Injection of Hsc/Hsp70 into the vitreous chamber of the eye protected retinal photoreceptors from bright light damage.[65,66] Application of exogenous Hsc/Hsp70 to the cut end of the sciatic nerve reduced cell death in sensory and motor neurons.[67,68] Tissue culture experiments demonstrated that glial cells release Hsp70 and that neurons take up Hsp70 from the medium, leading to greater resistance to cell death induced chemically or by lack of trophic factor.[69] Extracellular Hsp70 protected spinal cord motor neurons deprived of trophic support *in vitro* or undergoing cell death *in vivo*.[70] Thus, exogenous application of Hsps has potential as a therapeutic strategy for acute injury in the nervous system.

Hsps are released into the blood stream after stressful stimuli and this may represent an important feature of the stress response.[71] Exercise stress has been reported to induce the release of Hsp70 from the human brain into the blood stream *in vivo*.[72] The biological significance of this neural release is yet to be determined. For additional discussion of extracellular Hsps, see the article in this volume authored by Calderwood.

ACKNOWLEDGMENTS

The author holds a Canada Research Chair (Tier I) in Neuroscience. This work was supported by grants from NSERC Canada and the Canada Research Chair program.

REFERENCES

1. RAJDEV, S., K. HARA, Y. KOKUBO, *et al* 2000. Mice overexpressing rat heat shock protein 70 are protected against cerebral infarction. Ann. Neurol. **47:** 782–791.
2. HOEHN, B., T.M. RINGER, L. XU, *et al* 2001. Overexpression of HSP72 after induction of experimental stroke protects neurons from ischemic damage. J. Cereb. Blood Flow Meta. **21:** 1303–1309.
3. GIFFARD, R.G., L. XU, H. ZHAO, *et al* 2004. Chaperones, protein aggregation, and brain protection from hypoxic/ischemic injury. J. Exp. Biol. **207:** 3213–3220.

4. GARRIDO, C., S. GURBUXANI, L RAVAGNAN, et al 2001. Heat shock proteins: endogenous modulators of apoptotic death. Biochem. Biophys. Res. Commun. **286:** 433–442.
5. GIFFARD, R.G. & M.A. YENARI. 2004. Many mechanisms for Hsp70 protection from cerebral ischemia. J. Neurosurg. Anesthesiol. **16:** 53–61.
6. MOSSER, D.D. & R.I. MORIMOTO. 2004. Molecular chaperones and the stress of oncogenesis. Oncogenesis **23:** 2907–2918.
7. SUN, Y., Y. OUYANG, L. XU, et al. 2006. The carboxyl-terminal domain of inducible Hsp70 protects from ischemic injury *in vivo* and *in vitro*. J. Cereb. Blood Flow Meta. **26:** 937–955.
8. OUYANG, Y., L. XU, Y. SUN, et al. 2006. Overexpression of inducible heat shock protein 70 and its mutants in astrocytes is associated with maintenance of mitochondrial physiology during glucose deprivation stress. Cell Stress Chaperones **11:** 180–186.
9. YENARI, M.A., J. LIU, Z. ZHENG, et al. 2005. Anti-apoptotic and anti-inflammatory mechanisms of heat shock protein protection. Ann. N.Y. Acad. Sci. **1053:** 74–83.
10. SELKOE, D.J. 2004. Cell biology of protein misfolding: the examples of Alzheimer's and Parkinson's disease. Nat. Cell Biol. **6:** 1054–1061.
11. FORMAN, M.S., J.Q. TROJANOWSKI & V.M. LEE. 2004. Neurodegenerative diseases: a decade of discoveries paves the way for therapeutic breakthroughs. Nat. Med. **10:** 1055–1063.
12. MUCHOWSKI, P.J. & J.L. WALKER. 2005. Modulation of neurodegeneration by molecular chaperones. Nat. Rev. Neurosci. **6:** 11–22.
13. BROWN, I.R. 2007. Heat shock proteins and neurodegenerative diseases. *In* Cell Stress Proteins. S.K. Calderwood, Ed.: 396–421. Springer Science + Business Media, LLC. New York, NY.
14. MAGRANE, J., R.C. SMITH, K. WALSH, et al. 2004. Heat shock protein 70 participates in the neuroprotective response to intracellularly expressed beta-amyloid in neurons. J. Neurosci. **24:** 1700–1706.
15. SMITH, R.C., K.M. ROSEN, R. POLA, et al. 2005. Stress proteins in Alzheimer's disease. Int. J. Hyperthermia **21:** 421–431.
16. AULUCK, P.K., H.Y. CHAN, J.Q. TROJANOWSKI, et al. 2002. Chaperone suppression of alpha-synuclein toxicity in a Drosophila model of Parkinson's disease. Science **295:** 865–868.
17. MORIMOTO, R.I. 1998. Regulation of the heat shock transcriptional response: cross talk between a family of heat shock factors, molecular chaperones, and negative regulators. Genes Dev. **12:** 3788–3796.
18. WESTERHEIDE, S.D. & R.I. MORIMOTO. 2005. Heat shock response modulators as therapeutic tools for diseases of protein conformation. J. Biol. Chem. **280:** 33097–33100.
19. KIERAN D., B. KALMAR, J.R. DICK, et al. 2004. Treatment with arimoclomol, a co-inducer of heat shock proteins, delays disease progression in ALS mice. Nat. Med. **10:** 402–405.
20. BENN, S.C. & R.H. BROWN. 2004. Putting the heat on ALS. Nat. Med. **10:** 345–347.
21. KALMAR, B., D. KIERAN & L. GREENSMITH. 2005. Molecular chaperones as therapeutic targets in amyotrophic lateral scelerosis. Biochem. Soc. Trans. **33:** 551–552.
22. NIRMALANANTHAN, N. & L. GREENSMITH. 2005. Amyotrophic lateral sclerosis: recent advances and future therapies. Curr. Opin. Neurol. **18:** 712–719.

23. BATULAN, Z., G.A. SHINDER, S. MINOTTI, et al. 2003. High threshold for induction of the stress response in motor neurons is associated with failure to activate HSF1. J. Neurosci. **23:** 5789–5798.
24. ABBOTT, A. 2002. Neurologists strike gold in drug screen effort. Nature **417:** 109.
25. HEEMSKERK, J., A.J. TOBIN & L.J. BAIN. 2002. Teaching old drugs new tricks. Trends Neurosci. **25:** 494–496.
26. ZHOU, B.N. 1991. Some progress on the chemistry of natural bioactive terpenoids from Chinese medicinal plants. Mem. Inst. Oswaldo Cruz. **86**(Suppl. 2): 219–226.
27. PINNA, G.F., M. FIORUCCI, J.M. REIMUND, et al. 2004. Celastrol inhibits pro-inflammatory cytokine secretion in Crohn's disease biopsies. Biochem. Biophys. Res. Commun. **322:** 778–786.
28. TAO, X., J. YOUNGER, F.Z. FAN, et al. 2002. Benefit of an extract of Tripterygium Wilfordii Hook F in patients with rheumatoid arthritis: a double-blind, placebo-controlled study. Arthritis Rheum. **46:** 1735–1743.
29. HUANG, F.C., W.K. CHAN, K.J. MORIARTY, et al. 1998. Novel cytokine release inhibitors. Part I: Triterpenes. Bioorg. Med. Chem. Lett. **8:** 1883–1886.
30. CLEREN, C., N.Y. CALINGASAN, J. CHEN, et al. 2005. Celastrol protects against MPTP- and 3-nitropropionic acid-induced neurotoxicity. J. Neurochem. **94:** 995–1004.
31. KIAEI, M., K. KIPIANI, S. PETRI, et al. 2005. Celastrol blocks neuronal cell death and extends life in transgenic mouse model of amyotrophic lateral sclerosis. Neurodegener. Dis. **2:** 246–254.
32. WANG, J., S. GINES, M.E. MACDONALD, et al. 2005. Reversal of a full-length mutant hungtingtin neuronal phenotype by chemical inhibitors of polyglutamine-mediated aggregation. BMC Neurosci. **6:** 1–12.
33. WESTERHEIDE, S.D., J.D. BOSMAN, B.N. MBADUGHA, et al. 2004. Celastrols as inducers of the heat shock response and cytoprotection. J. Biol. Chem. **279:** 56053–56060.
34. MANZERRA, P., S.J. RUSH & I.R. BROWN. 1993. Temporal and spatial distribution of heat shock mRNA and protein (hsp70) in the rabbit cerebellum in response to hyperthermia. J. Neurosci. Res. **36:** 480–490.
35. FOSTER, J.A., S.J. RUSH & I.R. BROWN. 1995. Localization of constitutive and hyperthermia-inducible heat shock mRNAs (hsc70 and hsp70) in the rabbit cerebellum and brainstem by nonradioactive in situ hybridization. J. Neurosci. Res. **41:** 603–612.
36. DWYER, D.S., Y. LIU, S. MIAO, et al. 1996. Neuronal differentiation in PC12 cells is accompanied by diminished inducibility of Hsp70 and Hsp60 in response to heat and ethanol. Neurochem. Res. **21:** 659–666.
37. HATAYAMA, T., H. TAKAHASHI & N. YAMAGISHI. 1997. Reduced induction of HSP70 in PC12 cells during neuronal differentiation. J. Biochem. (Tokyo) **122:** 904–1010.
38. CHOW, A.M. & I.R. BROWN. 2007. Induction of heat shock proteins in differentiated human and rodent neurons by celastrol. Cell Stress Chaperones **12:** 237–244.
39. PARSIAN, A.J., J.E. SHEREN, T.Y. TAO, et al. 2000. The human Hsp70B gene at the HSPA7 locus of chromosome 1 is transcribed but non-functional. Biochim. Biophys. Acta **1494:** 201–205.
40. NOONAN, E.J., R.F. PLACE, R.J. RASOULPOUR, et al. 2007. Cell number-dependent regulation of Hsp70B' expression: evidence of an extracellular regulator. J. Cell Physiol. **210:** 201–211.

41. MORIMOTO, R.I., M.P. KLINE, D.N. BIMSTON, et al. 1997. The heat shock response: regulation and function of heat shock proteins and molecular chaperones. Essays Biochem. **32:** 17–29.
42. BARBE, M.F., M. TYTELL, D.J. GOWER, et al. 1988. Hyperthermia protects against light damage in the rat retina. Science **241:** 1817–1820.
43. WALSH, D.A., K. LI, J. SPEIRS, et al. 1989. Regulation of the inducible heat shock 71 genes in early neural development of cultured rat embryos. Teratology **40:** 321–334.
44. RORDORF, G., W.J. KOROSHETZ & J.V. BONTRE VEN. 1991. Heat shock protects cultured neurons from glutamate toxicity. Neuron **7:** 1043–1051.
45. MAYER, J. & I.R. BROWN. 1994. Heat shock proteins in the nervous system. 1–297. Academic Press London.
46. TYTELL, M., M.F. BARBE & I.R. BROWN. 1994. Induction of heat shock (stress) protein 70 and its mRNA in the normal and light-damaged rat retina after whole body hyperthermia. J. Neurosci. Res. **38:** 19–31.
47. YENARI, M.A. 2002. Heat shock proteins and neuroprotection. Adv. Exp. Med. Biol. **513:** 281–299.
48. MATTSON, M.P. & T. MAGNUS. 2006. Ageing and neuronal vulnerability. Nat. Rev. Neurosci. **7:** 278–294.
49. KARUNANITHI, S., J.W. BARCLAY, R.M. ROBERTSON, et al. 1999. Neuroprotection at Drosophila synapses conferred by prior heat shock. J. Neurosci. **19:** 4360–4369.
50. KARUNANITHI, S., J.W. BARCLAY, I.R. BROWN, et al. 2002. Enhancement of presynaptic performance in transgenic Drosophila over-expressing heat shock protein Hsp70. Synapse **44:** 8–14.
51. NEAL, S.J., S. KARUNANITHI, A. BEST, et al. 2006. Thermoprotection of synaptic transmission in a Drosophila heat shock factor mutant is accompanied by increased expression of Hsp83 and DNAJ-1. Physiol. Genomics **25:** 493–501.
52. KELTY, J.D., P.A. NOSEWORTHY, M.E. FEDER, et al. 2002. Thermal preconditioning and heat shock protein 72 preserve synaptic transmission during thermal stress. J. Neurosci. **22:** RC193, 1–6.
53. XIAO, C., V. MILEVA-SEITZ, L. SEROUDE, et al. 2007. Targeting HSP70 to motoneurons protects locomotor activity from hyperthermia in Drosophila. Dev. Neurobiol. **67:** 438–455.
54. BECHTOLD, D.A. & I.R. BROWN. 2000. Heat shock proteins Hsp27 and Hsp32 localize to synaptic sites in the rat cerebellum following hyperthermia. Mol. Brain Res. **75:** 309–320.
55. BECHTOLD, D.A., J.R. RUSH & I.R. BROWN. 2000. Localization of heat shock protein Hsp70 to the synapse following hyperthermic stress in the brain. J. Neurochem. **74:** 641–646.
56. CHEN, S., D. BAWA, S. BESSHOH, et al. 2005. Association of heat shock proteins and neuronal membrane components with lipid rafts from the rat brain. J. Neurosci. Res. **81:** 522–529.
57. MANZERRA, P.S., J. RUSH & I.R. BROWN. 1997. Tissue-specific differences in heat shock protein Hsc70 and Hsp70 in the control and hyperthermic rabbit. J. Cell. Physiol. **170:** 130–137.
58. CHEN, S. & I.R. BROWN. 2007. Translocation of constitutively expressed heat shock protein Hsc70 to synapse-enriched areas of the cerebral cortex after hyperthermic stress. J. Neurosci. Res. **85:** 402–409.

59. MANZERRA, P. & I.R. BROWN. 1992. Distribution of constitutive- and hyperthermia-inducible heat shock mRNA species (hsp70) in the Purkinje layer of the rabbit cerebellum. Neurochem. Res. **17:** 559–564.
60. MARCUCCILLI, C.J., S.K. MATHUR, R.I. MORIMOTO, et al. 1996. Regulatory differences in the stress response of hippocampal neurons and glial cells after heat shock. J. Neurosci. **16:** 478–485.
61. CHEN, S. & I.R. BROWN. 2007. Neuronal expression of constitutive heat shock proteins: implications for neurodegenerative diseases. Cell Stress Chaperones **12:** 51–58.
62. MERIIN, A.B. & M.Y. SHERMAN. 2005. Role of molecular chaperones in neurodegenerative disorders. Int. J. Hyperthermia **21:** 403–419.
63. TYTELL, M., S.G. GREENBERG & R.J. LASEK. 1986. Heat shock-like protein is transferred from glia to axon. Brain Res. **363:** 161–164.
64. TYTELL, M.. 2005. Release of heat shock proteins (Hsps) and the effects of extracellular Hsps on neural cells and tissues. Int. J. Hyperthermia **21:** 445–455.
65. TYTELL, M., M.F. BARBE & I.R. BROWN. 1993. Stress (heat shock) protein accumulation in the central nervous system- its relationship to cell stress and damage. In Neural injury and regeneration. F.J. Seil, Ed.: 293–303. Raven Press. New York, NY.
66. YU, Q, C.R. KENT & M. TYTELL. 2001. Retinal uptake of intravitreally injected Hsc/Hsp70 and its effect on susceptibility to light damage. Mol. Vision **7:** 48–56.
67. HOUENOU, L.J., L. LI, M. LEI, et al. 1996. Exogenous heat shock cognate protein Hsc70 prevents axotomy-induced death of spinal sensory neurons. Cell Stress Chaperones **1:** 161–166.
68. TIDWELL, J.L., L.J. HOUENOU & M. TYTELL. 2004. Administration of Hsp70 in vivo inhibits motor and sensory neuron degeneration. Cell Stress Chaperones **9:** 88–98.
69. GUZHOVA, I., K. KSENIA, O. MOSKALIOVA, et al. 2001. In vitro studies show that Hsp70 can be released by glia and that exogenous Hsp70 can enhance neuronal stress tolerance. Brain Res. **914:** 66–73.
70. ROBINSON, M.B., J.L. TIDWELL, T. GOULD, et al. 2003. Extracellular heat shock protein 70: a critical component for motoneuron survival. J. Neurosci. **25:** 9735–9745.
71. FLESHNER, M. & J.D. JOHNSON. 2005. Endogenous extra-cellular heat shock protein 72: releasing signal(s) and function. Int. J. Hyperthermia **21:** 457–471.
72. LANCASTER, G.I., K. MOLLER, B. NIELSEN, et al. 2004. Exercise induces the release of heat shock protein 72 from the human brain in vivo. Cell Stress Chaperones **9:** 276–280.

Heavy Metal Ions in Normal Physiology, Toxic Stress, and Cytoprotection

MICHAEL A. LYNES,[a] Y. JAMES KANG,[b] STEFANO L. SENSI,[c] GEORGE A. PERDRIZET,[d] AND LAWRENCE E. HIGHTOWER[a]

[a]*University of Connecticut, Storrs, Connecticut 06269, USA*

[b]*University of Louisville, Louisville, Kentucky 40202, USA*

[c]*University 'G. d'Annunizio', Chieti 66013, Italy*

[d]*Hartford Hospital, Hartford, Connecticut 06169, USA*

> ABSTRACT: As a group, heavy metals include both those essential for normal biological functioning (e.g., Cu and Zn), and nonessential metals (e.g., Cd, Hg, and Pb). Both essential and nonessential metals can be present at concentrations that disturb normal biological functions, and which evoke cellular stress responses. The cellular targets for metal toxicity include tissues of the kidney, liver, heart, and the immune response and nervous systems. Intriguingly, manipulations of specific metals, their reservoirs, and the cellular stress response can have therapeutic effects on certain diseases. In this minireview, we will consider both the biological responses to stressful levels of heavy metal cations, and experimental and clinical manipulations of these cations as a means to improve human health parameters.
>
> KEYWORDS: heavy metal; zinc; cadmium; tin; copper; cardiomyopathy; chelation; ischemia; inflammation; cytoprotection; heme-oxygenase; heat shock proteins; vascular surgery; immunomodulation; metallothionein; humoral immunity; neuronal injury; oxidative stress

INTRODUCTION

Heavy metals represent both essential components for the maintenance of normal biological functions, and toxic agents with damaging consequences when present in inappropriate amounts. One way to understand these agents as a group is to characterize the influences they have on sensitive biological systems. Metals such as copper (Cu) and zinc (Zn) play important roles as cofactors for normal enzyme functioning, and zinc contributes to the structural elements of some transcription factors. When in short supply, there are

Address for correspondence: Lawrence Hightower, Ph.D., Department of Molecular and Cell Biology, University of Connecticut, 91 N. Eagleville Road, Storrs, CT 06269. Voice: +1-860-486-4257; fax: +1-860-486-4331.

lawrence.hightower@uconn.edu

many biological processes that suffer, and the result can impede the normal roles played by many tissues, including those of the kidney, liver, heart, and the immune and nervous systems. Toxic heavy metals share some chemical similarities with the essential metals, and in excess can induce the production of reactive oxygen species (ROS) as well as interact with sulfhydryls, altering protein structure and function. The management of both essential and toxic heavy metals is in part accomplished by metallothionein (MT). In this article, we will consider the management of these heavy metals in several biological systems, and the effects of these agents on proper biological functioning.

METALLOTHIONEIN-MEDIATED IMMUNOMODULATION: NEW ROLES FOR AN OLD STRESS RESPONSE (M.A.L.)

MT serves as one of the points of intersection for many of the issues addressed in this review. Although this small stress response protein is not a member of the heat-shock protein (HSP) family, it serves many roles in both normal and stressed cells, acting as a reservoir of essential heavy metals (e.g., Cu^{2+} and Zn^{2+}), as a scavenger for both heavy metal toxicants (e.g., Hg^{2+}, Cd^{2+}) and free radicals, and as a regulator of transcription factor activity.[1] MT is induced by a range of different agents, including heavy metals, ROS, glucocorticoids, acute phase cytokines and interferon, and by endotoxin. Traditionally considered an intracellular protein, recent work has suggested important roles for MT, both in intracellular compartments and as an extracellular agent.[2] The protein is widely expressed in highly homologous forms, suggesting that the roles played by MT are essential to a variety of biological processes, and that the protein has a long evolutionary lineage.[3]

Metals have long been known to evoke changes in the immune response. These changes are dependent both on the specific metal and metal valency, as well as on the genetic makeup and physiological status of the exposed individual.[4] The changes induced in leukocyte behavior can result in immunosuppression and a consequent increase in susceptibility to infectious disease. These changes can also result in an autoreactive immune response that subsequently damages self tissues.[5] Our work has focused on the mechanisms of metal-mediated immunomodulation that are influenced by basal levels of MT, and by those levels of MT that are induced by inflammation, autoimmune disease, and toxic metal exposure. We have also explored the potential opportunities this immunomodulation provides both for the management of disease that results from toxic metal exposures, and for the management of diseases that are accompanied by MT synthesis as a consequence of cellular stresses that are associated with infection, inflammation or other biological stressors.

We have shown that exogenous MT can suppress elements of the humoral (T-dependent, antibody-mediated) response[6] in a manner consistent with the immune enhancements that have been reported in mice that carry targeted disruptions of the *Mt1* and *Mt2* genes.[7] Similarly, a monoclonal anti- MT antibody can enhance the T-dependent humoral response to antigen challenge.[8] Taken together, these results suggest that MT that is released from cells can act to diminish the normal immune response, and that this extracellular compartment of MT is a potential target for therapeutic manipulation. In addition, the results also suggest that MT that is released from stressed cells can interact with cells that participate in a humoral response, potentially via interactions with the membrane of the target cell population. There is one report that has described a putative MT receptor. El Refaey et al.[9] reported that astrocytes express a membrane-associated receptor that potentially regulates astrocyte function. We have observed MT binding on the plasma membranes of leukocytes cultured in the presence of MT,[10] on cells harvested from animals immunized in the presence of adjuvant,[8] and on cells harvested from animals that experience a chronic inflammatory disease.[11]

One of the biological processes activated by this extracellular pool of MT is leukocyte chemotaxis. We have shown that MT binds with surface molecules on many different subpopulations of leukocytes to initiate a chemotactic response.[12] This response can be inhibited by both pertussis toxin and cholera toxin, suggesting that one of the targets of MT binding is a G-coupled protein receptor (GCPR). This suggests that one of the ways that MT may influence immune activities is to initiate cell migration toward tissues that serve as the source of diffusible MT. In instances of metal intoxication, the organs where highest levels of this MT pool originate may become inflamed because of the accumulation of cells responding to the MT gradient, while, simultaneously, other wound sites may experience a less vigorous response due to a decreased immune response directed to those sites. In some instances where chemoattractants originating from the tissue wound overlap with a metal-initiated MT gradient, the resulting inflammatory response may be excessive. A thorough understanding of the role played by MT in these cases of metal intoxication will suggest novel therapeutic avenues, such as intentional induction of endogenous MT, or therapeutic administration of MT to change the course of disease. MT therapies have already been used experimentally in animals to alter the course of autoimmune disease. For example, injection of MT, or of Zn^{2+} at levels known to induce MT can diminish the severity of collagen-induced arthritis in experimental mice.[13] These observations correspond well with our work showing that combining a targeted disruption of the *Mt1* and *Mt2* genes with $Ptpn6^{me-v}/Ptpn6^{me-v}$ ("viable motheaten") results in a dramatic shortening of lifespan in these animals that display a congenital chronic inflammatory disease.[11] This suggests that the propensity to synthesize MT (and the additional propensity to release MT from stressed cells) may represent important susceptibility polymorphisms for certain types of disease in the human population.[14]

We have examined the humoral response in mice that carry an Mt1 transgene, and in mice with targeted disruption of the *Mt1* and *Mt2* genes. These experiments were done in the presence of Zn^{2+}, Cd^{2+}, or vehicle control at exposure levels that do not alter the response to antigen challenge in wild type mice. Our findings in these studies are intriguing: Cd^{2+}, but not Zn^{2+}, suppressed the humoral immune response to antigen challenge below vehicle control levels in the strain carrying the targeted *Mt* gene disruptions. In contrast, both Cd^{2+} and Zn^{2+} reduced the humoral response in the *Mt1* transgenic strain (unpublished data). We interpret these results to suggest that, in the absence of functional MT, the animal is susceptible to other mechanisms of Cd-mediated immunotoxicity. In the transgenic strain, both Cd^{2+} and Zn^{2+} can induce elevated levels of MT, and in both circumstances humoral immunity is diminished. These results suggest that the MT protein itself is an important component of the immunomodulatory effect, irrespective of the metal species that is complexed with the protein.

In recent experiments, we have explored the role of MT in the progression of an *in vivo* infection with *Listeria monocytogenes*, a model organism for intracellular infection. Our results illustrate that there can be dramatic differences in the ability of mice with different MT gene doses to respond to the infection (unpublished data). These data appear to indicate that MT synthesis influences the course of infection. In light of work from other laboratories that indicates critical influences of MT on various forms of tissue inflammation and ischemia, these results may indicate that preliminary testing of MT biosynthetic capacity in a patient population may represent a valuable prognostic for disease susceptibility and for predictions of the severity of disease.

COPPER METABOLIC DISORDER IN MYOCARDIAL PATHOGENESIS (Y.J.K.)

Cu is an essential mineral nutrient that participates in important cellular metabolism and function as a component of a number of cuproenzymes, an integrated structural element, and a regulatory agent.[15] However, Cu also catalyzes the production of highly ROS leading to oxidative damage of lipid, protein, DNA and other molecules.[16] Therefore, either Cu deficiency or excess can lead to diseases or affect the progression of diseases. There is virtually no free Cu in biological system under physiological conditions.[17] Cu intracellular movement is tightly controlled by Cu chaperons[18] and interorgan movement is mediated by different Cu carrier proteins.[19]

The current US–Canadian Recommended Dietary Allowance (RDA) for Cu is 0.9 mg/day for adults 19 years or older.[20] A debate continues regarding the appropriateness of the RDA for Cu.[21] A human study has examined potential adverse effect of marginal dietary Cu restriction on cardiovascular system.[22] Among 24 subjects consuming a typical American diet (1.03 mg Cu/day), one mildly obese subject sustained a myocardial infarction 4 weeks after consuming a starch-based diet and two subjects experienced severe tachycardia 7 and 10 weeks after consuming a fructose-based diet. Another subject experienced a type II, second-degree heart block 11 weeks after consuming the starch-based diet. It appears that preexisting cardiac conditions cannot be excluded for the outcome; however, the fact that these abnormalities appeared after the subjects were fed the controlled low Cu diet indicates the triggering effects of low levels of Cu. Moreover, these cardiac defects disappeared after the affected subjects were fed diets supplemented with 3.0 mg Cu/day.[22]

Cu supplementation would prevent or ameliorate cardiomyopathy if this pathogenesis is associated with Cu depression in the heart. We have observed[23] that dietary Cu supplementation (20 mg Cu/kg diet) can reverse an experimentally induced cardiac hypertrophy and improve heart contractile function in mice with established heart hypertrophy and dysfunction induced by ascending aortic constriction (AAC) if they were fed adequate Cu diet (6 mg Cu/kg). Moreover, Cu supplementation-induced recovery occurs in the presence of sustained pressure overload. We further found that AAC caused cardiac Cu depression, which can be corrected by dietary Cu supplementation.

Sco2 is an important Cu chaperone for cytochrome c oxidase [24] and mutations in Sco2 result in suppressed cytochrome c oxidase activity.[25,26] Patients with mutations in Sco2 developed severe hypertrophic cardiomyopathy.[25–27] An Sco2 patient with severe hypertrophic cardiomyopathy and heart dysfunction was treated with Cu-histidine. This Cu supplement therapy caused a reversal of the hypertrophic cardiomyopathy along with a significant improvement in all parameters of heart function and normalization of ECG signs and blood pressure.[28] A recent study with a small population of chronic heart failure patients has shown that dietary supplementation with micronutrients for 9 months increase left ventricle ejection and decreases left ventricle volume, along with an improvement of the quality of life.[29] Among the formulated micronutrients is Cu (1.2 mg/day).

Increased Cu concentrations and a high Cu/Zn ratio were found in patients with chronic rheumatic heart disease.[27] Some studies have shown that the plasma activity of semicarbazide-sensitive amine oxidase (SSAO), a Cu-containing protein, was elevated relative to the severity of diabetes mellitus and chronic heart failure.[30] Evidence obtained from epidemiological studies has suggested that serum ceruloplasmin, the main Cu-containing protein in plasma, is an important risk factor for myocardial infarction and cardiovascular disease because it is a potent catalyst of LDL oxidation *in vitro*.[31] We have observed that Cu concentrations in the plasma of streptozotocin (STZ)-induced diabetic mice significantly increased, along with a significant depression of Cu concentrations in the liver and a slight increase in the heart.[32–34]

A recent study has shown that Cu chelation using trientine substantially improves cardiomyocyte structure and contractile function and reverses left ventricular collagen deposition in diabetic patients and STZ-induced diabetic rat model.[35] It is unknown what forms of Cu interact with trientine *in vivo* and how trientine affects overall balance among different forms of Cu. However, the distinct effects of trientine in contrary to Cu supplementation described earlier may reflect the fundamental difference in Cu metabolic disorders between diabetic and pressure overload cardiomyopathy.

It is possible that systemic complications in diabetes lead to Cu accumulation in the blood due to disruption of Cu metabolism in the liver, thus increasing the risk of Cu toxicity to other organs including the heart. However, in pressure overload the heart is the primary organ of Cu metabolic disorder but the liver remains functional. Therefore, Cu functional deficiency would occur in the pressure overload heart. In this context, dietary Cu supplementation should reverse pressure-overload–induced hypertrophic cardiomyopathy, but Cu chelation should ameliorate diabetic cardiomyopathy, as observed in previous studies.[23,35] Although further studies are required to provide scientific understanding of Cu metabolic disorders under different disease conditions, differential manipulation of Cu metabolism under diabetic and pressure-overload conditions appears beneficial to patients with cardiomyopathy.

BRAIN INJURY AND OXIDATIVE STRESS: THINK ZINC! (S.L.S.)

Oxidative stress is considered one of the major triggers of neuronal degeneration in the brain. Zn^{2+} has been implicated in the regulation of many channels and receptors, but the cation can also act as a trigger for neuronal injury.[36] Zn^{2+} is coreleased with glutamate at many excitatory synapses, and synaptic Zn^{2+} can eventually enter neurons through channels associated with glutamatergic postsynaptic receptors such as NMDA and calcium-permeable AMPA/kainate receptors, or through voltage-sensitive calcium channels (VSCCs) and Zn^{2+} transporters. Neurons maintain relatively low levels of intracellular free Zn^{2+} ($[Zn^{2+}]_i$), through the coordinated activity of systems involving the extrusion, buffering, and sequestration of the cation. A key emerging concept is that many biological systems possess a finely tuned "Zn^{2+} set-point" and that deregulation of this set-point can have important consequences. Prenatal Zn^{2+} deficiency can impair brain development, resulting in serious cognitive deficits later in life. On the other hand, toxic $[Zn^{2+}]_i$ rises (resulting from either cation influx or its release from intracellular sites such as MTs and mitochondria) mediates toxic effects in a variety of pathological conditions, including cerebral ischemia, brain trauma and epilepsy.[36]

Interestingly, Zn^{2+} is also a strong inducer of oxidative stress. In neurons, Zn^{2+} can trigger ROS production through mitochondrial pathways by interfering with the activity of the electron transport chain (ETC).[37,38] Zn^{2+} can also modulate extramitochondrial pathways involved in ROS generation by promoting the increased activity of NADPH oxidase, protein kinase C (PKC) activation, as well as induction of neuronal nitric oxide synthase (nNOS), which together with superoxide can produce harmful peroxynitrite (ONOO–).

One of the major targets of ROS-dependent Zn^{2+} release involves the MTs. Recent findings strongly suggest that nitrosative stress can act as a critical trigger for $[Zn^{2+}]_i$ mobilization as Nitric Oxide (NO) or –ONOO interact preferentially with MT-3 and induce Zn^{2+} release from MTs both *in vitro* and *in vivo*.[39,40]

The facts that Zn^{2+} trigger ROS generation and cellular oxidation enhances $[Zn^{2+}]_i$ release set the stage for a dangerous feed-forward cycle, as the cation can elicit neuronal demise by activating multiple, intersecting necrotic and/or apoptotic pathways. Zn^{2+} can promote apoptosis via the induction of the mitochondrial permeability transition pore (mPTP) and release of mitochondrial pro-apoptotic factors,[41–43] but the cation has also been linked to necrotic disruption of neuronal metabolism, and biochemical studies have shown that it can inhibit key enzymes in the glycolytic pathway.

Zn^{2+}-dependent activation of both necrotic and apoptotic pathways may conceivably be linked to its intracellular mobilization upon oxidative stress. Findings in neurons indicate that cell oxidants can induce $[Zn^{2+}]_i$ rises that, in isolated mitochondria, are sufficient to cause a partial loss of mitochondrial membrane potential and trigger a multiconductance ion channel activity consistent with mPTP opening.[41–44] These ROS-induced $[Zn^{2+}]_i$ rises are also able to elicit activation of specific K^+ channels, leading to K^+ depletion, a key event in neuronal apoptosis.[45,46] Conversely, Zn^{2+}-induced generation of mitochondrial ROS might promote yet more Zn^{2+} release from the protein-bound pool. Thus, the two pathways appear to work synergistically to induce a self-perpetuating injurious cycle.

It is worth noting that the recent findings have prompted a very intriguing debate about the capability of Zn^{2+} to interplay with Ca^{2+} to promote excitotoxicity.[47–49] Although for at least two decades, excitotoxicity has been considered as a purely Ca^{2+}-dependent process, Ca^{2+} dependency has been substantiated mostly on experimental paradigms that employed either Ca^{2+} chelators or intracellular imaging with Ca^{2+}-sensitive fluorescent indicators. However, a potential caveat to this "Ca^{2+}-centered" view comes from the fact that both Ca^{2+}-sensitive fluorescent probes and chelators also bind Zn^{2+} with higher affinity. The confounding effect of Zn^{2+} on Ca^{2+} imaging has been so far neglected, moving from the assumption that $[Zn^{2+}]_i$ levels are negligible. On the other hand, as mentioned earlier, growing evidence has indicated that upon excitotoxic conditions, $[Zn^{2+}]_i$ increases can reach levels high enough to greatly interfere with fluorescent measurements of $[Ca^{2+}]_i$ and/or chelation by divalent chelators.[48,50,51]

The deleterious synergism between the two cations, though not yet fully explored, offers intriguing new perspectives on our understanding on the ionic determinants of many excitotoxic conditions. Given the emerging role of $[Zn^{2+}]_i$ release in neuronal death, and the fact that the cation is capable of triggering injury with greater potency compared to Ca^{2+}, Zn^{2+} is likely to emerge as a major mediator of excitotoxicity. For instance, recent data demonstrating that large $[Ca^{2+}]_i$ rises trigger important intracellular mobilization of Zn^{2+}, coupled with the likely probability that Ca^{2+}-induced mitochondrial ROS generation might also promote further Zn^{2+} release from MTs, suggest a more complex excitotoxic scenario. In such a model, glutamate-driven $[Ca^{2+}]_i$ rises might actually serve as a "partner in crime" to initiate the injurious mobilization of the main ionic mediator of neuronal death: Zn^{2+}.

CYTOPROTECTIVE METAL IONS (G.A.P AND L.E.H.)

All forms of surgical therapy are stressful and injurious for the patient. The majority of surgical and invasive medical procedures are performed in an elective fashion and thus provide the clinician an opportunity to preoperatively condition the patient to minimize iatrogenic tissue injury. Presently no preoperative clinical protocols exist that take advantage of intrinsic cellular mechanisms that have been shown to provide protection against iatrogenic ischemia–reperfusion (IR) and acute inflammatory injuries. We hypothesized that tissues could be protected from IR injury by pretreatment of the organism by using brief whole-body hyperthermia (HS, 42.5°C) or the systemic administration of metal ions ($SnCl_2$ or $ZnCl_2$) followed by a period of recovery (37°C, 6–8 hr) prior to major surgical procedures. We have successfully used HS to provide protection in diverse animal models of surgically induced IR, reviewed elsewhere.[52] The transient state of cytoprotection or protected phenotype is achieved via a complex adaptation in cellular metabolism analogous to that described for the thermotolerance phenomena.[53,54] The dominant metabolic change associated with hyperthermia-induced cytoprotection is the increased expression of the HS-related genes resulting in the rapid synthesis of HSPs.[55] The HSPs have been recently classified as molecular chaperones and, as such, contribute new insight into cellular resistance to and recovery from IR injury. In addition, heat-shock–related effects on acute inflammation, vascular biology, and non-HS gene expression have been described.[56,57] We wish to develop clinically relevant protocols for stress-conditioning humans prior to invasive surgical and medical procedures. Pharmacologic induction of the cellular stress response has been reported for a number of chemical agents. We have focused our work on the relatively nontoxic stannous and zinc metal ions as potential agents to develop pharmacologic stress-conditioning protocols.

It is known that the degree of acute inflammatory response to noxious agents can be radically altered by up- or downregulation of HSPs, specifically heme oxygenase–1 (HO-1, a.k.a. hsp32) and HSP 70 (hsp70).[58] Stannous chloride ($SnCl_2$), a tin salt, is a relatively nontoxic metal of the transition series of elements. Stannous chloride has been shown to be a potent inducer of HO-1 activity in many animal models[59,60] and of HO-1 and hsp70 mRNAs in human cells.[61] Acute inflammatory responses, such as acute lung injury associated

TABLE 1. Renal vascular resistance following warm ischemia

Experimental group	n	RVR (mmHg/mL/g) Mean ± SD	P vs. sham
Baseline/No ischemia	5	43.5 ± 10.1	Nd
Sham/NaCl	8	65 ± 10.3	Nd
$SnCl_2$	4	44.7 ± 13.8	0.016
$ZnCl_2$	5	48.3 ± 7.5	0.010

with the postoperative fat embolism syndrome, are observed in many medically relevant pathologic states.[62]

Ischemia–reperfusion and acute inflammation are fundamental mechanisms by which tissues can become injured during major surgical and invasive medical procedures. Four models of acute ischemia–reperfusion or acute inflammation have been used to test the pharmacologic stress-conditioning hypothesis: rodent renal artery occlusion (RAO), rabbit spinal cord ischemia, acute pulmonary inflammation (rabbit fat embolism syndrome by intravenous administration of oleic acid), and acute inflammation within the rodent mesenteric blood vessels.

HS-associated ischemic protection preserves renal microvascular integrity and is dependent on new HS-gene expression.[63] Sprague–Dawley rats (200–250 g, male) were divided into four pretreatment groups. Animals that received no pretreatment and no ischemia provided baseline renal vascular resistance data. Sham control animals received normal saline injections (0.9% NaCl) at 16 h prior to surgery. Tin- and zinc-pretreated animals received 0.15 mg/kg of the metal solution by the subcutaneous route 16 h prior to RAO. Bilateral RAO was performed in hydrated (3 mL, 0.9% NaCl, intravenous), anesthetized (5 mg pentobarbital/100 g body weight, IP), animals for 60 min at 37°C. Immediately following RAO, all kidneys were perfused *in situ* via the renal artery with phosphate buffered saline (PBS, 27°C) at a constant flow rate (0.76 mL/min). Perfusion pressures were measured (mmHg) using an in-line pressure monitor (Tektronix) and renal vascular resistance was calculated (RVR, mmHg/mL/g). Kidneys pretreated with either tin or zinc demonstrate a significantly reduced RVR following 60 min of *in situ* warm ischemia (TABLE 1).

HS-associated ischemic protection can also preserve neurologic function in the rabbit model of acute aortic occlusion (AAO).[64] Three groups of New Zealand White rabbits were subjected to 20 min of infra-renal aortic occlusion to induce acute spinal cord ischemia, and divided into three pretreatment groups (TABLE 2). Sham control animals received normal saline injections (0.9% NaCl, intravenous) 16 h prior to AAO. Unstressed control animals received no pretreatment or handling prior to AAO. The tin-pretreated animals received one injection of $SnCl_2$ (0.15 mg/kg, subcutaneous) followed by 16 h of recovery. Following 20 min of AAO, all animals were followed with daily neurological exams for 2 days and then euthanized. Intraoperative hemodynamic parameters were not significantly different between groups. Paralysis

TABLE 2. Neurologic function following acute spinal cord ischemia

Experimental group (n)	% Normal	% Paretic	% Paralysis
Sham/NaCl (7)	14	72	14
Control (8)	0	12	88
SnCl$_2$ (4)	100	0	0*

*$P < 0.001$ SnCl$_2$ vs control.

TABLE 3. Lung function and inflammation following oleic acid injection

Experimental group	n	Oxygenation (mm Hg)	WBC count (10^6 cells/mL)
Sham/NaCl	13	47.7 ± 11.5	1.13 ± 0.32
SnCl$_2$	11	65.2 ± 20.8	0.81 ± 0.9
P value		0.013	0.04

developed in seven of eight animals in the unstressed control group but was not seen in the SnCl$_2$ group (none in four, $P < 0.001$). The sham group demonstrated an intermediate level of protection: one animal developed paralysis, five animals developed paresis (weakness), and one animal had normal function.

Similarly, HS could influence the progression of acute pulmonary inflammation induced by fat embolism syndrome in the rabbit (TABLE 3). New Zealand White rabbits (2–3 kg) received a single dose of oleic acid (OA, 0.025 mL/kg, intravenous) 16 h after randomization to two pretreatment groups. Group 1 ($n = 13$, Sham/NaCl) controls received an injection of normal saline (0.9% NaCl, subcutaneous) and Group 2 ($n = 11$, SnCl$_2$) an injection of SnCl$_2$ (0.15 mg/kg, subcutaneous). Pulmonary oxygen transfer and bronchoalveolar lavage for cell infiltrate were performed 24 h after OA administration. Oxygen transfer was significantly better in animals following pretreatment with SnCl$_2$ compared to sham-treated animals. The improvement in pulmonary function is associated with a significant reduction in WBC infiltration into the alveolar space.

We have also evaluated the ability of tin to moderate acute inflammation within rodent mesenteric blood vessels that is induced by an injected inflammatory agent (TABLE 4).[65] Adult male Wistar rats received normal saline or tin pretreatments 16 h prior to intraarterial suffusion of the proinflammatory agent formyl-methionyl-leucyl-phenylalanine (FMLP, 10^{-7} M). Leukocyte adherence to mesenteric vascular endothelium was determined by intravital microscopy techniques and reported as cell number per 100 μm of vessel length. Tin pretreatment significantly reduced the leukocyte adhesion to rodent mesenteric blood vessels induced by FMLP treatment.

The systemic pretreatment of rodents with stannous chloride or zinc chloride can provide tissue-level protection against acute ischemia and reperfusion injury. The cytoprotected state is associated with a reduction in leukocyte

TABLE 4. Leukocyte–endothelial adhesion following FMLP

Experimental group	n	Baseline*	FMLP treatment*
Sham/NaCl	42	3.1 ± 0.4	6.4 ± 0.6
SnCl$_2$	28	2.5 ± 0.3	2.5 ± 0.2
P value		Ns	<0.05

*Units are cell number per 100 μm of vessel length.

adherence to FMLP-primed endothelium and enhanced expression of hsp70 within vascular tissue. The low toxicity profile of these metal salts makes these agents ideal phramacologic candidates for future clinical trials targeting preoperative stress conditioning in the human.

In a recent study of human hsp72 and hsp70B′ induction in colon cell lines by ZnCl^2, several cell-specific effects were found, indicating a potentially useful selectivity in human cellular responses to zinc ions. Hsp72 was inducible in HT-29 human colon carcinoma cells but not SW-480 colon cancer cells or CRL-1807 human colonocyte line. Hsp70B′ was induced in HT-29 and CRL-1807 cells but not SW-480 cells.[66]

ACKNOWLEDGMENTS

The work described was supported in part by NIH grants ES07408 and AI46790 (to M.A.L.), PRIN 2004 and 2006, and FIRB 2003 (to S.L.S.), and by NIH grants HL63760 and HL59225 (to Y.J.K).

REFERENCES

1. VASAK, M. 2005. Advances in metallothionein structure and functions. J. Trace Elem. Med. Biol. **19:** 13–17.
2. LYNES, M.A. *et al.* 2006. The physiological roles of extracellular metallothionein. Exp. Biol. Med. (Maywood). **231:** 1548–1554.
3. BINZ, P.A. & J.H.R. KAGI. 1999. Metallothionein: molecular evolution and classification. *In* MT IV. C.D. Klaassen, Ed.: 7–14. Birkhauser Verlag. Basel.
4. HEMDAN, N.Y. *et al.* 2006. The *in vitro* immune modulation by cadmium depends on the way of cell activation. Toxicology **222:** 37–45.
5. HESS, E.V. 2002. Environmental chemicals and autoimmune disease: cause and effect. Toxicology **181–182:** 65–70.
6. LYNES, M.A. *et al.* 1993. Immunomodulatory activities of extracellular metallothionein. I. Metallothionein effects on antibody production. Toxicology **85:** 161–177.
7. CROWTHERS, K.C. *et al.* 2000. Augmented humoral immune function in metallothionein-null mice. Toxicol. Appl. Pharmacol. **166:** 161–172.

8. CANPOLAT, E. & M.A. LYNES. 2001. *In vivo* manipulation of endogenous metallothionein with a monoclonal antibody enhances a T-dependent humoral immune response. Toxicol. Sci. **62:** 61–70.
9. EL REFAEY, H. *et al.* 1997. Identification of metallothionein receptors in human astrocytes. Neurosci. Lett. **231:** 131–134.
10. BORGHESI, L.A. *et al.* 1996. Interactions of metallothionein with murine lymphocytes: plasma membrane binding and proliferation. Toxicology **108:** 129–140.
11. LYNES, M.A., C.A. RICHARDSON, R. MCCABE, *et al.* 1999. Metallothionein-mediated alterations in autoimmune disease processes. *In* MT IV. C.D. Klaassen, Ed.: 437–444. Birkhauser Verlag. Basel.
12. YIN, X., D.A. KNECHT & M.A. LYNES. 2005. Metallothionein mediates leukocyte chemotaxis. BMC Immunol. **6:** 21.
13. YOUN, J. *et al.* 2002. Metallothionein suppresses collagen-induced arthritis via induction of TGF-beta and down-regulation of proinflammatory mediators. Clin. Exp. Immunol. **129:** 232–239.
14. LYNES, M.A. *et al.* 2006. Gene expression influences on metal immunomodulation. Toxicol. Appl. Pharmacol. **210:** 9–16.
15. TAPIERO, H., D.M. TOWNSEND & K.D. TEW. 2003. Trace elements in human physiology and pathology. Copper. Biomed. Pharmacother. **57:** 386–398.
16. GALHARDI, C.M. *et al.* 2004. Toxicity of copper intake: lipid profile, oxidative stress and susceptibility to renal dysfunction. Food Chem. Toxicol. **42:** 2053–2060.
17. RAE, T.D. *et al.* 1999. Undetectable intracellular free copper: the requirement of a copper chaperone for superoxide dismutase. Science **284:** 805–808.
18. FIELD, L.S., E. LUK & V.C. CULOTTA. 2002. Copper chaperones: personal escorts for metal ions. J. Bioenerg. Biomembr. **34:** 373–379.
19. PENA, M.M., J. LEE & D.J. THIELE. 1999. A delicate balance: homeostatic control of copper uptake and distribution. J. Nutr. **129:** 1251–1260.
20. TRUMBO, P. *et al.* 2001. Dietary reference intakes: vitamin A, vitamin K, arsenic, boron, chromium, copper, iodine, iron, manganese, molybdenum, nickel, silicon, vanadium, and zinc. J. Am. Diet. Assoc. **101:** 294–301.
21. KLEVAY, L.M. & D.M. MEDEIROS. 1996. Deliberations and evaluations of the approaches, endpoints and paradigms for dietary recommendations about copper. J. Nutr. **126:** 2419S–2426S.
22. REISER, S. *et al.* 1985. Indices of copper status in humans consuming a typical American diet containing either fructose or starch. Am. J. Clin. Nutr. **42:** 242–251.
23. JIANG, Y., *et al.* 2007. Dietary copper supplementation reverses hypertrophic cardiomyopathy induced by chronic pressure overload in mice. J. Exp. Med. **204:** 657–666.
24. HAMZA, I. & J.D. GITLIN. 2002. Copper chaperones for cytochrome c oxidase and human disease. J. Bioenerg. Biomembr. **34:** 381–388.
25. JAKSCH, M. *et al.* 2000. Mutations in SCO2 are associated with a distinct form of hypertrophic cardiomyopathy and cytochrome c oxidase deficiency. Hum. Mol. Genet. **9:** 795–801.
26. PAPADOPOULOU, L.C. *et al.* 1999. Fatal infantile cardioencephalomyopathy with COX deficiency and mutations in SCO2, a COX assembly gene. Nat. Genet. **23:** 333–337.
27. VESELA, K. *et al.* 2004. Clinical, biochemical and molecular analyses of six patients with isolated cytochrome c oxidase deficiency due to mutations in the SCO2 gene. Acta Paediatr. **93:** 1312–1317.

28. FREISINGER, P. *et al.* 2004. Reversion of hypertrophic cardiomyopathy in a patient with deficiency of the mitochondrial copper binding protein Sco2: is there a potential effect of copper? J. Inherit. Metab. Dis. **27:** 67–79.
29. WITTE, K.K. *et al.* 2005. The effect of micronutrient supplementation on quality-of-life and left ventricular function in elderly patients with chronic heart failure. Eur. Heart J. **26:** 2238–2244.
30. BOOMSMA, F. *et al.* 2000. Plasma semicarbazide-sensitive amine oxidase (SSAO) is an independent prognostic marker for mortality in chronic heart failure. Eur. Heart J. **21:** 1859–1863.
31. FOX, P.L. *et al.* 2000. Ceruloplasmin and cardiovascular disease. Free Radic. Biol. Med. **28:** 1735–1744.
32. CAI, L. *et al.* 2005. Inhibition of superoxide generation and associated nitrosative damage is involved in metallothionein prevention of diabetic cardiomyopathy. Diabetes **54:** 1829–1837.
33. SONG, Y. *et al.* 2005. Cardiac metallothionein synthesis in streptozotocin-induced diabetic mice, and its protection against diabetes-induced cardiac injury. Am. J. Pathol. **167:** 17–26.
34. WANG, J. *et al.* 2006. Cardiac metallothionein induction plays the major role in the prevention of diabetic cardiomyopathy by zinc supplementation. Circulation **113:** 544–554.
35. COOPER, G.J. *et al.* 2004. Regeneration of the heart in diabetes by selective copper chelation. Diabetes **53:** 2501–2508.
36. FREDERICKSON, C.J., J.Y. KOH & A.I. BUSH. 2005. The neurobiology of zinc in health and disease. Nat. Rev. Neurosci. **6:** 449–462.
37. SENSI, S.L., H.Z. YIN & J.H. WEISS. 2000. AMPA/kainate receptor-triggered Zn^{2+} entry into cortical neurons induces mitochondrial Zn^{2+} uptake and persistent mitochondrial dysfunction. Eur. J. Neurosci. **12:** 3813–3818.
38. BROWN, A.M. *et al.* 2000. Zn^{2+} inhibits alpha-ketoglutarate-stimulated mitochondrial respiration and the isolated alpha-ketoglutarate dehydrogenase complex. J. Biol. Chem. **275:** 13441–13447.
39. AIZENMAN, E. *et al.* 2000. Induction of neuronal apoptosis by thiol oxidation: putative role of intracellular zinc release. J. Neurochem. **75:** 1878–1888.
40. FREDERICKSON, C.J. *et al.* 2002. Nitric oxide causes apparent release of zinc from presynaptic boutons. Neuroscience **115:** 471–474.
41. BONANNI, L. *et al.* 2006. Zinc-dependent multi-conductance channel activity in mitochondria isolated from ischemic brain. J. Neurosci. **26:** 6851–6862.
42. JIANG, D. *et al.* 2001. $Zn(2+)$ induces permeability transition pore opening and release of pro-apoptotic peptides from neuronal mitochondria. J. Biol. Chem. **276:** 47524–47529.
43. MALAIYANDI, L.M. *et al.* 2005. Direct visualization of mitochondrial zinc accumulation reveals uniporter-dependent and -independent transport mechanisms. J. Neurochem. **93:** 1242–1250.
44. SENSI, S.L. *et al.* 2003. Modulation of mitochondrial function by endogenous Zn^{2+} pools. Proc. Natl. Acad. Sci. USA **100:** 6157–6162.
45. MCLAUGHLIN, B. *et al.* 2001. p38 activation is required upstream of potassium current enhancement and caspase cleavage in thiol oxidant-induced neuronal apoptosis. J. Neurosci. **21:** 3303–3311.
46. PAL, S. *et al.* 2003. Mediation of neuronal apoptosis by Kv2.1-encoded potassium channels. J. Neurosci. **23:** 4798–4802.

47. SENSI, S.L. & J.M. JENG. 2004. Rethinking the excitotoxic ionic milieu: the emerging role of Zn(2+) in ischemic neuronal injury. Curr. Mol. Med. **4:** 87–111.
48. BOSSY-WETZEL, E. et al. 2004. Crosstalk between nitric oxide and zinc pathways to neuronal cell death involving mitochondrial dysfunction and p38-activated K+ channels. Neuron **41:** 351–365.
49. FRAZZINI V. et al. 2007. Mild acidosis enhances AMPA receptor-mediated intracellular zinc mobilization in cortical neurons. Mol. Med. In press.
50. DEVINNEY, M.J., 2ND, I.J. REYNOLDS & K.E. DINELEY. 2005. Simultaneous detection of intracellular free calcium and zinc using fura-2FF and FluoZin-3. Cell Calcium **37:** 225–232.
51. STORK, C.J. & Y.V. LI. 2006. Intracellular zinc elevation measured with a "calcium-specific" indicator during ischemia and reperfusion in rat hippocampus: a question on calcium overload. J. Neurosci. **26:** 10430–10437.
52. PERDRIZET, G.A. 1997. Heat Shock Response and Organ Preservation: Models of Stress Conditioning. RG Landes Co, and Chapman & Hall. Georgetown, TX, and New York, NY.
53. KREGEL, K.C. 2002. Heat shock proteins: modifying factors in physiological stress responses and acquired thermotolerance. J. Appl. Physiol. **92:** 2177–2186.
54. LEPOCK, J.R. 2005. How do cells respond to their thermal environment? Int. J. Hyperthermia **21:** 681–687.
55. MARTIN, J. 2000. Group II chaperonins as mediators of cytosolic protein folding. Curr. Protein Pept. Sci. **1:** 309–324.
56. RINALDI, B. et al. 2006. Inflammatory events in a vascular remodeling model induced by surgical injury to the rat carotid artery. Br. J. Pharmacol. **147:** 175–182.
57. SCHELL, M.T. et al. 2005. Heat shock inhibits NF-kB activation in a dose- and time-dependent manner. J. Surg. Res. **129:** 90–93.
58. WILLIS, D. et al. 1996. Heme oxygenase: a novel target for the modulation of the inflammatory response. Nat. Med. **2:** 87–90.
59. KAPPAS, A. & M.D. MAINES. 1976. Tin: a potent inducer of heme oxygenase in kidney. Science **192:** 60–62.
60. NEIL, T.K. et al. 1995. Differential heme oxygenase induction by stannous and stannic ions in the heart. J. Cell Biochem. **57:** 409–414.
61. MITANI, K. et al. 1993. The role of inorganic metals and metalloporphyrins in the induction of haeme oxygenase and heat-shock protein 70 in human hepatoma cells. Biochem. J. **290**(Pt. 3): 819–825.
62. BULGER, E.M. et al. 1997. Fat embolism syndrome. A 10-year review. Arch. Surg. **132:** 435–439.
63. GARCIA, J.C. et al. 1997. Heat shock protects rodent kidneys from acute ischemia/reperfusion injury. Surg. Forum **48:** 381–382.
64. LENA, C.J., D.S. SHAPIRO & G.A. PERDRIZET. 1996. Heat shock and recovery prevents paraplegia in a model of acute spinal cord ischemia. Surg. Forum **47:** 270–272.
65. HOUSE, S.D. et al. 2001. Effects of heat shock, stannous chloride, and gallium nitrate on the rat inflammatory response. Cell Stress Chaperones **6:** 164–171.
66. NOONAN, E.J., R.F. PLACE, C. GIARDINA & L.E. HIGHTOWER. 2007. Hsp70B' Regulation and Function. Cell Stress Chaperones. Early edition DOI: 10.1379/CSC-278.

Interleukin-1 System in CNS Stress

Seizures, Fever, and Neurotrauma

TAMAS BARTFAI,[a] MANUEL SANCHEZ-ALAVEZ,[a]
SIV ANDELL-JONSSON,[b] MARIANNE SCHULTZBERG,[b]
ANNAMARIA VEZZANI,[c] ERIK DANIELSSON,[d] AND BRUNO CONTI[a]

[a]*The Scripps Research Institute, La Jolla, California, USA*

[b]*Stockholm University, Stockholm, Sweden*

[c]*M. Negri Institute, Milano, Italy*

[d]*Malmö Almänna Sjukhus, Sweden*

ABSTRACT: Proteins of the interleukin-1 (IL-1) system include the secreted agonist IL-1β, and the receptor antagonist IL-1ra, both competing for binding to the IL-1 receptor (IL-1R). IL-1β and IL-1ra are highly inducible under different forms of stress, such as excitatory neurotransmitter excess occurring during seizures, in infection and inflammation, and during neurotrauma. In each of these conditions induction of IL-1β precedes that of IL-1ra, resulting in up to 10–20-fold elevation of IL-1β concentrations. Consequently, IL-1β induces the elevation of other proinflammatory molecules, including IL-6, IL-1R1, COX2, and iNOS, as well as of IL-1ra. Elevation of IL-1ra is of key importance for quenching the inflammatory response at the IL-1R1 as part of an autoregulatory loop. In seizures, IL-1ra is a strong anticonvulsant and in IL-1β-dependent fever, a powerful antipyretic. In traumatic brain injury (TBI), the ability of patients to mount an IL-1ra response, as measured in the CSF, strongly correlated with the neurological outcome. Selective induction or pharmacological application of IL-1ra may be sparing neurons in seizures and neurotrauma.

KEYWORDS: interleukin-1β; interleukin-1 receptor antagonist; neurotrauma; seizure; fever; anticonvulsant

INTRODUCTION

The cytokines interleukin 1β (IL-1β) and interleukin 1 receptor antagonist (IL-1ra) can be synthesized in the central nervous system (CNS) by neurons,

Address for correspondence: Tamas Bartfai, Ph.D., Molecular and Integrative Neurosciences Department, The Scripps Research Institute, 10550 North Torrey Pines Road, La Jolla, CA, USA. Voice: 858-7848404; fax: 858-7849099.
tbartfai@scripps.edu

TABLE 1. GFAP-IL-1ra overexpressor mice in studies on CNS IL-1 signaling

Effects	Reference
Reduced fever response	10
Elevated seizure threshold	5
Improved recovery neurotrauma	14
Influence on behavior, corticosterone, and brain monoamines	15
Antagonism of morphine analgesia	16

microglia, and infiltrating macrophages.[1] The healthy brain uses IL-1β signaling in synaptic modulation.[2] Neurons in the hippocampus and hypothalamus express functional IL-1 receptor type I (IL-1R1) that upon activation signals through the transcription-dependent toll-type signaling and through a recently discovered nontranscription-dependent signaling involving ceramide-mediated activation of c-Src and PI3K/Akt.[3] Through IL-1RI, IL-1β can affect IPSPs,[1] and LTP in the hippocampus.[4] During stressful conditions including excitatory, infectious–inflammatory, or neurotrauma, the production of IL-1β is greatly stimulated. Elevation of IL-1β is associated with an elevation of other proinflammatory molecules in the central nervous system (CNS) including the interleukin-6 (IL-6), and the anti-inflammatory cytokine IL-1ra. IL-1β-dependent stimulation of IL-1ra is an important component of the autoregulatory network controlling inflammatory response. Thus, the level of occupancy of IL-1R1 and the ratio of IL-1β and IL-1ra or that with other cytokines are key determinants in neuronal responses to stress in the brain. In this communication we present examples of the role of IL-1R1 signaling in seizures, fever, and neurotrauma and comment on the prognostic and therapeutic value of measuring and modulating the IL-1 system in the brain through enhanced IL-1ra levels.

RESULTS AND DISCUSSION

IL-1β effects on neurons have been studied most extensively in the hippocampus in the context of long-term potentiation (LTP) and in seizure.[2,4,5] We have shown that IL-1β is a strong proconvulsant in kainate-induced seizure, and that IL-1ra is a potent anticonvulsant agent.[5] Hypothalamic application of IL-1β causes sickness syndrome, anorexia,[6] and fever of rapid and sustained phase.[3] The central effects of leptin on appetite and energy balance and febrile response are mediated by hypothalamic IL-1 signaling.[7]

In view of the rapid rise of IL-1β under different stress conditions, the use of transgenic animals with null mutations of IL-1RI, IL-1β, or IL-1ra[8,9] or overexpressing IL-1ra[10] can inform about the importance of the IL-1β/IL-ra ratio. We generated mice with specific overexpression of the IL-1ra cDNA in astrocytes by driving its expression with the glial fibrillary acidic protein

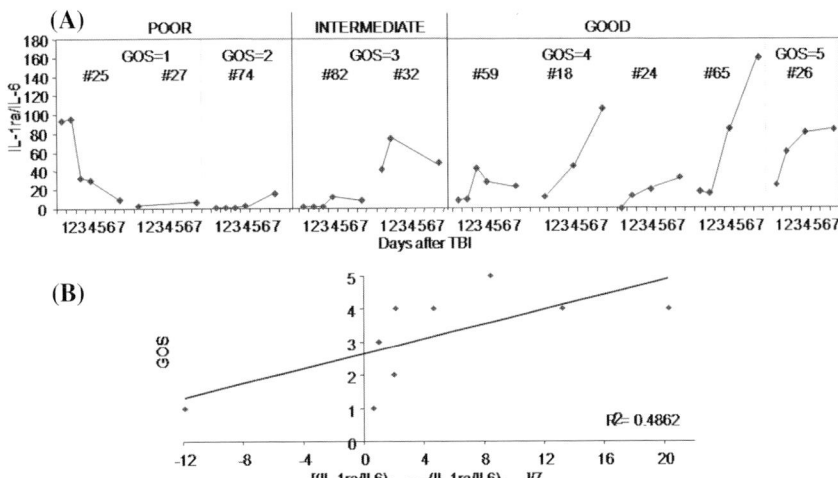

FIGURE 1. (A) Profiles of the ratio between the CSF concentration of IL-1ra and IL-6 during the first 7 days after injury of 10 patients hospitalized with traumatic brain injury (TBI). The profiles are organized in three groups depending on the clinical outcome classified as good (GOS 4–5), intermediate (GOS 2–3), or poor (GOS 1) neurological outcome. The distribution clearly showed a positive correlation between good outcome and rising levels of IL-1ra/IL-6. IL-1ra = inteleukin-1 receptor antagonist; IL-6 = interleukin-6; GOS = Glasgow Outcome Scale, TBI = traumatic brain injury; # indicates the arbitrary ID number of a subject. **(B)** The correlation between the molar ratios of CSF IL-1ra and /IL-6 levels during the first week after injury can be described by the formula ([IL-1ra/IL-6] day 7 – (IL-1ra/IL-6) day 1)/7.

promoter.[10] This transgenic mouse line with intact IL-1 signaling in the periphery, but 3–8 fold overexpression of IL-1ra in the CSF, has been shown very useful in delineating the effects of IL-1β/IL-1ra changes in the brain in a number of studies summarized in TABLE 1. This model has also inspired experiments with viral expression of IL-1ra[11] and pharmacological studies with recombinant IL-1ra in stroke models[12] and in stroke patients in a phase 2 trial.[13]

Work on animal models indicated that local inflammatory processes occurring after traumatic brain injury (TBI) contribute to secondary neuronal damage. In a specific study, the cerebrospinal fluid (CSF) was collected from 10 TBI patients, sampled daily for seven consecutive days after hospitalization. CSF levels of IL-6 and IL-1ra were correlated with the clinical outcome at 3 months after injury as measured with the Glasgow outcome scale (GOS). Individual responses in CSF levels of IL1-ra and IL-6 varied dramatically and no correlation was found between the absolute levels of IL-6 or IL-1ra and the GOS. However, the trend of the molar ratio of IL-1ra/IL-6 over 7 days was a good index of the clinical outcome (FIG. 1). Subjects with steadily rising

IL-ra/IL-6 values during the 7 days after TBI had a significantly better neurological outcome than those who failed to muster such a response. These results provide a possible measure for the clinical evolution of TBI patients and suggest that elevating IL-1ra or decreasing IL-6 levels in the first week following trauma may provide effective measures to contain neurological damage and improve the clinical outcome.

As IL-6 can be induced by IL-1 and TNF-α it provides a general measure of the degree of inflammation. IL-1ra provides an anti-inflammatory response of importance in order to control inflammation and seizures following neurotrauma. Thus, we chose the IL-1ra/IL-6 ratio to follow the patients. The present data suggest that it may be a useful parameter to follow in TBI and stroke. Since there are important polymorphisms in the IL-1ra gene, it may be necessary to pharmacologically assist the patients with genetically lower IL-1ra biosynthesis rates, in order to improve neurological outcome in TBI.

REFERENCES

1. XIA, Y. *et al.* 1999. IL-1β and IL-6 excite neurons and suppress nicotinic and noradrenergic neurotransmission in guinea pig enteric nervous system. J. Clin. Invest. **103**: 1309–1316.
2. VIVIANI, B. *et al.* 2003. Interleukin-1β enhances NMDA receptor-mediated intracellular calcium increase through activation of the Src family of kinases. J. Neurosci. **23**: 8692–8700.
3. SANCHEZ-ALAVEZ, M. *et al.* 2006. Ceramide mediates the rapid phase of febrile response to IL-1β. Proc. Natl. Acad. Sci. USA **103**: 2904–2908.
4. KATSUKI, H. *et al.* 1990. Interleukin-1 beta inhibits long-term potentiation in the CA3 region of mouse hippocampal slices. Eur. J. Pharmacol. **181**: 323–326.
5. VEZZANI, A. *et al.* 2000. Powerful anticonvulsant action of IL-1 receptor antagonist on intracerebral injection and astrocytic overexpression in mice. Proc. Natl. Acad. Sci. USA **97**: 11534–11539.
6. DANTZER, R. *et al.* 1998. Cytokines and sickness behavior. Ann. N. Y. Acad. Sci. **840**: 586–590.
7. LUHESHI, G.N. *et al.* 1999. Leptin actions on food intake and body temperature are mediated by IL-1. Proc. Natl. Acad. Sci. USA **96**: 7047–7052.
8. ALHEIM, K. & T. Bartfai. 1998. The interleukin-1 system: receptors, ligands, and ICE in the brain and their involvement in the fever response. Ann. N. Y. Acad. Sci. **840**: 51–58.
9. NICKLIN, M.J. *et al.* 2000. Arterial inflammation in mice lacking the interleukin 1 receptor antagonist gene. J. Exp. Med. **191**: 303–312.
10. LUNDKVIST, J. *et al.* 1999. Acute-phase responses in transgenic mice with CNS overexpression of IL-1 receptor antagonist. Am. J. Physiol. **276**: R644.
11. TSAI, T.H. *et al.* 2003. Gene treatment of cerebral stroke by rAAV vector delivering IL-1ra in a rat model. Neuroreport **14**: 803–807.
12. MULCAHY, N.J. *et al.* 2003. Delayed administration of interleukin-1 receptor antagonist protects against transient cerebral ischaemia in the rat. Br. J. Pharmacol. **140**: 471–476.

13. EMSLEY, H.C. *et al.* 2005. A randomised phase II study of interleukin-1 receptor antagonist in acute stroke patients. J. Neurol. Neurosurg. Psychiatry **76:** 1366–1372.
14. TEHRANIAN, R. *et al.* 2002. Improved recovery and delayed cytokine induction after closed head injury in mice with central overexpression of the secreted isoform of the interleukin-1 receptor antagonist. J. Neurotrauma **19:** 939–951.
15. OPRICA, M. *et al.* 2005. Transgenic overexpression of interleukin-1 receptor antagonist in the CNS influences behaviour, serum corticosterone and brain monoamines. Brain Behav. Immun. **19:** 223–234.
16. SHAVIT, Y. *et al.* 2005. Interleukin-1 antagonizes morphine analgesia and underlies morphine tolerance. Pain **115:** 50–59.

Chaperonopathies by Defect, Excess, or Mistake

ALBERTO J. L. MACARIO AND EVERLY CONWAY DE MACARIO

University of Maryland Biotechnology Institute, Center of Marine Biotechnology, Baltimore, Maryland 21202, USA

ABSTRACT: The stress response, stress proteins, heat-shock genes and proteins, molecular chaperone genes and proteins, and a number of closely related molecules and cellular processes have been studied over the last few decades. A huge amount of information has accumulated that is scattered in printed and electronic literature and databases. Most of this information constitutes the subject matter of the science of chaperonology. More recently, the concept of chaperone pathology, sick chaperones, has evolved since various pathological conditions have been identified in which defective chaperones play an etiologic role. These conditions are the chaperonopathies. Recent findings on chaperonopathies are briefly discussed in this article. Chaperonopathies occur at all ages; as a rule the genetic cases have an early clinical onset while the acquired chaperonopathies become manifest in the elderly and/or in association with other diseases. Other fields of chaperonology, which will most likely be expanded in the near future, are the study of extracellular chaperones, chaperone networks, the therapeutic use of chaperones (i.e., chaperonotherapy) to manage chaperonopathies and to improve cell performance in the face of stress, the evaluation of chaperones as diagnostic markers and as prognostic indicators, and the development of antichaperone agents to suppress chaperone-gene expression or inhibit chaperone function when chaperones contribute to disease rather than the opposite.

KEYWORDS: genetic chaperonopathies; acquired chaperonopathies; chaperonology; sick chaperones; chaperonotherapy; antichaperone agents

INTRODUCTION

The study of molecular chaperones has progressed steadily since they came into prominence in the early 1970s. Many aspects of their role in cell survival in the face of stress and of their physiological functions, including assisting

Address for correspondence: Alberto J. L. Macario, M.D., University of Maryland Biotechnology Institute, Center of Marine Biotechnology, 701 E. Pratt Street, Baltimore, MD 21202. Voice: 410-234-8849; fax: 410-234-8896.
macario@umbi.umd.edu

protein maturation and translocation have been elucidated in a few species, particularly bacterial and eukaryotic model organisms. The molecular mechanisms involved in chaperoning client polypeptides, the structure and evolution of chaperones, and the participation of chaperones in various cellular processes other than protein folding have also been actively investigated in the last decades. The information is available in a great number of original articles and reviews and constitutes the subject matter of a scientific discipline we call chaperonology.

Also within the realm of chaperonology is the study of chaperone pathology. The idea that chaperones could participate in pathology developed since the early times of chaperonology.[1,2] More recently the notion that chaperones may themselves be etiological factors in disease and ageing has emerged in full view.[3,4] Pathological conditions in which a defective chaperone is an etiological factor have been grouped and classified as chaperonopathies, syndromes, and diseases that form a definite area of general pathology and medicine.[5] A few recent findings pertinent to chaperonopathies will be briefly discussed in this article.

GENETIC CHAPERONOPATHIES

Some recent developments within the area of genetic chaperonopathies pertain to Bardet–Biedl syndrome (BBS), neuropathies associated with sHsp mutations, Williams syndrome (WS), dilated cardiomyopathies (DCs), ataxia of Charlevoix–Saguenay (ARSACS), and endoplasmic reticulum (ER) pathology.

Bardet–Biedl Syndrome

This is a genetically heterogeneous developmental disorder characterized by obesity, polydactyly, retinopathy, mental retardation, renal and cardiac anomalies, hypertension, and diabetes. It is considered to be essentially a ciliopathy but many genes with various prospective roles seem to be implicated. Until recently, nine BBS loci had been identified, one of which, *BBS6*, encodes a chaperonin II–like subunit. It was later established that this gene is the same as the *MKKS*, involved in McKusick Kaufman syndrome (MKKS) and known to be related with CCT subunit 8. *BBS10* and *BBS11* were recently added to the list of BBS loci, both encoding proteins that would be involved in protein maturation and processing. *BBS10* encodes a chaperonin II–like molecule,[6] while *BBS11* encodes an E3 ubiquitin ligase,[7] a component of the ubiquitin–proteasome system, which plays a major role in protein degradation, and thus in protein homeostasis, together with chaperones.[5]

Interestingly, *BBS6* (i.e., *MKKS*) SNPs were found to be associated with common obesity, and also tended to associate with dyslipidemia and hyperglycemia, in the absence of other manifestations of BBS.[8]

Neuropathies Associated with sHsp Mutations

A key property of chaperones, including sHsp, is the ability to interact with other molecules: client polypeptide in need of assistance for folding or refolding, another chaperone molecule of the same team or subunit of the same chaperoning complex, another chaperone of another team to build a chaperoning network, and components of the ubiquitin–proteasome system.[9] These interactions are mediated by defined domains in the chaperone molecule. If any of these domains is structurally modified because of a gene mutation or an aberrant posttranslational modification, the respective interaction may be impaired or abolished, resulting in a defective chaperoning pathway. To understand the mechanisms involved in the pathogenesis of neurological diseases associated with mutations of the sHsp Hsp22(HSPB8), two mutants, K141E and K141N, were investigated.[10] These mutants are associated with inherited peripheral motor neuron disorders such as distant hereditary motor neuropathy type II and axonal Charcot–Marie–Tooth disease.[11] It is known that Hsp22 forms homodimers, heterodimers with other sHsps, and large oligomers; Fontaine *et al.* investigated the ability of each mutant to interact with itself, with wild type Hsp22, and with other sHsps that are abundant in neurons: alphaB-crystallin, Hsp20, and Hsp27.[10] It was found that the mutants displayed increased interaction with alphaB-crystallin and with Hsp27, but not with Hsp20. Likewise, a mutant Hsp27 (S135F) that is also associated with distant hereditary motor neuropathy type II and axonal Charcot–Marie–Tooth disease, showed increased interaction with wild-type Hsp22. The measurements performed demonstrated that each of the Hsp22 mutants investigated has its distinctive pattern of interaction with the various molecules tested, indicating that one can expect clinicopathological variations depending on the mutant involved. The data also indicated that the mutants' exaggerated tendency to aggregate could contribute to defective chaperoning, and to the formation of pathological protein precipitates.

Williams Syndrome

Whether WS is in fact a chaperonopathy has not yet been established. However, there is evidence suggesting that hemideletion of a gene encoding FK506-binding protein 6 (*FKBP6*) is associated with WS.[12] FKBP6 belongs to the PPIase (peptidyl–prolil *cis–trans* isomerase) group of chaperones that isomerize Thr/Ser–Pro motifs, and of which three major families have been studied:

cyclophilins, FKBP6, and parvulins.[13] WS is characterized by a distinctive facial appearance, growth retardation, cardiovascular anomalies, hypercalcemia, impaired glucose tolerance, constipation, scoliosis, and a series of neuropsychiatric manifestations, including mild mental retardation or learning difficulties. Genetically, WS has been attributed to a deletion in chromosome 7q11.23 encompassing about 1.6 megabases with about 28 genes, one of which is *FKBP6*. A recent study of new cases has shown that mental retardation does not occur if the FKB6 gene is present.[12]

Dilated Cardiomyopathy

DC is the most common among the nonischemic cardiopathies and is characterized by enlargement of the ventricular cavity, weakening of ventricular walls, and conduction abnormalities. In a recent paper it was reported that mitochondrial Hsp40 (DnaJ3, or Tid1) is differentially expressed during cardiac development and also in pathologic cardiac hypertrophy.[14] It was demonstrated, using transgenic mice lacking *DnaJ3*, that absence of DnaJ3 resulted in the development of DC followed by early death. The mitochondrial respiratory chain and DNA were deficient and decreased, respectively, in the affected mice. It was found in addition that a prominent client for DnaJ3 was the alpha subunit of DNA polymerase gamma (Polga). The data considered together indicate that Hsp40 (DnaJ3) is necessary for mitochondrial biogenesis, most likely through its interaction with Polga. Thus, DC can be considered a mitochondriopathy due, at least in part, to a quantitative defect of Hsp40, and one may speculate that this disease may occur in humans as a chaperonopathy due to a genetic defect in, or absence of, the mitochondrial *hsp40* gene.

Mutations in the gene encoding alphaB-crystallin (*CRYAB*) have been implicated in DC. It is known that mutations in this gene cause cataracts and cardioskeletal myopathies,[11] but their participation in the pathogenesis of DC is not completely elucidated. An overview of the literature revealed that the phenotypes associated with mutations in *CRYAB* are varied, including cataract only or combined with cardioskeletal anomalies, and DC only.[15] The association of *CRYAB* mutation with DC is difficult to search for because there is no certain clinical marker that is specific for the mutation; nonetheless, the search is justified because although rare the association with DC does occur even in the absence of cataract or skeletal myopathy.

Ataxia of Charlevoix–Saguenay

The autosomal recessive spastic ataxia of Charlevoix-Saguenay (ARSACS) is characterized, as suggested by the name, by ataxia (of early onset) and spasticity, which are accompanied by manifestations of peripheral neuropathy, with finger and foot deformities. Usually there is no mental deterioration. The disease was described in some inhabitants of the Quebec region in Canada. The major pathological indicators are hypermyelination of the retinal nerve fibers, atrophy of the upper vermis and loss of Purkinje cells in the cerebellum, and degeneration of lateral corticospinal tracts. A gene involved in the disease of patients from Quebec encodes sacsin, which shares some structural features with chaperones.[11] Recently, ARSACS-like conditions were found in five Japanese families.[16] Interestingly, some of the clinical features of the disease in the Japanese patients differed from those described for the Canadian patients, for example, there was a later onset of ataxia in the Japanese compared with the Canadian patients. In addition, the Japanese patients did not show hypermyelination of retinal nerve fibers or spasticity, and had mental deterioration. It is increasingly clear that one can expect various types or ARSACS, depending on the population examined. The disease's heterogeneity may be attributed to a variety of factors, one of which could be the type of alteration in the sacsin molecule. This protein has a predicted molecular mass of 437 kDa, and a pI of 6.85. The molecule is made of 3,829 amino acids, encompasses two leucine zippers, three coiled coils, and seven nuclear localization signals, and does not show significant extensive similarity with any known molecule. However, the C-terminal portion of sacsin contains a DnaJ motif. Furthermore, sacsin also has two domains similar to the N-terminal portion of Hsp90. These features shared with molecular chaperones have been the basis for inclusion of ARSACS among the candidates for chaperonopathies,[11] and should stimulate research on the functions of sacsin *in vitro* and *in vivo*, focusing on the possible chaperoning properties of the molecule in normal individuals and their alterations in ARSACS patients.

Pathology of Protein Transport into the Endoplasmic Reticulum

Several pathological conditions have been identified that show alteration of the mechanism translocating proteins into the ER.[17] Among these conditions, one in particular, autosomal dominant polycystic liver disease (ADPLD),[18] qualifies for inclusion in the chaperonopathy class. Another one would be Marinesco–Sjodren syndrome, associated with mutation of a nucleotide exchange factor gene, *SIL1* (discussed previously[9]). ADPLD is characterized, histopathologically, by biliary epithelial liver cysts, which become bigger and more numerous with time, leading to organ failure. One of the genes involved encodes SEC63, which is mutated in some cases of ADPLD.[18] SEC63 seems to be crucial for folding and/or import into the ER of one or more proteins participating in the development of cells in the biliary ducts and in the proliferation of those cells. In the absence of a functional SEC63, the proteins that regulate cell growth and multiplication do not translocate and fail to reach their place of residence and work and, as a consequence, the affected cells multiply excessively. Noteworthy in this regard is that the *SEC63* gene has been found mutated in certain types of cancers (e.g., hereditary nonpolyposis colorectal cancer–associated small-bowel cancer) that are very malignant tumors characterized by extremely excessive cell multiplication.[19]

ACQUIRED CHAPERONOPATHIES

Failure of Inducible Chaperones and Disease

The distribution of Hsp73 (constitutive) and Hsp72 (heat-inducible) was examined in the neuromelanin-containing neurons by applying immunohistochemistry.[20] Brain samples were obtained within 10 h after death from patients who had Parkinson's disease (PD) and from matched controls without the disease. In these controls, Hsp72 was undetectable and Hsp73 was present at low levels in cytosol and the nucleus. In PD samples, Hsp73 was high in the nucleus, in which Hsp72 was undetectable. Lewy bodies were specifically examined, and Hsp73 was present in a subset of them, many more than the bodies showing Hsp72. The immunoreactivity of this latter chaperone was considerably stronger in nonmelanized neurons than in the melanized ones. These results suggest that the expected increase of Hsp72 in response to increased need of assistance for the defective peptides characteristic of PD, and to the cellular stress caused by misfolded and aggregated peptides, does not occur. This lack of reaction by the Hsp72 chaperoning pathway may contribute to pathogenesis in PD by allowing accumulation–aggregation of misfolded peptides.

Age-Related Chaperone Deficiency

The effect of stress preconditioning on the subsequent response to stress was studied in human dermal fibroblasts (HDF) from primary cultures.[21] HDF were obtained from young (15–28 years of age) and from old (61–77 years of age) individuals to assess the consequences of age in determining the outcome of stress preconditioning. The latter preconditioning was done by incubating cultured cells at 42°C for 1 h while the tested stressor was applied to the preconditioned cells at 1, 2, or 20 h of recovery at 37°C after the 1-h-long heat shock. Both preconditioned and nonstressed cells were submitted to hypoxic and oxidative stress by treating them with cyanide-m-chlorophenylhydrazone or hydrogen peroxide, respectively, for 1 h. Western blot was used to assess the levels of Hsp70 in cell lysates, and lactic dehydrogenase (LDH) was quantified in the used culture medium to assess cell damage. Hsp70 levels increased in young HDF by nearly 100%, 200%, and more than 200% at 1, 2, and 20 h after the preconditioning stress, respectively. In contrast, the increases in aged HDF were considerably lower, that is, nearly 30, 60, and less than 30% at the same intervals. Therefore, aged cells responded poorly to the preconditioning heat stress by comparison with young counterparts. Furthermore, a decrease in cell damage caused by the secondary stressor (hypoxia or oxidative) likely attributable to the preconditioning stress was evident only in young HDF (LDH release after secondary stress was decreased only in used media from the young cells that had been preconditioned 20 h before). It was concluded that aged cells respond poorly to hypoxic and oxidative stresses even after preconditioning, and that the deficient response to the secondary stress is associated with a decreased response to the preconditioning stress measured by Hsp70 levels. Aged cells fail to increase the Hsp70 levels after heat shock by comparison with young counterparts, and this failure may contribute to their poor response to a secondary stress.

Basal levels of Hsps during senescence in the absence of deliberate or known induction by stressors could be assumed, at least as a working hypothesis, to be good indicators of how ageing affects antistress mechanisms and protein homeostasis. However, information on basal levels of Hsps and chaperones in humans of any age is scarce. To address this lack of data on what appears to be an important aspect of the ageing process, the expression patterns of various Hsps were studied in human peripheral blood cells from individuals of old age and younger controls.[22] Twelve women and five men (mean age 36.4 ± SD 8.1 years) were compared to six women and 12 men with an average age of 75.0 ± SD 6.3 years. In both groups all participants were healthy, with no detectable pathology. In addition, patients with acute infection (C-reactive serum protein levels greater than 10 mg/L) were also studied divided into two groups: "young" comprised 10 women and 6 men, mean age 37.9 ± SD 6.7 years, and "old" comprised 11 women and 4 men, mean age 82.7 ± SD 7.2 years. An increase in the basal levels of Hsp32, Hsp70, and Hsp90 was evident with age in the healthy set of individuals but not in the patients with infection, although the latter patients had overall higher levels of the Hsps than the healthy counterparts. The data also indicated that Hsp70 levels in patients with infection parallel the levels of serum C-reactive protein and cytokines.

Posttranslational Modification of PDIase and Protein Misfolding Disease

A chaperone that is part of the cellular antistress mechanisms residing in the ER is protein-disulphide isomerase (PDIase). This chaperone assists folding and transport of secretory proteins via a mechanism that involves catalysis of thiol disulfate exchange, which leads to disulphide bond formation and conformational rearrangements. In neurodegenerative diseases and cell stress due to ischemia–hypoxia, a stress response is mounted in the ER with participation of PDIase in response to accumulation of misfolded proteins. However, this important function of PDIase may be compromised in some neurodegenerative disorders. For example, it was found that the chaperone is S-nitrosylated (a nitric oxide [NO] group is transferred to a key cysteine thiol) in brain cells from individuals with Alzheimer's and Parkinson's diseases.[23] S-nitrosylation has a negative impact on the chaperone and on the cell: it inhibits the PDIase's enzymatic activity and promotes accumulation of polyubiquitinated peptides with induction of the ER stress response that may proceed to apoptosis and cell death. As a consequence of S-nitrosylation the cytoprotective capacity of PDIase is cancelled. This is a documented example on the molecular and pathological consequences of an acquired chaperonopathy (posttranslational modification) that add up on top of those because of the intrinsic molecular abnormalities, that is, proteinopathies, occurring in neurodegenerative disorders.

Physical Association of Chaperones with Precipitates of Abnormal Proteins in Neurodegenerative Disorders

Presence of chaperones and components of the ubiquitin–proteasome system in lesions characteristic of neurodegenerative disorders has been reported on numerous occasions. A recent article shows that sHsp are present in lesions typical of Alzheimer's disease (AD).[24] AD shows various pathological signatures: extracellular senile plaques (SP), intracellular neurofibrillary tangles (NFT), and cerebral amyloid angiopathy (CAA). All of them are constituted of defective proteins with a tendency to misfold, aggregate, and precipitate, thus forming the three kinds of deposits mentioned earlier. Senile plaques and CAA are chiefly made of A-beta amyloid, while the main component of NFT is tau. The article discussed here reports that SP was accompanied by Hsp20, Hsp27, and HspB2 in the extracellular space.[24] In addition, CAA was accompanied by extracellular HspB2. In contrast, NFT did not contain any of these sHsps. What one can infer from these results is that the various types of sHsp play differential roles in the pathogenesis of the different types of lesions that are characteristic of AD. Therapeutic strategies based on the use of chaperones as therapeutic agents (chaperonotherapy), or based on the use of other agents targeted to chaperones, must take into account the findings just discussed, that is, not all sHsp have the same role and not all participate in every type of lesion. See also the section on chaperonopathies by "mistake" further ahead.

Sequestration of Chaperone Molecules in Protein Deposits May Cause Chaperoning Deficiency

A quantitative deficiency of chaperones can conceivably occur if they become trapped into precipitates, which will subtract them from the pool of useful molecules. This phenomenon is not only possible in theory but it is likely to occur also in real life. Indeed, chaperones are found in protein deposits of various misfolding disorders, including neurodegenerative diseases. The question then is, are the quantities subtracted from the useful pool big enough to cause a deficiency in the overall chaperoning capacity of the cell? It has recently been reported that alphaB-crystallin and Hsp27 are abundant in Rosenthal fibers.[25] These fibers are protein deposits typically found in astrocytes from patients with Alexander disease, and contain mainly glial fibrillary acidic protein (GFAP), which occurs almost exclusively in astrocytes. In most cases of Alexander disease studied, mutations of GFAP have been documented. One of the most frequent mutations, R416W, was found to disrupt intermediate filament assembly: the filaments do form but they tend to aggregate. The mutant also induces the generation of intracellular precipitates in many cell types, including astrocytes, if the mutant gene is expressed in transfected host cells. These precipitates experimentally induced in transfected cells share properties with Rosenthal fibers; for example, alphaB-crystallin and Hsp27 are present in the precipitates together with the mutant GFAP. Analysis of the mechanism of formation of the experimentally induced aggregates showed that the mutant GFAP and the two chaperones associate simultaneously, during the formation of the aggregates. Interestingly, Hsp70 does not coaggregate, suggesting that aphaB-crystallin and Hsp27 have a measure of specificity for aggregate formation with mutant GFAP. Furthermore, it was demonstrated by immunofluorescence with a monoclonal antibody for R416W GFAP that the latter protein is present in Rosenthal fibers.

Inactivation of Chaperones by Exogenous Toxins

A newly identified type of chaperonopathy is caused by inactivation of a chaperone by a bacterial toxin.[26] A toxin from a highly virulent strain of *Escherichia coli*, termed AB5 subtilase, was found capable of cutting the Hsp70 chaperone that resides in the ER, BiP, between two amino acids at positions 416 and 417. In this manner, the substrate-binding and the ATPase domains (for structural details see Ref. [9]) are severed from one another, causing inactivation of BiP. This chaperone plays a central role in the cell, not only in protein folding but also in various other related functions (e.g., translocation of extending polypeptide chains into the ER lumen, degradation of damaged or misfolded proteins, sealing of inactive translocons, maintenance of calcium concentration in the ER, and triggering the unfolded protein response in response to ER stress caused by accumulation of damaged and misfolded polypeptides). Because of these roles, inactivation of BiP is lethal. One can envisage that similar mechanisms still to be unveiled may be involved in the pathogenesis of many other bacterial infections and conditions due to toxic compounds from a variety of sources. In this regard, one may assume that even if a chaperone is not irreversibly destroyed as in the case of AB5 subtilase described earlier but just partially inactivated or blocked, pathological consequences will ensue, the extent and severity of which will depend on the role of the chaperone affected.

Quantitative Changes of Hsp47 and Fibrotic Disorders

Collagen, the most abundant protein in the body of mammals, is an important component of many tissues providing a matrix that supports other molecules and endowing these tissues, for example, skin and cartilage, with elasticity. Collagen synthesis is a complex process that involves a number of enzymes and also the molecular chaperones PDIase and Hsp47. The first set of steps occur inside the ER, in which the procollagen molecules form trimers assisted by PDIase and then fold with the help of Hsp47 to build stable triple-helical procollagen molecules. These triplets are then transported to the extracellular space via Golgi, in which parts of the procollagen molecules are cleaved and thus proceed to the formation of the collagen fibrils. Structurally, Hsp47 is characterized by a signal sequence at the N-terminus that directs it to the ER and by an ER-retention signal sequence (made of the amino acids Arg, Asp, Glu, and Leu), which makes Hsp47 stay in the ER. When Hsp47 binds the nascent procollagen chains, the ER-retention signal sequence is cleaved and the chaperone dissociates from its client peptide, which is then exported to the extracellular space. In view of these mechanisms with central participation of Hsp47, it has recently been proposed that this chaperone might play a role in pathological disorders characterized by accumulation of fibrotic material made of collagen.[27] Fibrosis occurs, for example, in skin keloids, pulmonary fibrosis, and renal glomerular and tubulointerstitial sclerosis. In experimental models of these disorders it has been observed that Hsp47 is increased, sometimes quite markedly. It is not completely clear whether this increase is primary (primary quantitative chaperonopathy), namely, whether it precedes the fibrosis and causes it, or rather the reverse, increase in collagen production forces the increase of Hsp47 (secondary quantitative chaperonopathy), since the chaperone is required for the synthesis–assembly of collagen, as seen earlier. The fact that experimental blockage of Hsp47 activity reduces collagen accumulation would tend to indicate, although it does not prove, that upregulation of the chaperone is a causative factor rather than a response to a disorder affecting other genes-molecules, such as collagen.

CHAPERONOPATHIES BY MISTAKE OR COLLABORATIONISM AND THE NEED FOR ANTI-CHAPERONE AGENTS

Typically, chaperones are considered to be devoted to the maintenance of healthy, normal cells and organisms. However, in some instances chaperones are recruited into pathogenic pathways and thus contribute to disease rather than the opposite.

In principle, one can envisage at least two types of pathogenic pathways involving chaperones: (1) the pathway is normal, part of the cell physiology, and includes the participation of one or more chaperones, but for some reason the pathway becomes part of a pathologic process; and (2) the pathway is not part of the normal physiology of the cell but it becomes active in pathologic situations, involving chaperones. In both instances 1 and 2, chaperones contribute to a pathogenic pathway by "mistake" as it were. This scenario must be viewed from the standpoint of the affected cell and also from the standpoint of the tissue or organism of which the affected cell is part. The pathway can be deleterious for the affected cell, and thus the chaperone would be promoting cell pathology and death, and disease of the tissue and organism. An example would be chaperones that fail to prevent protein precipitation when they are overwhelmed by an excess of misfolded peptides and coprecipitate with them, as it occurs in some neurodegenerative disorders of the elderly in which protein deposits are important components of the histopathological picture.[4,5,24,28]

On the other hand, if the pathway in which a chaperone participates leads to the malignant transformation of a cell and is required for the tumor cell to grow and proliferate, the chaperone while beneficial to the tumor cell is pathogenic for the organism. Examples of the conditions in which chaperones do promote cell growth but are nevertheless pathogenic are found in certain types of malignancies, in which chaperones (e.g., Hsp70, Hsp60, and Hsp10) are elevated and are required for tumor-cell proliferation and growth.[29–31] It has also been shown that increased expression of Hsp90 predicts lower survival in breast cancer.[32] The mechanisms by which chaperones could contribute to tumorigenesis are not yet fully understood, but it is thought that the impact that some chaperones have on the cell cycle, and also the antiapoptotic effect of some chaperones, could be key components of these mechanisms.[31,33] For example, protection of tumor cells from apoptosis would favor tumor progression.

In situations 1 and 2 described earlier, if the levels of the pathogenic chaperones are elevated, the pathologic entity could be classified as a dysregulatory, quantitative chaperonopathy.[5] However, elevated levels of the pathogenic chaperones may not be essential for the pathogenic pathway to proceed and cause disease, in which case normal chaperones at normal levels become involved in the causation of disease. Whether the chaperones are increased or not, treatment of these conditions would benefit from antichaperone agents that would suppress their genes and/or would block their function. It is clear then that the development of antichaperone agents ought to be a priority in the research agenda of chaperonologists for the next decade.

CONCLUSIONS AND PERSPECTIVE

Chaperonology is an emerging science born after many years of work that has unveiled a variety of chaperones with distinct evolution, structure, function, pathology, and potential applications. It is now the time for progress in the understanding of the chaperonopathies, and in the refinement and standardization of chaperonotherapy, namely the use of chaperone genes and proteins to treat chaperonopathies and other conditions, such as cancer, in which chaperones have shown promise to help in the induction of antitumor immunity.[34] Also, very promising is the study of chaperone networks to achieve a more complete understanding of protein homeostasis.[35,36] This strategy will help elucidate the interactions between the various chaperoning systems and between chaperones and other molecules, and will also help to unveil pathological disruptions of these interactions and predict their pathogenetic impact in what regards, for example, ageing and associated diseases. It can be envisaged that other fields of chaperonology will soon expand considerably, such as (a) the use of chaperones, genes or proteins, to make cells more resistant to stress in clinical and biotechnological settings. (e.g., in preparation for surgical interventions with expected general hypoxia; in bioreactors in which productive cells must be protected from a variety of stressors that come with the influent or are generated during fermentation; and in industry focused on the health and productivity of plants and animals that need to be made resistant to environmental stressors [e.g., water pollutants in aquaculture tanks]); (b) the evaluation of chaperones as diagnostic markers and prognostic indicators; (c) the elucidation of the role of extracellular chaperones and their alterations during development, ageing, and disease; (d) the optimization of chaperone-based agents for the treatment of cancer; and (e) the development of antichaperone agents for prevention and treatment of conditions in which chaperones play an active pathogenetic role.

ACKNOWLEDGMENTS

This is Publication No. 07-173 from the Center of Marine Biotechnology.

REFERENCES

1. WELCH, W.J. 1992. Mammalian stress response: cell physiology, structure/function of stress proteins, and implications for medicine and disease. Physiol. Rev. **72:** 1063–1081.
2. MACARIO, A.J.L. 1995. Heat-shock proteins and molecular chaperones: implications for pathogenesis, diagnostics, and therapeutics. Intl. J. Clin. Lab. Res. **25:** 59–70.
3. MACARIO, A.J.L. & E. CONWAY DE MACARIO. 2002. Sick chaperones and ageing: a perspective. Ageing Res. Rev. **1:** 295–311.
4. NARDAI, G., P. CSERMELY & C. SOTI. 2002. Chaperone function and chaperone overload in the aged. A preliminary analysis. Exp. Gerontol. **37:** 1257–1262.
5. MACARIO, A.J.L. & E. CONWAY DE MACARIO. 2005. Sick chaperones, cellular stress and disease. N. Engl. J. Med. **353:** 1489–1501.
6. STOETZEL, C., V. LAURIER, E.E. DAVIS, *et al.* 2006. *BBS10* encodes a vertebrate-specific chaperonin-like protein and is major BBS locus. Nat. Genet. **38:** 521–524.

7. CHIANG, A.P., J.S. BECK, H.-J. YEN, et al. 2006. Homozygosity mapping with SNP arrays identifies *TRIM32*, an E3 ubiquitin ligase, as a Bardet-Biedl syndrome gene (*BBS11*). Proc. Natl. Acad. Sci. USA **103:** 6287–6292.
8. BENZINOU, M., A. WALLEY, S. LOBBENS, et al. 2006. Bardet-Biedl syndrome gene variants are associated with both childhood and adult common obesity in French Caucasians. Diabetes **55:** 2876–2882.
9. MACARIO, A.J.L. & E. CONWAY DE MACARIO. 2007. Molecular chaperones: multiple functions, pathologies, and potential applications. Front. Biosci. **12:** 2588–2600. To see abstract, figures, and tables go to: http://www.bioscience.org/ click on "Journal" then click on "Volume 12" and find "Macario".
10. FONTAINE, J.M., X. SUN, A.D. HOPPE, et al. 2006. Abnormal small heat shock protein interactions involving neuropathy-associated HSP22 (HSPB8) mutants. FASEB J. **20:** 2168–2170.
11. MACARIO, A.J.L., T.M. GRIPPO & E. CONWAY DE MACARIO. 2005. Genetic disorders involving molecular-chaperone genes: a perspective. Genet. Med. **7:** 3–12.
12. MEYER-LINDENBERG, A., C.B. MERVIS & K.F. BERMAN. 2006. Neural mechanisms in Williams syndrome: a unique window to genetic influences on cognition and behaviour. Nat. Rev. Neurosci. **7:** 380–393.
13. MARUYAMA, T., R. SUZUKI & M. FURUTANI. 2004. Arachael peptidyl-prolyl *cis-trans* isomerases (PPIases) update 2004. Front. Biosci. **9:** 1680–1700. To see abstract, figures and tables go to http://www.bioscience.org/2004/v9/af/1361/fulltext.htm.
14. HAYASHI, M., K. IMANAKA-YOSHIDA, T. YOSHIDA, et al. 2006. A crucial role of mitochondrial Hsp40 in preventing dilated cardiomyopathy. Nat. Med. **12:** 128–132.
15. PILOTTO, A., N. MARZILIANO, M. PASOTTI, et al. 2006. Alpha-beta-crystallin mutation in dilated cardiomyopathies: low prevalence in a consecutive series of 200 unrelated probands. Biochem. Biophys. Res. Commun. **346:** 115–117.
16. TAKIYAMA, Y. 2006. Autosomal recessive spastic ataxia of Charlevoix-Saguenay. Neuropathology **26:** 368–375.
17. ZIMMERMANN, R., L. MUELLER & B. WULLICH. 2006. Protein transport into the endoplasmic reticulum: mechanisms and pathologies. Trends Mol. Med. **12:** 567–573.
18. DAVILA, S., L. FURU, A.G. GHARAVI, et al. 2004. Mutations in SEC63 cause autosomal dominant polycystic liver disease. Nat. Genet. **36:** 575–577.
19. SCHULMANN, K., Y. MORI, V. CROOG, et al. 2005. Molecular phenotype of inflammatory bowel disease-associated neoplasms with microsatellite instability. Gastroenterology **129:** 74–85.
20. ANDRINGA, G., J.G.J.M. BOL, X. WANG, et al. 2006. Changed distribution pattern of the constitutive rather than the inducible HSP70 chaperone in neuromelanin-containing neurones of the Parkinsonian midbrain. Neuropathol. Appl. Neurobiol. **32:** 157–169.
21. TANDARA, A.A., O. KLOETERS, I. KIM, et al. 2006. Age effect on HSP70: decreased resistance to ischemic and oxidative stress in HDF. J. Surg. Res. **132:** 32–39.
22. NJEMINI, R., M. LAMBERT, C. DEMANET, et al. 2007. Basal and infection-induced levels of heat shock proteins in human aging. Biogerontology **8:** 353–364.
23. UEHARA, T., T. NAKAMURA, D. YAO, et al. 2006. S-Nitrosylated protein-disulphide isomerase links protein misfolding to neurodegeneration. Nature **441:** 513–517.

24. WILHELMUS, M.M.M., O. HOELLER, P. WESSELING, et al. 2006. Specific association of small heat shock proteins with the pathological hallmarks of Alzheimer's disease brains. Neuropathol. Appl. Neurobiol. **32:** 119–130.
25. PERNG, M.D., M. SU, S.F. WEN, et al. 2006. The Alexander disease-causing glial fibrillary acidic protein mutant, R416W, accumulates into Rosenthal fibers by a pathway that involves filament aggregation and the association of alpha-beta-crystallin and HSP27. Am. J. Hum. Gen. **79:** 197–213.
26. PATON, A.W., T. BEDDOE, C.M. THORPE, et al. 2006. AB5 subtilase cytotoxin inactivates the endoplasmic reticulum chaperone BiP. Nature **443:** 548–552.
27. TAGUCHI, T. & M.S. RAZZAQUE. 2007. The collagen-specific molecular chaperone HSP47: is there a role in fibrosis? Trends Mol. Med. **13:** 45–53.
28. KAWAMOTO, Y., I. AKIGUCHI, Y. SHIRAKASHI, et al. 2007. Accumulation of Hsc70 and Hsp70 in glial cytoplasmic inclusions in patients with multiple system atrophy. Brain Res. **1136:** 219–227.
29. CAPPELLO, F., M. BELLAFIORE, S. DAVID, et al. 2003. Ten kilodalton heat shock protein (HSP10) is overexpressed during carinogenesis of large bowel and uterine exocervix. Cancer Lett. **196:** 35–41.
30. CAPPELLO, F., S. DAVID, F. RAPPA, et al. 2005. The expression of HSP60 and HSP10 in large bowel carcinomas with lymph node metastase. BMC Cancer **5:** 139 (10 pages).
31. ROHDE, M., M. DAUGARD, M.H. JENSEN, et al. 2005. Members of the heat-shock protein 70 family promote cancer cell growth by distinct mechanisms. Genes Dev. **19:** 570–582.
32. PICK, E., Y. KLUGER, J.M. GILTNANE, et al. 2007. High HSP90 expression is associated with decreased survival in breast cancer. Cancer Res. **67:** 2932–2937.
33. CZARNECKA, A.M., C. CAMPANELLA, G. ZUMMO & F. CAPPELLO. 2006. Heat shock protein 10 and signal transduction: a "capsula eburnea" of carcinogenesis? Cell Stress Chap. **11:** 287–294.
34. LEE, K.P., L.E. RAEZ & E.R. PODACK. 2006. Heat-shock protein-based cancer vaccines. Hematol. Oncol. Clin N. Am. **20:** 637–659.
35. NARDAI, G., E.M. VEGH, Z. PROHASZKA & P. CSERMELY. 2006. Chaperone-related immune dysfunction: an emergent property of distorted chaperone networks. Trends Immunol. **27:** 74–79.
36. SOTI, C. & P. CSERMELY. 2007. Aging cellular networks: chaperones as major participants. Exp. Gerontol. **42:** 113–119.

Heat Shock Proteins in Cancer

MICHAEL SHERMAN[a] AND GABRIELE MULTHOFF[b]

[a]Department of Biochemistry, Boston University School of Medicine, Boston, Massachusetts, USA

[b]Department of Radiotherapy and Radiooncology, Klinikum rechts der Isar, Technische Universität München, Munich, Germany, and GSI Institute of Pathology

> ABSTRACT: Heat shock proteins (Hsps) are highly conserved and inhabit nearly all subcellular locations where they perform a variety of chaperoning functions including folding and unfolding of nascent polypeptides, proteins, transport of proteins, and support of antigen presentation processes. Apart from their intracellular location Hsps with a molecular weight of 70 kDa (Hsp70) also have been found on the plasma membrane of malignantly transformed cells, on virally/bacterial infected cells and in the extracellular space. Depending on their intra- and extracellular location Hsps exert either protection against environmental stress or act as potent stimulators of the immune response. In this review we address the dual function of intracellular and extracellular located small Hsps and members of the Hsp70 family and its immunological consequences for cancer immunity.
>
> KEYWORDS: heat shock proteins; adaptive/innate immune response; cancer

INTRODUCTION

The major groups of molecular chaperones have been implicated in cancer development. Here we will focus on the role of Hsp70 and small heat shock proteins (Hsps) family members in cancer, because Hsp90 and ER chaperone Grp78, Grp74 (gp96) are subjects of different symposia.

There are many reports indicating that the major inducible chaperones Hsp70 (member of the Hsp70 family) and Hsp27 (member of the small Hsps family) are present at elevated levels in various human tumors, especially of epithelial origin or gliomas. For example, these Hsps are expressed at high levels in a large fraction (up to 50%) of breast, endometrial, lung, prostate, and other types of tumor biopsies,[1–5] and their expression often correlates with increased cell proliferation, lymph node metastases, poor response to chemotherapy, and

Address for correspondence: Gabriele Multhoff, Ph.D., Department of Radiotherapy and Radioon-cology, Klinikum rechts der Isar, Technische Universität München, Ismaningerstr.22, D-81675 Munich, Germany. Voice: +49-89-4140-4299; fax: +49-89-4140-4299.
gabriele.multhoff@lrz.tu-muenchen.de

poor survival (Refs. 4,6,7 for review). In addition to their intracellular location Hsps are also found on the plasma membrane and in the extracellular space where they can activate the immune system.

Beside correlation studies with human tumor biopsies, there have been many works that attempt to address the role of these Hsps in cancer using cell culture and animal experiments. Overall these studies indicate that Hsp70 and Hsp27 have dual effects on cancer—(1) they promote cancer development by suppression of various anticancer mechanisms, like apoptosis and senescence, as well as by facilitating expression of metastatic genes; and (2) they facilitate tumor rejection by immune system. Accordingly, there have been attempts to inhibit or downregulate intracellular Hsps to facilitate apoptosis or senescence of cancer cells. On the other hand, extracellular and membrane-associated Hsps have been used for cancer immunotherapy. Here, we will address both opposing activities of Hsp70 and small Hsp family members in cancer.

Intracellular Hsp70 Family Members in Cancer

Earlier experiments with permanent overproduction of Hsp70 in several cell types supported the idea that Hsp70 facilitates cell's tumorigenicity, including the ability to form tumors in nude mice,[8] formation of foci, anchorage-independent growth, and formation of tumors in mice xenografts,[9] as well as development of multiple lymphomas.[10] Furthermore, in human breast cancer MCF-7 cells overproduction of recombinant Hsp70 led to a strong acceleration of cell growth by shortening of G0/G1 phases.[11] This effect could be related to stabilization of the cyclin D1 upon overproduction of Hsp70.[12]

On the other hand, experiments with a specific depletion of Hsp70 in cancer cells indicated that Hsp70 is important for the tumor cell survival. For example, depletion of Hsp70 led to an apoptosis-like death of a variety of tumor cell types, including human oral carcinoma cells isolated from primary tumors, HSC-2, MCF-7, Molt-4, PC-3, and others.[13–15] On the other hand, untransformed cells (e.g., primary fibroblasts, Rat1, CM, MIN-6) did not lose their viability upon Hsp70 depletion[15,16] suggesting that, in contrast to untransformed cell types, many tumor cell lines cannot survive endogenously activated death if they lack Hsp70.

An interesting extension of these studies was a recent finding reported in this symposium that depletion of Hsp70 leads to rapid premature senescence in several cancer cell lines.[17] Originally cellular senescence was described as a limit to a number of divisions that a normal cell can undergo. For example, normal fibroblasts can undergo about 60 divisions in culture before acquiring a specific "flat" morphology and becoming permanently growth arrested.[18] While originally it was thought that the replicative senescence is an ultimate result of the telomers shortening, at present it is commonly accepted that senescence could be triggered by the cell cycle inhibitors p16 and p21 (see Ref. 19 for review). Senescence is a very complex program with multiple end points that include not only growth arrest, but also enlargement of cells, extensive vacuolization, repression, and derepression of certain sets of genes, secretion of various signaling molecules, inhibition of the heat shock response, and other manifestations.

The senescence program seems to represent one of the major breaks on cancer emergence at the cellular level in addition to activation of apoptosis. Indeed, limiting cell divisions seems to be a perfect way of preventing tumor growth, and mammalian cells appear to use both apoptotic and senescence programs to counteract tumor-inducing action of the major oncogenes. In fact, surprisingly, overexpression of major oncogenes could either activate apoptosis, as seen with myc or E1A,[20–24] or trigger senescence as seen with Ras, Her-2, PTEN, Raf, and other oncogenes of the Ras pathway.[25–27] Under these conditions, both apoptosis and senescence are associated with activation of the p53 pathway.[27–29] If a p53 target gene p21 is induced, senescence is initiated, however, if p21 induction is abrogated, for example, by myc, apoptosis is activated.[28–29] Importantly, in recent experiments where senescence of cancer cells was seen upon depletion of Hsp70, a strong activation of p53 and p21 was observed under these conditions. Furthermore, induction of p21 was dependent on p53. These data indicate that endogenous high levels of Hsp70 in these cancer lines are critical for control of the p53 pathway, and thus for cell proliferation.

Irina Guzhova and her colleagues (Institute of Cytology RAS, St. Petersburg, Russia) report in this symposium that Hsp70 can also suppress apoptosis stimulated by myc oncogene. In fact, they observed that introduction of v-myc into U-937 cells or activation of myc in Rat1 cells sensitizes them to camptothecin and etoposide by enhancing apoptosis. Overexpression of Hsp70 suppressed the myc-induced sensitization, which represents the first example when Hsp70 interferes with myc-facilitated apoptosis. Interestingly, myc can promote apoptosis via stimulation of p53,[30] and furthermore, both drugs used in this study also activate p53. Therefore, it is likely that there is a unifying mechanism of suppression of both apoptosis and senescence in cancer cells by Hsp70, which involves downregulation of the p53 pathway, and the choice of a specific pathway of cell demise appears to be dependent on the ability of p53 to induce p21.

These data suggest a simple model that explains overexpression of Hsp70 (as well as Hsp27, see below) in various cancers. In fact, since oncogenes initiate apoptosis or senescence, emergence of cancer clones indicates that they somehow bypass these breaks on cancer development. One of the mechanisms is accumulation of mutations in the p53 pathway, which was reported for many cancers. On the other hand, there must be alternative mechanisms of suppression of the p53-dependent cell demise, which take over in cancers with normal p53 pathway. Hsps appear to be ideal candidate factors that can provide such a suppression, which can explain their accumulation in cancers.

Interestingly, as reported in the symposium by Marja Jäättelä (Institute of Cancer Research, Copenhagen, Denmark), another member of the Hsp70 family, a less abundant but ubiquitously expressed protein Hsp70–2, which is essential for testis development, also protects from senescence as seen with various cancer cell lines.[31] Depletion of this protein using siRNA approach caused upregulation of p53 followed by permanent G1 arrest, cell enlargement, and flattening typical for senescent cells.[31] Therefore, as with Hsp70, the primary effect of Hsp70–2 depletion seems to involve p53 activation, which in turn triggers the senescence program. Accordingly, Hsp70–2 seems to play a role in keeping p53 pathway suppressed, and Hsp70–2 depletion leads to the abrogation of this control and reactivation of the default senescence program.

In addition to controlling the p53 system Hsp70–2 controls lysosome stability and a lysosome-dependent apoptosis-like process. This control appears to involve expression of a number of genes, including a lens epithelium-derived growth factor (LEDGF).[32] Accordingly, downregulation of either Hsp70–2 or LEDGF led to lysosomal destabilization and a cathepsin-dependent cell death. Interestingly, overexpression of LEDGF increased the tumorigenic potential of human cancer cells, and both LEDGF and Hsp70–2 are found at high levels in breast and bladder carcinomas.[32] Therefore, it appears that Hsp70–2 has multiple tumor-promoting activities, including suppression of p53, induction of LEDGF, and possibly others.

In promoting cancer development, Hsp70 family members may cooperate with cofactors. For example, Vince Guerriero and his colleagues (University of Arizona, Tucson) report in the symposium that HspBP1, a co-chaperone with nucleotide exchange activity that binds to and regulates Hsp70, is highly overexpressed in a variety of tumors. Although this group does not identify a role of HspBP1 in cancer, they found an interesting correlation between expressions of HspBP1 and Hsp70, where the ratio between these proteins is constant in various biopsies. These data suggest that HspBP1 cooperates with Hsp70 in promoting tumors.

Intracellular Small Hsps in Cancer

Hsp27, the major member of the small Hsps family, is overexpressed in a variety of cancers, and was suggested to play a major role in tumor development. These findings are discussed in details in recent reviews by Calderwood and Ciocca, and Arrigo. In part implication of small Hsps in cancer is based on the well-described antiapoptotic activity of Hsp27, which upon overexpression blocks apoptosis caused by a variety of stimuli (see Refs. 4, 7, 33 for review). There are multiple steps in the apoptotic process that can be controlled by Hsp27, but this discussion is beyond the scope of this article.

In addition to suppression of apoptosis, Hsp27 can suppress the senescence program, as reported by Michael Sherman and his colleagues (Boston

University, USA) at this symposium. In fact, depletion of Hsp27 in highly transformed cells caused activation of the p53 pathway, induction of p21, and expression of typical signs of cell senescence. On the other hand, overexpression of Hsp27 in immortalized mammary epithelium cells caused suppression of the senescence program activated by genotoxic drugs and oxidants. Therefore, the function of Hsp27 in cancer appears to be suppression of the default senescence program that involves p53. In that sense the role of Hsp27 is similar to that of Hsp70. These two proteins may serve as two alternative factors responsible for suppression of the oncogene-induced senescence in cells with normal p53 pathway. Furthermore, these data suggest that suppression of p53 may be a novel factor in the antiapoptotic activity of Hsp27.

In addition to antiapoptotic and antisenescence activity of Hsp27, this chaperone appears to play a major role in cell migration and metastases. These functions appear to be unrelated to p53, since they were detected in cell lines with defective p53. In part, the role of Hsp27 in cell migration can be explained by its interaction with the actin cytoskeleton.[34,35] In line with these findings, recently it was reported that a novel inhibitor of phosphorylation of Hsp27 suppresses tumor cell migration and invasion.[36] On the other hand, Hsp27 appears to be critical for expression of metalloproteases MMP2 and MMP9 that are essential for metastases.[37] In fact, it was found that in PC-3 cells activation of metastases by TGF-α required Hsp27, where this Hsp serves as a regulator of transcription of the metalloproteases.[37] In this pathway, TGF-α activates p38 kinase, which in turn activates the MAPKAP-2 kinase, which phosphorylates and activates Hsp27.

Surprisingly, a close homolog of Hsp27 αB-crystallin appears to be sufficient for cancer transformation.[38] Vince Cryns and his colleagues (Northwestern University, Chicago, USA) report at this symposium that overexpression of αB-crystallin transforms immortalized human mammary epithelial cells. It induces EGF- and anchorage-independent growth, increases cell migration and invasion, and constitutively activates the MAPK kinase/ERK (MEK/ERK) pathway. The transformed phenotype required activity of this signaling pathway. In addition, cells overexpressing αB-crystallin formed invasive mammary carcinomas in nude mice that recapitulated aspects of human basal-like breast tumors. These results indicate that αB-crystallin is a novel oncoprotein.

Extracellular and Membrane-Bound Hsps in Cancer

Apart from their intracellular location Hsps with molecular weights of 60, 70 (Hsp70, Hsc70), and 90 (gp96) kDa have been found on the plasma membrane of malignantly transformed cells[39,40] and in the extracellular milieu[41] (Triantafilou, University of Sussex, UK). Pioneering work of the group of Srivastava[42] demonstrated that Hsp90 as well as Hsp70 peptide complexes are potent stimulators of the adaptive immune system. Hsp-chaperoned peptides are taken up by professional and nonprofessional antigen-presenting cells including monocytes, macrophages, dendritic cells (DC), and B cells via receptor-mediated endocytosis and thus become cross-presented as classical antigens for CD8-positive cytotoxic T cells on MHC class I molecules. Presently, the α 2 macroglobulin receptor CD91, toll-like receptors 2 and 4, the LPS receptor CD14,[43,44] scavenger receptors (SR-A-H)[45] including lectin-like oxidized LDL receptor LOX-1, SR expressed by endothelial cells (SREC-1), SR-H member FEEL-1 (identical to CLEVER-1/stabilin-1), B cell receptor CD40, CD36, chemokine receptor CCR5, as well as C-type-lectin receptors NKG2A, NKG2C in association with CD94, and NKG2D[46,47] are discussed as potential mediators of binding and uptake of members of the Hsp70 and Hsp90 families into antigen-presenting cells.

Our group (Multhoff, Technische Universität München, Germany) demonstrated a tumor-specific cell surface location of Hsp70, the major stress-inducible member of the Hsp70 family by selective cell-surface iodination and by flow cytometry using the mAb cmHsp70.1 multimmune GmbH, Munich), detecting membrane-bound Hsp70 in the plasma membrane of viable tumor cells.[48,49] Even in the absence of chaperone-bound peptides, plasma membrane-bound Hsp70 has been determined as a tumor-specific recognition structure for preactivated NK cells. Following incubation with the Hsp70-peptide TKDNNLLGRFELSG plus proinflammatory cytokines NK cells acquire migratory capacity and efficiently kill Hsp70 membrane-positive tumor cells. TKD peptide represents a 14-mer amino acid sequence of the C-terminal domain, which is exposed to the extracellular milieu of tumor cells and contains the recognition epitope of the Hsp70 mAb cmHsp70.1.[50]

In addition to Hsp70, extracellular located Hsp60, Hsp90, Hsp110, and the ER chaperone glucose-related protein 170 (Grp170) have been found to exhibit proinflammatory and proimmune activities. However, the species from which the Hsps or Grps are derived from and the concentration can affect immune function. In contrast to stress-inducible Hsp70, BiP an ER-residing Hsp70 member as well as Grp78 and Grp74 (gp96), and members of the small Hsp families, including Hsp27 and cpn10, have been found to exhibit anti-inflammatory (Pockley, University of Sheffield, UK) immunoregulatory activities toward the adaptive immune system.[51,52] These data indicate that despite their homologies, members of different Hsp families can mediate divergent functions on the innate and adaptive immune system.

Although the immunological consequences of membrane-bound and extracellular located Hsps are obvious, the mechanism of anchorage, export, and uptake of Hsps remains to be elucidated. Evidence is accumulating that similar to IL-1-β Hsps lacking a membrane translocation and anchorage sequence are released in an ER-Golgi independent manner via a nonclassical lysosomal pathway.[53] Anchorage of Hsp70 in the plasma membrane has been found to be associated with a lipid–protein rather than a receptor–protein interaction since high salt and pH changes did not affect the Hsp70 membrane expression (unpublished). In contrast, methyl-β-cyclodextrin, a cholesterol-depleting agent, is highly efficient in diminishing the amount of membrane-bound Hsp70.[45] These data might provide a first hint that Hsp70 is associated with lipid rafts in the plasma membrane of tumor cells. Another question that is presently a matter of debate is related to the physiological function of membrane-bound Hsp70 in lipid rafts and in the extracellular space. Are membrane-bound and extracellular Hsps still capable to act as chaperones and if so which cochaperones are associated with Hsps. Last but not least it is still not clear as to whether extracellular Hsps can gain access to the plasma membrane from outside and thus might be reintegrated into the plasma membrane as already shown much earlier by Tytell and Hightower[54,55] for neurological cells.

Although many questions still remain open, it is obvious that Hsps possess dual roles depending on their intra- and extracellular location and depending on the species from which they are derived of. As potent stimulators of the adaptive and innate immune response presently Hsps with molecular weight of 70 and 90 kDa strongly increase the interest in using them as immune modulators in cancer therapy, as mediators of tolerance small Hsps and BiP might be useful for the cure of autoimmune diseases and as intracellular molecular chaperones they might support survival of normal cells from apoptotic cell death following environmental stress.

REFERENCES

1. NANBU, K. *et al.* 1998. Prognostic significance of heat shock proteins Hsp70 and Hsp90 in endometrial carcinomas. Cancer Detect. Prev. **22:** 549–555.
2. COSTA, M.J.M. *et al.* 1997. Expression of heat shock protein 70 and P53 in human lung cancer. Oncol. Reports **4:** 1113–1116.
3. VARGASROIG, L.M. *et al.* 1997. Heat shock proteins and cell proliferation in human breast cancer biopsy samples. Cancer Detect. Prev. **21:** 441–451.
4. JÄÄTTELÄ, M. 1999. Escaping cell death: Survival proteins in cancer [Review]. Exp. Cell Res. **248:** 30–43.
5. CIOCCA, D.R. *et al.* 1993. Heat shock protein hsp70 in patients with axillary lymph node-negative breast cancer: prognostic implications. J. Natl Cancer Inst. **85:** 570–574.
6. CIOCCA, D.R. & S.K. CALDERWOOD. 2005. Heat shock proteins in cancer: diagnostic, prognostic, predictive, and treatment implications. Cell Stress Chaperones **10:** 86–103.
7. CALDERWOOD, S.K. *et al.* 2006. Heat shock proteins in cancer: chaperones of tumorigenesis. Trends Biochem. Sci. **31:** 164–172.
8. JÄÄTTELÄ, M. 1995. Over-expression of hsp70 confers tumorigenicity to mouse fibrosarcoma cells. Int. J. Cancer **60:** 689–693.
9. VOLLOCH, V.Z. & M.Y. SHERMAN. 1999. Oncogenic potential of Hsp72. Oncogene **18:** 3648–3651.
10. SEO, J. *et al.* 1996. T cell lymphoma in transgenic mice expressing the human Hsp70 gene. Biochem. Biophys. Res. Comm. **218:** 582–587.

11. BARNES, J.A. et al. 2001. Expression of inducible Hsp70 enhances the proliferation of MCF-7 breast cancer cells and protects against the cytotoxic effects of hyperthermia. Cell Stress Chaperones **6:** 316–325.
12. DIEHL, J.A. et al. 2003. Hsc70 regulates accumulation of cyclin D1 and cyclin D1-dependent protein kinase. Mol. Cell. Biol. **23:** 1764–1774.
13. KAUR, J. & R. RALHAN. 2000. Induction of apoptosis by abrogation of HSP70 expression in human oral cancer cells. Int. J. Cancer **85:** 1–5.
14. WEI, Y.Q. et al. 1995. Inhibition of proliferation and induction of apoptosis by abrogation of heat-shock protein (Hsp) 70 expression in tumor cells. Cancer Immunol. Immunother. **40:** 73–78.
15. NYLANDSTED, J., K. BRAND & M. JÄÄTTELÄ. 2000. Heat shock protein 70 is required for the survival of cancer cells. Ann. N. Y. Acad. Sci. **926:** 122–125.
16. GABAI, V.L. et al. 2000. Suppression of stress kinase JNK is involved in HSP72-mediated protection of myogenic cells from transient energy deprivation. HSP72 alleviates the stress-induced inhibition of JNK dephosphorylation. J. Biol. Chem. **275:** 38088–38094.
17. YAGLOM, J.A., V.L. GABAI & M.Y. SHERMAN. 2007. High levels of heat shock protein Hsp72 in cancer cells suppress default senescence. Cancer Res. **67:** 2373–2381.
18. HAYFLICK, L. 1979. Cell biology of aging. Fed. Proc. **38:** 1847–1850.
19. BRAIG, M. & C.A. SCHMITT. 2006. Oncogene-induced senescence: putting the brakes on tumor development. Cancer Res. **66:** 2881–2884.
20. PELENGARIS, S., B. RUDOLPH & T. LITTLEWOOD. 2000. Action of Myc in vivo—proliferation and apoptosis [Review]. Opin. Gen. Dev. **10:** 100–105.
21. PRENDERGAST, G.C. 1999. Mechanisms of apoptosis by c-Myc [Review]. Oncogene **18:** 2967–2987.
22. PEREZ-SALA, D. & A. REBOLLO. 1999. Novel aspects of Ras proteins biology: regulation and implications [Review]. Cell Death Diff. **6:** 722–728.
23. WHITE, E. 1998. Regulation of apoptosis by adenovirus E1A and E1B oncogenes [Review]. Semin. Virol. **8:** 505–513.
24. BLYTH, K. et al. 2000. Sensitivity to myc-induced apoptosis is retained in spontaneous and transplanted lymphomas of CD2-mycER (TM) mice. Oncogene **19:** 773–782.
25. FERBEYRE, G. et al. 2002. Oncogenic ras and p53 cooperate to induce cellular senescence. Mol. Cell. Biol. **22:** 3497–3508.
26. TROST, T.M. et al. 2005. Premature senescence is a primary fail-safe mechanism of ERBB2-driven tumorigenesis in breast carcinoma cells. Cancer Res. **65:** 840–849.
27. CHEN, Z. et al. 2005. Crucial role of p53-dependent cellular senescence in suppression of Pten-deficient tumorigenesis. Nature **436:** 725–730.
28. NILSSON, J.A. & J.L. CLEVELAND. 2003. Myc pathways provoking cell suicide and cancer. Oncogene **22:** 9007–9021.
29. EISCHEN, C.M. et al. 1999. Disruption of the ARF-Mdm2-p53 tumor suppressor pathway in Myc-induced lymphomagenesis. Genes Dev. **13:** 2658–2669.
30. YU, K. et al. 1997. Regulation of Myc-dependent apoptosis by P53, C-Jun N-terminal kinases stress-activated protein kinases, and Mdm-2. Cell Growth Diff. **8:** 731–742.
31. ROHDE, M. et al. 2005. Members of the heat-shock protein 70 family promote cancer cell growth by distinct mechanisms. Genes Dev. **19:** 570–582.

32. DAUGAARD, M. et al. 2007. Lens epithelium-derived growth factor is an Hsp70-2 regulated guardian of lysosomal stability in human cancer. Cancer Res. **67:** 2559–2567.
33. GARRIDO, C. et al. 2003. HSP27 and HSP70: potentially oncogenic apoptosis inhibitors. Cell Cycle **2:** 579–584.
34. LANDRY, J. & J. HUOT. 1995. Modulation of actin dynamics during stress and physiological stimulation by a signaling pathway involving p38 MAP kinase and heat-shock protein 27. Biochem. Cell Biol. **73:** 703–707.
35. PIOTROWICZ, R.S., E. HICKLEY & E.G. LEVIN. 1998. Heat shock protein 27 kDa expression and phosphorylation regulates endothelial cell migration. FASEB J. **12:** 1481–1490.
36. SHIN, K.D. et al. 2005. Blocking tumor cell migration and invasion with biphenyl isoxazole derivative KRIBB3, a synthetic molecule that inhibits Hsp27 phosphorylation. J. Biol. Chem. **280:** 41439–41448.
37. XU, L., S. CHEN & R.C. BERGAN. 2006. MAPKAPK2 and HSP27 are downstream effectors of p38 MAP kinase-mediated matrix metalloproteinase type 2 activation and cell invasion in human prostate cancer. Oncogene **25:** 2987–2998.
38. MOYANO, J.V. et al. 2006. Alpha B-crystallin is a novel oncoprotein that predicts poor clinical outcome in breast cancer. J. Clin. Invest. **116:** 261–270.
39. SHIN, B.K. et al. 2003. Global profiling of the cell surface proteome of cancer cells uncovers an abundance of proteins with chaperone function. J. Biol. Chem. **278:** 7607–7616.
40. MULTHOFF, G. et al. 1995. A stress-inducible 72 kDa heat shock protein (Hsp72) is expressed on the surface of human tumor cells, but not on normal cells. Int. J. Cancer **61:** 272–279.
41. TRIANTAFILOU, M. & K. TRIANTAFILOU. 2004. HSP70 and HSP90 associate with Toll-like receptor 4 in response to bacterial lipopolysaccharide. Biochem. Soc. Trans. **32:** 636–639.
42. SUTO, R. & P.K. SRIVASTAVA. 1995. A mechanism for the specific immunogenicity of heat shock protein-chaperoned peptides. Science **269:** 1585–1588.
43. ASEA, A. et al. 2000. Hsp70 stimulates cytokine production through a CD14-dependent pathway, demonstrating its dual role as a chaperone and cytokine. Nat. Med. **6:** 435–442.
44. VABULAS, R.M. et al. 2002. The endoplasmatic reticulum-resident heat shock protein gp96 activates dendritic cells via the Toll-like receptor 2/4 pathway. J. Biol. Chem. **277:** 20847–20853.
45. TRIANTAFILOU, M. et al. 2006. Membrane sorting of TLR-2/6 and TLR2/1 heterodimers at the cell surface determines heterotypic associations with CD36 and intracellular targeting. J. Biol. Chem. **281:** 31002–31011.
46. CALDERWOOD, S.K. et al. 2007. Extracellular HSP in cell signaling. FEBS Lett. **581:** 3689–3694.
47. THERIAULT, J.R. et al. 2006. Role of scavenger receptors in the binding and internalization of Hsp70. J. Immunol. **177:** 8604–8611.
48. GROSS, C. et al. 2003. Cell surface-bound Hsp70 mediates perforin-independent apoptosis by specific binding and uptake of granzyme B. J. Biol. Chem. **278:** 41173–41181.
49. SCHMITT, E. et al. 2007. Intracellular and extracellular functions of heat shock proteins: repercussion in cancer therapy. Leuco. Biol. **81:** 15–27.
50. MULTHOFF, G. et al. 2001. A 14-mer Hsp70 peptide stimulates natural killer (NK) cell activity. Cell Stress Chaperones **6:** 337–344.

51. PANYANI, G.S. & V.M. CORRIGAL. 2006. BiP regulates autoimmune inflammation and tissue damage. Autoimmun. Rev. **5:** 140–142.
52. POCKLEY, A.G. 2002. Heat shock proteins, inflammation, and cardiovascular disease. Circulation **105:** 1012–1017.
53. MAMBULA, S.S. & S.K. CALDERWOOD. 2006. Hsp70 is secreted from tumor cells by a non-classical pathway involving lysosomal endosomes. J. Immunol. **177:** 7849–7857.
54. TYTELL, M. *et al*. 1986. Heat-shock like protein is transferred from glia to axon. Brain Res. **363:** 161–164.
55. HIGHTOWER, L.E. & P.T. GUIDON JR. 1989. Selective release form cultured mammalian cells of heat- shock (stress) proteins that resemble glia-axon transfer proteins. J. Cell. Physiol. **138:** 257–266.

Drugging the Cancer Chaperone HSP90

Combinatorial Therapeutic Exploitation of Oncogene Addiction and Tumor Stress

PAUL WORKMAN,[a] FRANCIS BURROWS,[b] LEN NECKERS,[c] AND NEAL ROSEN[d]

[a]*Cancer Research UK, Centre for Cancer Therapeutics, The Institute of Cancer Research, Sutton, Surrey, United Kingdom*

[b]*Biogen IDEC, San Diego, California, USA*

[c]*Urologic Oncology Branch, Center for Cancer Research, National Cancer Institute, Bethesda, Maryland, USA*

[d]*Memorial Sloan-Kettering Cancer Center, New York, New York, USA*

ABSTRACT: The molecular chaperone HSP90 has emerged as an exciting target for cancer treatment. We review the potential advantages of HSP90 inhibitors, particularly the simultaneous combinatorial depletion of multiple oncogenic "client" proteins, leading to blockade of many cancer-causing pathways and the antagonism of all of the hallmark pathological traits of malignancy. Cancer selectivity is achieved by exploiting cancer "dependencies," including oncogene addiction and the stressed state of malignant cells. The multiple downstream effects of HSP90 inhibitors should make the development of resistance more difficult than with agents having more restricted effects. We review the various classes of HSP90 inhibitor that have been developed, including the natural products geldanamycin and radicicol and also the purine scaffold and pyrazole/isoxazole class of synthetic small molecule inhibitors. A first-in-class HSP90 drug, the geldanamycin analog 17-AAG, has provided proof of concept for HSP90 inhibition in patients at well tolerated doses and therapeutic activity has been seen. Other inhibitors show promise in preclinical and clinical development. Opportunities and challenges for HSP90 inhibitors are discussed, including use in combination with other agents. Most of the current HSP90 inhibitors act by blocking the essential nucleotide binding and ATPase activity required for chaperone function. Potential new approaches are discussed, for example, interference with cochaperone binding and function in the superchaperone complex. Biomarkers for use with HSP90 inhibitors are described.

Address for correspondence: Paul Workman, Ph.D., Cancer Research UK Centre for Cancer Therapeutics, The Institute of Cancer Research, Haddow Laboratories, 15 Cotswold Road, Sutton, Surrey. UK. Voice: +44-208-722-4301; fax: +44-208-722-4324.

paul.workman@icr.ac.uk

We stress how basic and translational research has been mutually beneficial and indicate future directions to enhance our understanding of molecular chaperones and their exploitation in cancer and other diseases.

KEYWORDS: HSP molecular chaperone; inhibitors; cancer therapy

INTRODUCTION: SETTING THE SCENE

We are in a very exciting era of cancer drug discovery in which effective mechanism-based therapies are being designed and developed, which act on specific oncogenic proteins and pathways that are hijacked by pathological genetic and epigenetic changes to bring about the initiation of cancer and subsequent malignant progression.[1–3] The success of the approach of targeting individual oncogenic proteins is exemplified by the clinical activity and regulatory approval of drugs like trastuzumab, imatinib, gefitinib, erlotinib, and others.[2] Therapeutic selectivity for tumor versus normal cells is achieved by taking advantage of various "dependencies" that develop during the induction of cancers. These dependencies have been usefully categorized[4] as involving oncogene addiction,[5] lineage addiction, tumor–host cell and tumor microenvironment factors, and genetic abnormalities offering opportunities for synthetic lethality.

Although significant success has been achieved with cancer therapeutics designed to act on individual molecular targets, there are considerable challenges that indicate that optimal clinical efficacy will require combinatorial approaches. Firstly, evidence from human cancer genome-sequencing efforts indicates the existence of a very large number of potential cancer genes and the presence of multiple driver mutations within any one individual cancer.[6,7] Secondly, resistance is very common and is associated not only with mutations in target kinases, for example, but also with feedback loops and activation of alternative pathways.[2,8,9] A combinatorial attack on cancer can be achieved by using specific agents in rational combinations. An alternative strategy is to identify molecular targets that when inhibited pharmacologically result in simultaneous, powerful, pleiotropic downstream effects on many oncogenic proteins and pathways. Proof of principle for such an approach is the clinical activity of the proteasome inhibitor bortezomib in multiple myeloma.[10] In recent years, the molecular chaperone HSP90 has emerged as a leading example of such a therapeutic target. In this article, we summarize current views relating to HSP90 as a cancer drug target and describe recent research on the discovery and development of HSP90 inhibitors, focusing on work carried out in the authors' laboratories. For more detailed background the reader is referred to several review articles.[11–17]

HSP90 AS AN ATTRACTIVE CANCER DRUG TARGET

HSP90 is a highly abundant 90-kDa protein and a member of the group of heat shock proteins (HSPs) that were first reported in 1962 as undergoing increased expression as a protective response to elevated temperature.[18,19] The HSP90 family comprises HSP90α and HSP90β that are mainly cytoplasmic, GRP94 in the endoplasmic reticulum, and TRAP1 in the mitochondria. We will focus here on HSP90α and β, which are the best understood at this time.

HSP90 acts as a molecular chaperone to regulate the conformation, activation, function, and stability of so-called "client proteins." The structure and function of HSP90 have been elucidated in considerable detail, particularly with the aid of X-ray crystallography.[20] HSP90 is made up of three functional domains: an NH_2-terminal ATP/ADP-binding domain, a middle domain involved in client protein binding, and a COOH-terminal dimerization domain.[20] The chaperone activity of HSP90 depends on an orchestrated series of conformational changes and interactions with client proteins and accessory proteins or cochaperones, operating in a cycle that is driven by nucleotide binding and ATP hydrolysis.

There are over 100 known HSP90 client proteins (for an updated list see: htpp://www.picard.ch/downloads/HSP90interactors.pdf). These comprise a variety of functional types, but feature a high proportion of signal transduction proteins, particularly kinases and transcription factors. The rules for what confers client protein status on a protein are not clear, although it has been suggested that surface charge and hydrophobicity determine binding of the kinase ERBB2 to HSP90.[21]

Until recently HSP90 was perhaps not seen as an obvious cancer drug target, most likely due to concerns about possible toxicity but also since it is not apparently mutated or amplified in malignant cells. HSP90 has, however, been shown to support malignant transformation, is overexpressed in cancer cells, and is associated with decreased survival in breast cancer.[22]

A particular attraction of HSP90 as a cancer drug target is that a large number of client proteins are *bona fide* oncoproteins. These include many kinases, such as ERBB2, EGFR, CDK4, CRAF, BRAF, AKT, MET, and BCR-ABL, as well as transcription factors, such as estrogen and androgen receptors, HIF-1α and p53. Other cancer-relevant client proteins include survivin and the catalytic subunit of telomerase hTERT. Inhibition of HSP90 leads to simultaneous combinatorial depletion (via the ubiquitin–proteasome pathway) of a wide range of client proteins, thereby attacking addiction to multiple oncogenes, exploiting the various cancer dependencies (see Introduction) and antagonizing all of the hallmark pathological traits of malignant cells, that is, self-sufficiency in growth signals, insensitivity to growth suppression signals, evasion of apoptosis, and acquisition of limitless replicative potential, together with sustained angiogenesis, invasion, and metastasis[11,13,16,23] (FIG. 1).

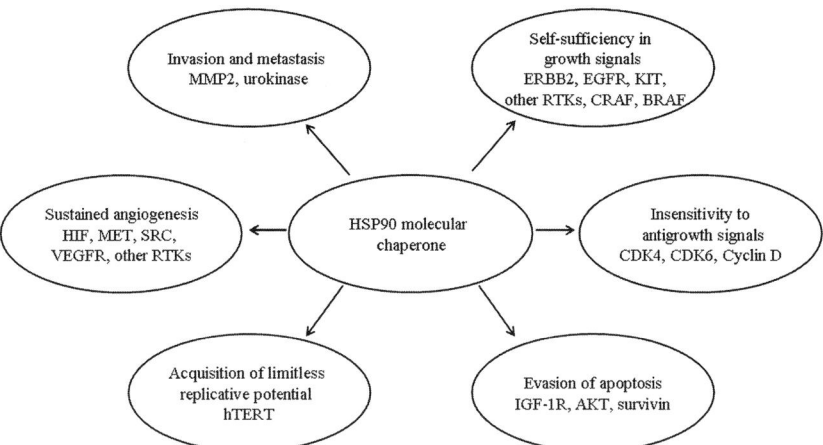

FIGURE 1. Schematic illustration of the central role of HSP90 in supporting all of the hallmark traits of cancer cells by chaperoning the various client proteins shown, as well as others. See also references 11,13,14,16, and 23.

It is important also to emphasize the potential contribution of cellular stress to the mechanism of action of HSP90 inhibitors. The increased levels of HSP90 and other HSPs in cancers may be seen as a response to the stresses experienced by malignant cells, due not only to the mutation and deregulation of cancer genes (see above), but also to the hostile tumor microenvironment, including factors such as hypoxia, nutrient deprivation, and acidosis.[24,25] The induction of HSPs is controlled by the transcription factor heat shock factor 1 (HSF1) and also by oncogenic signaling pathways.[24,25]

It seems likely that cancer cells have a higher dependence on stress response pathways compared to normal cells. An important finding in this regard was the observation that in malignant cells HSP90 is present in an activated superchaperone complex that is hypersensitive to HSP90 inhibition, whereas in normal cells HSP90 is predominantly uncomplexed and less sensitive to pharmacologic inhibitors.[26] This most likely contributes to the accumulation of HSP90 inhibitors seen in malignant cells and tissues and may represent an operative "HSP90 addiction" in some cancers.

A final point to make with respect to the sensitivity of cancer cells to HSP90 inhibition concerns the ability of the molecular chaperone to buffer potentially unstable mutant proteins. It has been demonstrated that a number of oncoproteins, including, for example, BRAF and EGFR, are much more dependent on HSP90 than the corresponding wild-type proteins.[27–29] There is a clear analogy here with the role played by HSP90 in development and evolution in model organisms, acting as a buffer or capacitor for metastable mutant proteins.[30,31] It is clear, however, that we have considerably more to learn about the consequences of HSP90 inhibition in various types of cells and organisms, and also on the impact of genetic and epigenetic differences. Nevertheless, it seems clear that we know enough to appreciate that therapeutic selectivity can be achieved against cancer versus normal cells.

Taken together, the above findings form a powerful case for targeting the HSP90 molecular chaperone for cancer treatment. Indeed, HSP90 has emerged from being considered a somewhat unlikely molecular target to one that has allowed us to attack the multiple genetic drivers, biochemical pathways, signaling networks, and malignant traits of cancer.

DISCOVERY OF HSP90 INHIBITORS

In the last few years we have seen a considerable increase in the discovery of HSP90 inhibitors, progressing from the original natural products to derivatives of these and to fully synthetic small molecule agents.[12,15,16,23,32] The chemical structures of some representative HSP90 inhibitors are shown in FIGURE 2.

The earliest inhibitors of HSP90 were the natural products related to geldanamycin and radicicol.[33,34] As is often the case for natural products, these acted as pathfinder molecules, helping us to understand the biology of HSP90 as well as to explore the consequences of HSP90 inhibition and to understand the issues in translating HSP90 biology into HSP90 drugs.[35] X-ray crystallography studies showed that geldanamycin and radicicol act as a nucleotide mimetics, binding within the deep ATP/ADP pocket of the NH_2-terminal domain of HSP90.[36] This nucleotide mimicry has provided the basis for the subsequent design of a range of HSP90 inhibitors.[32] Furthermore, the unusual structure of the nucleotide-binding site, requiring the adoption of a folded C-shape by both the nucleotide ligands and the natural products, provides a clear explanation for the selectivity of HSP90 inhibitors versus most other ATPases, as well other purine-binding proteins, such as kinases.[37]

The prototype benzoquinone ansamycin geldanamycin was shown to deplete client proteins and exhibit anticancer effects (e.g., cell cycle arrest and apoptosis) but was too toxic for clinical development.[38] For reasons that are not entirely clear, the analog 17-AAG exhibited a greater therapeutic index in animal models of cancer and has progressed to clinical trials.[15,39–41] The more soluble and orally bioavailable derivative 17-DMAG[42] is also in clinical development, as is the hydroquinone of 17-AAG, known as IPI-504.[43] The hydroquinone appears to be active in its own right, as well as acting as a soluble prodrug of the parent quinone. The quinone moiety remains as a possible liability in the geldanamycin series, having potential for adduct formation and redox metabolism, most likely contributing to the liver toxicity seen with these agents.[39–41] The 17-AAG quinone is metabolized by the oxidoreductase NQO1/DT-diaphorase; this gives rise to the variable activity of 17-AAG with high NQO1-expressing cancer cells being much more sensitive to the drug than are low expressing cells.[44] 17-AAG is also a substrate for

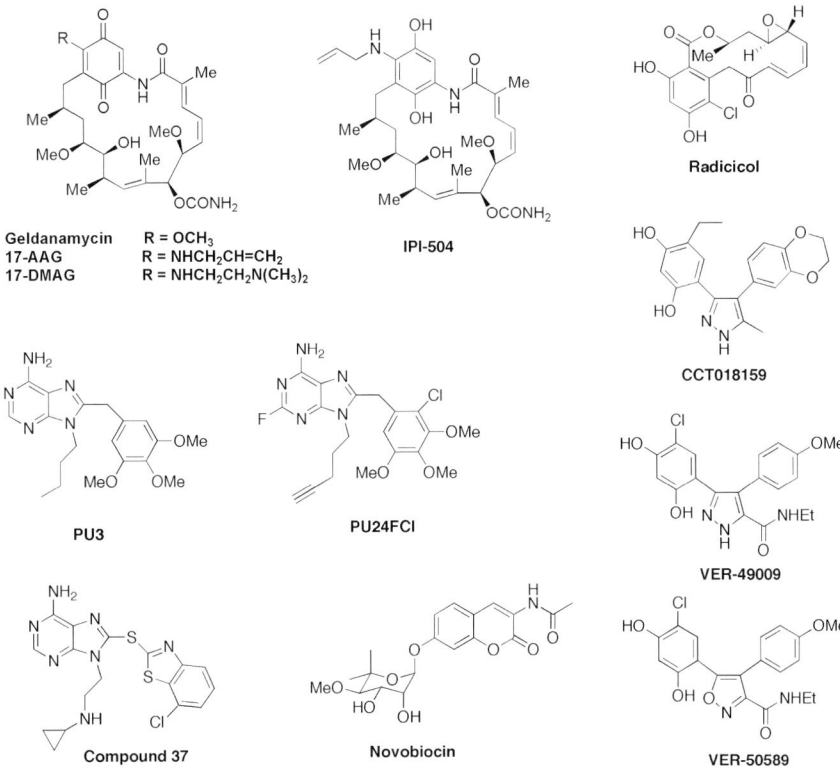

FIGURE 2. Chemical structures of representative HSP90 inhibitors.

polymorphic cytochrome P450 CYP3A4 and the P-glycoprotein efflux pump.[44] Since HSP90 functions as a dimer, dimeric geldanamycin derivatives have also been explored as anticancer agents.[45,46] The dimeric ansamycin CF237 displays enhanced duration of binding and client protein degradation relative to 17-AAG *in vitro* and *in vivo*, although its large size (∼900 kDa) and pharmaceutical limitations may preclude clinical development.

Like the geldanamycins, the radicicol structure also contains chemically reactive features, probably contributing to its disappointing *in vivo* activity. The more stable oxime derivatives showed activity in animal cancer models.[47] However, these have not progressed to clinical trials, possibly due to reported toxicity to the eye.[48]

The first series of synthetic small molecule HSP90 inhibitors were based on a purine-scaffold and were conceived by structure-based modeling.[49,50] The purines are exemplified by the initial lead PU3 and the more potent PU24FCl. An understanding of the precise HSP90-binding mode,[51] together with optimization of potency and oral bioavailability,[52,53] led to compounds, such as the 8-arylsulfanyl adenine analog 37,[53] which has oral activity in animal cancer models. An oral purine-scaffold drug CNF2024 (structure not disclosed) has entered clinical trials.

Containing the resorcinol anchoring unit that is also critical for binding in radicicol, the diarylpyrazole-scaffold HSP90 inhibitor CCT018159 was identified by high-throughput screening.[54,55] Structure-based design led to the discovery of the more potent resorcinylic pyrazole amide VER-49009[56] and the even more potent resorcinylic isoxazole amide VER-50589.[57] The latter compound is the most potent small molecule HSP90 inhibitor yet reported (Kd 5nM) and has shown activity in an animal cancer model.[57] These compounds are soluble and avoid liabilities due to P450s, NQO1, and P-glycoprotein.

The amino-coumarin-containing antibiotic novobiocin was found to inhibit HSP90, apparently by binding to a proposed second ATP-binding site in the C-terminal domain.[58,59] More potent analogs have been synthesized[60] and derivatives that show selectivity for HSP90 versus the original novobiocin target DNA gyrase have been obtained.[61]

Results for a range of other HSP90 inhibitors that are emerging have been presented at scientific meetings, but the chemical structures have not yet been disclosed. It is likely that most of them at this stage will be N-terminal ATP-site inhibitors identified by high-throughput screening and structure-based design.

An interesting development has been the demonstration that inhibitors of histone deacteylase (HDAC) are able to reduce HSP90 activity by increasing the acetylation of the molecular chaperone.[62,63] HDAC6 is implicated in HSP90 deacetylation and a specific acetylation site (K294 in HSP90α) has been shown to play an important role in the chaperone cycle.[64]

Additional putative HSP90 inhibitors have been identified by screening for compounds that reproduce the established gene expression profile associated with HSP90 inhibition.[65,66] In this way, celestrol and gedunin were revealed as natural products that inhibit HSP90 by an as yet unclear mechanism that does not involve direct ATP-site antagonism.

An antibody has been developed against the yeast homolog of HSP90 that is intended for antifungal therapy, but which may have potential for cancer treatment.[67]

Thus it can be seen that a variety of approaches are being taken to inhibit HSP90. Indeed, the complex structure and mechanism action of the HSP90 superchaperone complex suggest numerous alternative opportunities for pharmacologic intervention, in addition to ATP-site mimetics.

CLINICAL DEVELOPMENT

A considerable amount of understanding has been gained from the clinical experience to date with 17-AAG [15,39–41] (see also htpp://www.clinicals.gov). Use of a validated biomarker signature of HSP90 inhibition[68] (see also next section on biomarkers) showed clearly that HSP90 could be inhibited in solid tumors at doses that were quite well tolerated by patients.[39] To date, evidence of clinical activity has been reported for 17-AAG in melanoma, multiple myeloma, prostate, and breast cancer.[39–41] In addition to difficulties with the initially cumbersome and DMSO-containing formulation, the main toxicities have involved liver and gastrointestinal effects.

17-DMAG has followed 17-AAG into the clinic, offering greater solubility and oral bioavailability. The 17-AAG hydroquinone IPI-504, now in clinical trials, has similar advantages and a purine-scaffold compound CNF2024 has also begun clinical evaluation.

It is likely that a range of other new inhibitors of various chemical types will enter the clinic over the next couple of years. Experience with these agents will help us to understand the nature of on-target and off-target effects that may be contributing to efficacy and toxicity. This information, together with the emerging basic science, will likely form the basis for the design of a further generation of HSP90 inhibitors.

BIOMARKERS FOR HSP90 INHIBITORS

Rational drug development requires the use of biomarkers to obtain proof of concept for the correct molecular mechanism of action and potentially for patient selection prior to therapy. The molecular signature of depletion of client proteins and induction of HSP72 has been used frequently to show pharmacodynamic effects of HSP90 inhibition in peripheral blood mononuclear cells and tumor biopsies from treated patients.[39,68] Two new biomarkers of target inhibition, insulin-like growth factor-binding protein-2 and ERBB2 extracellular domain, are readily detected in patient sera by ELISA.[69] Gene expression microarray and proteomic profiling of cells treated with HSP90 inhibitors have identified additional potential biomarkers of HSP90 modulation.[65,70]

Minimally invasive molecular and functional imaging methods have advantages. A positron emission tomography technique for detecting depletion of ERBB2 shows promise.[71] A magnetic resonance spectroscopy signature of HSP90 inhibition also offers potential.[72]

The above methods all provide valuable pharmacodynamic biomarkers of HSP90 inhibition. Further studies are required to identify biomarkers that will predict sensitivity and allow the identification of patients who are most likely to respond to HSP90 inhibitors. These might be client proteins (e.g., ERBB2 in breast cancer), antiapoptotic proteins (e.g., HSP72), or unknown gene products.

CONCLUDING REMARKS AND FUTURE PERSPECTIVE

In this review we have emphasized the promising therapeutic potential of inhibiting the HSP90 molecular chaperone for cancer treatment. The powerful combinatorial effects on multiple client proteins, signaling pathways, and hallmark cancer traits provide the potential for an attractive therapeutic profile across a range of malignant diseases where several oncogenic drivers often conspire together to accelerate the malignant process. This feature should also reduce the likelihood of resistance developing.

There is much more to be discovered about the basic biology and genetics of HSP90. In addition to traditional hypothesis-testing approaches, genome-wide strategies are revealing the complexity of the HSP90 interaction network.[65,70,73] HSP90 has genetic and physical interactions with up to 10% of the yeast proteome.[73] The complicated relationship between the HSP90 chaperone complex, chromatin remodeling, and transcriptional control is especially intriguing and has therapeutic significance.[17,30,70,73]

There is much more to learn about the effects of HSP90 inhibition on client proteins and their cognate biochemical pathways. For example, loss of HSP90 association upregulates SRC-dependent ERBB2 activity.[74] HSP90 inhibition leads to transient activation of SRC, AKT, and ERK.[75] The significance of these effects remains to be determined. There are also additional potential downsides of HSP90 inhibitors. The tumor-suppressor protein LKB1 is a client protein[76,77] and depletion of this and any other wild-type tumor suppressors could have undesirable consequences if not counteracted by beneficial effects. A role of HSP90 in the maturation of hERG channels[78] has suggested at least the potential for cardiac toxicity with HSP90 inhibitors, and particular care may be required when these are used in combination with doxorubicin.[79] Effects on the heart and other organs will be determined during toxicology studies with clinical candidates. It will also be important to assess possible adverse effects in patients very carefully, including determination of the consequences of long-term administration, particularly given the ability of HSP90 to buffer mutations. One study has shown that 17-AAG can increase osteoclast formation and bone metastasis.[80] However, this can be prevented by alendronate (Neckers, L, unpublished data). In addition, a decrease in metastasis to other sites may be expected with HSP90 inhibitors. Any possible negative effects will need to be considered in the light of the many potential advantages of HSP90 inhibition.

The beneficial effects of HSP90 inhibitors on tumor angiogenesis require more detailed study, particularly given the modulation of multiple proangiogenic factors.[81,82]

The effects of combining HSP90 inhibitors with cytotoxic agents, as well as the newer molecular cancer therapeutics, are already being evaluated in the clinic. Combinations must be designed carefully as there is potential for antagonism as well as additivity or synergy. Of particular interest is the combination of HSP90 inhibitors with proteasome inhibitors, such as bortezomib, where endoplasmic reticulum vacuolization and protein relocalization is seen, contributing to cytotoxic consequences.[83] The combination of 17-AAG and taxanes is promising in breast cancer[84] and also in PI3K-activated ovarian cancer.[85]

A general characteristic feature of HSP90 inhibitors is that they activate the heat shock response, mediated by HSF-1.[17] Induction of antiapoptotic HSPs, such as HSP72, HSP27, and HSP90 itself, represents protective mechanisms against HSP90 inhibition and other therapies.[17] It is unclear what the balance of effects of this will be in cancer versus normal cells and hence on therapeutic index *in vivo*. One approach will be to modulate the expression or function of the HSPs alongside HSP90 inhibition. Another is to explore the therapeutic potential of compounds that selectively induce client protein degradation for cancer therapy, as well as those that favor cytoprotective pathways for the treatment of conditions like Alzheimer's and Huntingdon's Disease.[86]

As mentioned earlier, most research to identify HSP90 inhibitors has focused on ligands acting at the N-terminal ATP-binding site. There is considerable potential for interfering with the superchaperone complex in other ways, for example, by affecting cochaperone interactions. Blockade of CDC37 could produce specific depletion of kinase clients. Modulation of the HSP90 activator AHA1 may be beneficial in cystic fibrosis.[87] A recent cell-based phenotypic screen identified compounds that inhibit HSP90 by a novel mechanism, as well as others that block its ATPase activity directly.[88]

Work with HS90 inhibitors as potential cancer drugs has opened up a potentially bright future, which extends beyond HSP90 into other molecular chaperones and protein quality control targets, as well into other diseases in which molecular chaperones and related proteins may play a pathological role.

REFERENCES

1. WORKMAN, P. 2005. Drugging the cancer kinome: progress and challenges in developing personalized molecular cancer therapeutics. Cold Spring Harb. Symp. Quant. Biol. **70:** 499–515.
2. COLLINS, I & P. WORKMAN. 2006. New approaches to molecular cancer therapeutics. Nat. Chem. Biol. **2:** 689–700.
3. ROSEN, N & Q.B. SHE. 2006. AKT and cancer-is it all mTOR? Cancer Cell. **10:** 254–256.
4. BENSON, J.D., Y.N. CHEN, S.A. CORNELL-KENNON, *et al*. 2006. Validating cancer drug targets. Nature **441:** 451–456.
5. WEINSTEIN, I.B. 2002. Cancer. Addiction to oncogenes—the Achilles heal of cancer. Science **297:** 63–64.
6. SJOBLOM, T., S. JONES, L.D. WOOD, *et al*. 2006. The consensus coding sequences of human breast and colorectal cancers. Science **314:** 268–274.
7. GREENMAN, C., P. STEPHENS, R. SMITH, *et al*. 2007. Patterns of somatic mutation in human cancer genomes. Nature **446:** 153–158.
8. O'REILLY, K.E., F. ROJO, Q.B. SHE, *et al*. 2006. mTOR inhibition induces upstream receptor tyrosine kinase signaling and activates Akt. Cancer Res. **66:** 1500–1508.
9. SHE, Q.B., D.B. SOLIT, Q. YE, *et al*. 2005. The BAD protein integrates survival signaling by EGFR/MAPK and PI3K/Akt kinase pathways in PTEN-deficient tumor cells. Cancer Cell **8:** 287–297.

10. ROCCARO, A.M., T. HIDESHIMA, P.G. RICHARDSON, *et al*. 2006. Bortezomib as an antitumor agent. Curr. Pharm. Biotechnol. **7:** 441–448.
11. WORKMAN, P. 2004. Combinatorial attack on multistep oncogenesis by inhibiting the Hsp90 molecular chaperone. Cancer Lett. **206:** 149–157.
12. SHARP, S. & P. WORKMAN. 2006. Inhibitors of the HSP90 molecular chaperone: current status. Adv. Cancer Res. **95:** 323–348.
13. ZHANG, H. & F. BURROWS. 2004. Targeting multiple signal transduction pathways through inhibition of Hsp90. J. Mol. Med. **82:** 488–499.
14. XU, W. & L. NECKERS. 2007. Targeting the molecular chaperone heat shock protein 90 provides a multifaceted effect on diverse cell signaling pathways of cancer cells. Clin. Cancer Res. **13:** 1625–1629.
15. SOLIT, D.B. & N. ROSEN. 2006. Hsp90: a novel target for cancer therapy. Curr. Top. Med. Chem. **6:** 1205–1214.
16. CHIOSIS, G., M. VILENCHIK, J. KIM & D. SOLIT. 2004. Hsp90: the vulnerable chaperone. Drug Discov. Today. **9:** 881–888.
17. WHITESELL, L. & S. LINDQUIST. 2005. HSP90 and the chaperoning of cancer. Nat. Rev. Cancer **5:** 761–772.
18. RITOSSA, F. 1962. A new puffing pattern induced by temperature shock and DNP in Drosophila. Experentia **19:** 571–573.
19. RITOSSA, F. 1996. Discovery of the heat shock response. Cell Stress Chaperones **1:** 97–98.
20. PEARL, L.H. & C. PRODROMOU. 2006. Structure and mechanism of Hsp90 molecular chaperone machinery. Ann. Rev. Biochem. **75:** 271–294.
21. XU, W., X. YUAN, Z. XIANG, *et al*. 2005. Surface charge and hydrophobicity determine ErbB2 binding to the Hsp90 chaperone complex. Nat. Struct. Mol. Biol. **12:** 120–126.
22. PICK, E., Y. KLUGER, J.M. GILTNANE, *et al*. 2007. High HSP90 expression is associated with decreased survival in breast cancer. Cancer Res. **67:** 2932–2937.
23. NECKERS, L. 2006. Using natural product inhibitors to validate HSP90 as a molecular target in cancer. Curr. Med. Chem. **6:** 1163–1171.
24. WHITESELL, L., R. BAGATELL & R. FALSEY. 2003. The stress response: implications for the development of Hp90 inhibitors. Curr. Cancer Drug Targets **3:** 349–358.
25. CALDERWOOD, S.K., M.A. KHALEQUE, D.B. SAWYER & D.R. CIOCCA. 2006. Heat shock proteins in cancer: chaperones of tumorigenesis. Trends Biochem. Sci. **31:** 164–172.
26. KAMAL, A., L. THAO, J. SENSINTAFFAR, *et al*. 2003. A high-affinity conformation of Hsp90 confers tumour selectivity on Hsp90 inhibitors. Nature **425:** 407–410.
27. DA ROCHA DIAS, S., F. FRIEDLOS, Y. LIGHT, *et al*. 2005. Activated B-RAF is an Hsp90 client protein that is targeted by the anticancer drug 17-allylamino-17-demethoxygeldanamycin. Cancer Res. **65:** 10686–10691.
28. GRBOVIC, O.M., A.D. BASSO, A. SAWAI, *et al*. 2006. V600E B-Raf requires the Hsp90 chaperone for stability and is degraded in Response to Hsp90 inhibitors. Proc. Natl. Acad. Sci. USA. **103:** 57–62.
29. SHIMAMURA, T., A.M. LOWELL, J.A. ENGELMAN & G.I. SHAPIRO. 2005. Epidermal growth factor receptors harboring kinase domain mutations associate with the heat shock protein 90 chaperone and are destabilized following exposure to geldanamycins. Cancer Res. **65:** 6401–6408.

30. RUTHERFORD, S.L. & S. LINDQUIST. 1998. Hsp90 as a capacitor for morphological evolution. Nature **396:** 336–342.
31. YEYATI, P.L., R.M. BANCEWICZ, J.M. MAULE & V. VAN HEYNINGEN. 2007. Hsp90 selectively modulates phenotype in vertebrate development. PLoS Genetics **43:** 431–447.
32. MCDONALD, E., P. WORKMAN & K. JONES. 2007. Inhibitors of the HSP90 molecular chaperone: attacking the master regulator in cancer. Curr. Top. Med. Chem. **6:** 1091–1107.
33. WHITESELL, L., E.G. MIMNAUGH, B. DE COSTA, et al. 1994. Inhibition of heat shock protein HSP90-pp60v-src heteroprotein complex formation by benzoquinone ansamycins: essential role for stress proteins in oncogenic transformation. Proc. Natl. Acad. Sci. USA **91:** 8324–8328.
34. SHARMA, S.V., T. AGATSUMA & H. NAKANO. 1998. Targeting of the protein chaperone HSP90, by the transformation suppressing agent, radicicol. Oncogene **16:** 2639–2645.
35. WORKMAN, P. 2003. Overview: translating Hsp90 biology into Hsp90 drugs. Curr. Cancer Drug Targets **3:** 297–300.
36. ROE, S.M., C. PRODROMOU, R. O'BRIEN, et al. 1999. Structural basis for inhibition of the Hsp90 molecular chaperone by the antitumor antibiotics radicicol and geldanamycin. J. Med. Chem. **42:** 260–266.
37. CHENE, P. 2002. ATPases as drug targets: learning from their structure. Nat. Rev. Drug Discov. **1:** 665–673.
38. SUPKO J.G., R.L. HICKMAN, M.L. GREVER & L. MALSPEIS 1995. Preclinical pharmacologic evaluation of geldanamycin as an antitumor agent. Cancer Chemother. Pharmacol. **36:**305–315.
39. BANERJI, U., A O'DONNELL, M. SCURR, et al. 2005. Phase I pharmacokinetic and pharmacodynamic study of 17-allylamino, 17-demethoxygeldanamycin in patients with advanced malignancies. J. Clin. Oncol. **23:** 4152–4161.
40. SOLIT, D.B., S.P. IVY, C. KOPIL, et al. 2007. Phase I trial of 17-allylamino-17-demethoxygeldanamycin in patients with advanced cancer. Clin. Cancer Res. **13:** 1775–1782.
41. PACEY, S., U. BANERJI, I. JUDSON & P. WORKMAN. Hsp90 inhibitors in the clinic. Hand. Exp. Pharmacol. **172:** 331–358.
42. BURGER A.M., H.H. FIEBIG, S.F. STINSON & E.A. SAUSVILLE. 2004. 17-(Allylamino)-17-demethoxygeldanamycin activity in human melanoma models. Anticancer Drugs **15:** 377–387.
43. SYDOR J.R., E. NORMANT, C.S. PIEN, et al. 2006. Development of 17-allylamino-17-demethoxygeldanamycin hydroquinone hydrochloride (IPI-504), an anticancer agent directed against Hsp90. Proc. Natl. Acad. Sci. USA **103:** 17408–17413.
44. KELLAND, L.R., S.Y. SHARP, P.M. ROGERS, et al. 1999. DT-Diaphorase expression and tumor cell sensitivity to 17-allylamino, 17-demethoxygeldanamycin, an inhibitor of heat shock protein 90. J. Natl. Cancer Inst. **91:** 1940–1949.
45. ZHENG, F.F., S. KUDUK, G. CHIOSIS, et al. 2000. Identification of a geldanamycin dimer that induces the selective degradation of HER-family tyrosine kinases. Cancer Res. **15:** 20904–2094.
46. ZHANG, H., Y.C. YANG, L. ZHANG, et al. 2006. Dimeric ansamycins—a new class of antitumor Hsp90 modulators with prolonged inhibitory activity. Int. J. Cancer **120:** 918–926.

47. SOGA, S., Y. SHIOTSU, S. AKINAGA & S.V. SHARMA. 2003. Development of radicicol analogues. Curr. Cancer Drug Targets **3:** 359–369.
48. JANIN, Y.L. 2005. Heat shock protein 90 inhibitors. A text book example of medicinal chemistry? J. Med. Chem. **48:** 7503–7512.
49. CHIOSIS, G., M.N. TIMAUL, B. LUCAS, et al. 2001. A small molecule designed to bind to the adenine nucleotide pocket of Hsp90 causes Her2 degradation and the growth arrest and differentiation of breast cancer cells. Chem. Biol. **8:** 289–299.
50. CHIOSIS, G. 2006. Discovery and development of purine-scaffold Hsp90 inhibitors. Curr. Top. Med. Chem. **6:** 1183–1191.
51. WRIGHT, L., X. BARRIL, B. DYMOCK, et al. 2004. Structure-activity relationships in purine-based inhibitor binding to HSP90 isoforms. Chem. Biol. **11:** 775–785.
52. LLAUGER, L., H. HE, J. KIM, et al. 2005. Evaluation of 8-arylsulfanyl, 8-arylsulfoxyl, and 8-arylsulfonyl adenine derivatives as inhibitors of the heat shock protein 90. J. Med. Chem. **48:** 2892–2905.
53. ZHANG, L., J. FAN, K. VU, et al. 2006. 7′-Substituted benzothiazolothio- and pyridinothiazolothio-purines as potent heat shock protein 90 inhibitors. J. Med. Chem. **49:** 5352–5362.
54. CHEUNG, K.M., T.P. MATTHEWS, K. JAMES, et al. 2005. The identification, synthesis, protein crystal structure and in vitro biochemical evaluation of a new 3,4-diarylpyrazole class of Hsp90 inhibitors. Bioorg. Med. Chem. Lett. **15:** 3338–3343.
55. SHARP, S.Y., K. BOXALL, M. ROWLANDS, et al. 2007. In vitro biological characterization of a novel, synthetic diaryl pyrazole resorcinol class of heat shock protein 90 inhibitors. Cancer Res. **67:** 2206–2216.
56. DYMOCK, B.W., X. BARRILL, P.A. BROUGH, et al. 2005. Novel, potent small-molecule inhibitors of the molecular chaperone Hsp90 discovered through structure-based design. J. Med. Chem. **48:** 4212–4215.
57. SHARP, S.Y, C. PRODROMOU, K. BOXALL, et al. 2007. Inhibition of the heat shock protein 90 molecular chaperone in vitro and in vivo by novel, synthetic, potent resorcinylic pyrazole/isoxazole amide analogues. Mol. Cancer Ther. **6:** 1198–1211.
58. MARCU, M.G., A. CHADLI, I. BOUHOUCHE, et al. 2000. The heat shock protein 90 antagonist novobiocin interacts with a previously unrecognized ATP-binding domain in the carboxyl terminus of the chaperone. J. Biol. Chem. **275:** 37181–37186.
59. MARCU, M.G., T.W. SCHULTE & L. NECKERS. 2000. Novobiocin and related coumarins and depletion of heat shock protein 90-dependent signaling proteins. J. Natl. Cancer Inst. **92:** 242–248.
60. YU, X.M., G. SHEN, L. NECKERS, et al. 2005. Hsp90 inhibitors identified from a library of novobiocin analogues. J. Am. Chem. Soc. **127:**12778–12779.
61. BURLISON, J.A., L. NECKERS, A.B. SMITH, et al. 2006. Novobiocin: redesigning a DNA gyrase inhibitor for selective inhibition of hsp90. J. Am. Chem. Soc. **128:** 15529–15536.
62. YU, X., M.G. GUO, L. NECKERS, D.M. NQUYEN, et al. 2002. Modulation of p53, ErbB1, ErbB2, and Raf-1 expression in lung cancer cells by depsipeptide FR901228. J. Natl. Cancer Inst. **94:** 504–513.
63. BALI P, M. PRANPAT, J. BRADNER, et al. 2005. Inhibition of histone deacetylase 6 acetylates and disrupts the chaperone function of heat shock protein 90: a novel

basis for antileukemia activity of histone deacetylase inhibitors. J. Biol. Chem. **280:** 26729–26734.
64. SCROGGINS, B.T., K. ROBZYK, D. WANG, et al. 2007. An acetylation site in the middle domain of Hsp90 regulates chaperone function. Mol. Cell. **25:** 151–159.
65. CLARKE, P.A., I. HOSTEIN, U. BANERJI, et al. 2000. Gene expression profiling of human colon cancer cells following inhibition of signal transduction by 17-allylamino-17-demethoxygeldanamycin, an inhibitor of the hsp90 molecular chaperone. Oncogene **19:** 4125–4133.
66. HIERONYMUS, H., J. LAMB, K.N. ROSS, et al. 2006. Gene expression signature-based chemical genomic prediction identifies a novel class of HSP90 pathway modulators. Cancer Cell **10:** 321–330.
67. MATTHEWS, R.C., C. RIGG, S. HODGETTS, et al. 2003. Preclinical assessment of the efficacy of mycograb, a human recombinant antibody against fungal HSP90. Antimicrob. Agents Chemother. **47:** 2208–2216.
68. BANERJI, U., M. WALTON, F. RAYNAUD, et al. 2005. Pharmacokinetic-pharmacodynamic relationships for the heat shock protein 90 molecular chaperone inhibitor 17-allylamino, 17-demethoxygeldanamycin in human ovarian cancer xenograft models. Clin. Cancer Res. **11:** 7023–7032.
69. ZHANG, H., D. CHUNG, Y.C. YANG, et al. 2006. Identification of new biomarkers for clinical trials of Hsp90 inhibitors. Mol. Cancer Ther. **5:** 1256–1264.
70. MALONEY, A., P.A. CLARKE, S. NAABY-HANSEN, et al. 2007. Gene and protein expression profiling of human ovarian cancer cells treated with the heat shock protein 90 inhibitor 17-allylamino-17-demethoxygeldanamycin. Cancer Res. **67:** 3239–5323.
71. SMITH-JONES, P.M., D. SOLIT, F. AFROZE, et al. 2006. Early tumor response to Hsp90 therapy using HER2 PET: Comparison with 18F-FDG PET. J. Nucl. Med. **47:** 793–796.
72. CHUNG, Y.L., H. TROY, U. BANERJI, et al. 2003. Magnetic resonance spectroscopic pharmacodynamic markers of the heat shock protein 90 inhibitor 17-allylamino,17-demethoxygeldanamycin (17AAG) in human colon cancer models. J. Natl. Cancer Inst. **95:** 1624–1633.
73. ZHAO, R., M. DAVEY, H. YA-CHIEH, et al. 2005. Navigating the chaperone network: an interactive map of physical and genetic interactions mediated by the Hsp90 chaperone. Cell **120:** 715–727.
74. XU, W., X. YUAN, K. BEEBE, et al. 2007. Loss of Hsp90 association up-regulates Src-dependent ErbB2 activity. Mol. Cell Biol. **27:** 220–228.
75. KOGA, F., W. XU, T.S. KARPOVA, et al. 2006. Hsp90 inhibition transiently activates Src kinase and promotes Src-dependent Akt and Erk activation. Proc. Natl. Acad. Sci. USA **103:** 11318–1122.
76. BOUDEAU, J., M. DEAK, M.A. LAWLOR, et al. 2003. Heat shock protein 90 and Cdc37 interact with LKB1 and regulate its stability. Biochem. J. **370:** 849–857.
77. NONY, P., H. GAUDE, M. ROSSEL, et al. 2003. Stability of the Peutz-Jeghers syndrome kinase LKB1 requires its binding to the molecular chaperones Hsp90/Cdc37. Oncogene **22:** 9165–9175.
78. FICKER, E, A.T. DENNIS, L. WANG, et al. 2003. Role of the cytosolic chaperones Hsp70 and Hsp90 in maturation of the cardiac potassium channel hERG. Circ. Res. **92**:87–100.
79. GABRIELSON, K., D. BEDJA, S. PIN, et al. 2007. Heat shock protein 90 and ErbB2 in the cardiac response to doxorubicin injury. Cancer Res. **67:** 1436–1441.

80. PRICE, J.T., J.M.W. QUINN, N.A. SIMS, *et al.* 2005. Heat shock protein 90 inhibitor, 17-allylamino-17-demethoxygeldanamycin, enhances osteoclast formation and potentiates bone metastasis of a human breast cancer cell line. Cancer Res. **65:** 4929–4938.
81. SANDERSON, S, M. VALENTI, S. GOWAN, *et al.* 2006. Benzoquinone ansamycin heat shock protein 90 inhibitors modulate multiple functions required for tumor angiogenesis. Mol. Cancer Ther. **5:** 522–532.
82. ISAACS, J.S., Y.J. JUNG, E.G. MIMNAUGH, *et al.* 2002. Hsp90 regulates a von Hippel Lindau-independent hypoxia-inducible factor-1 alpha-degradative pathway. J. Biol. Chem. **277:** 29936–29944.
83. MIMNAUGH, E.G., W. XU, M. VOS, *et al.* 2006. Endoplasmic reticulum vacuolization and valosin-containing protein relocalization result from simultaneous hsp90 inhibition by geldanamycin and proteasome inhibition by velcade. Mol. Cancer Res. **4:** 667–681.
84. BASSO, A., D. SOLIT, L. NORTON, *et al.* 2001. Modulation of Hsp90 function by ansamycins sensitizes breast cancer cells to chemotherapy-induced apoptosis in an RB- and schedule-dependent manner. Clin. Cancer Res. **7:** 2228–2236.
85. SAIN, N, B. KRISHNAN, M.G. ORMEROD, *et al.* 2006. Potentiation of paclitaxel activity by the HSP90 inhibitor 17-allylamino-17-demethoxygeldanamycin in human ovarian carcinoma cell lines with high levels of activated AKT. Mol. Cancer Ther. **5:** 1197–1208.
86. DICKEY, C.A., A. KAMAL, K. LUNDGREN, *et al.* 2007. A high-affinity HSP90-CHIP complex recognizes and selectively degrades phosphorylated tau client proteins. J. Clin. Invest. **117:** 648–658.
87. WANG, X., J. VENABLE, P. LAPOINTE, *et al.* 2006. Hsp90 cochaperone Aha1 down-regulation rescues misfolding of CFTR in cystic fibrosis. Cell **127:** 803–815.
88. HARDCASTLE, A., P. TOMLIN, C. NORRIS, *et al.* 2007. A duplexed phenotypic screen for the simultaneous detection of inhibitors of the molecular chaperone heat shock protein 90 and modulators of cellular acetylation. Mol. Cancer Ther. **6:** 1112–1122.

Stress, Heat Shock Proteins, and Autoimmunity

How Immune Responses to Heat Shock Proteins Are to Be Used for the Control of Chronic Inflammatory Diseases

WILLEM VAN EDEN,[a] GEORGE WICK,[b] SALVATORE ALBANI,[c] AND IRUN COHEN[d]

[a]*Department of Infectious Diseases and Immunology, Faculty of Veterinary Medicine, Yalelaan, Utrecht, the Netherlands*

[b]*Division of Experimental Pathophysiology and Immunology, Biocenter, Innsbruck Medical University, Innsbruck, Austria*

[c]*Department of Medicine and Pediatrics, University of California, San Diego, California, USA*

[d]*Department of Immunology, The Weizmann Institute of Science, Rehovot, Israel*

ABSTRACT: Especially since the (re-)discovery of T cell subpopulations with specialized regulatory activities, mechanisms of anti-inflammatory T cell regulation are studied very actively and are expected to lead to the development of novel immunotherapeutic approaches, especially in chronic inflammatory diseases. Heat shock proteins (Hsp) are possible targets for regulatory T cells due to their enhanced expression in inflamed (stressed) tissues and the evidence that Hsp induce anti-inflammatory immunoregulatory T cell responses. Initial evidence for an immunoregulatory role of Hsp in chronic inflammation was obtained through analysis of T cell responses in the rat model of adjuvant arthritis and the findings that Hsp immunizations protected against the induction of various forms of autoimmune arthritis in rat and mouse models. Since then, immune reactivity to Hsp was found to result from inflammation in various disease models and human inflammatory conditions, such as rheumatoid arthritis (RA), type 1 diabetes, and atherosclerosis. Now, also in the light of a growing interest in T cell regulation, it is of interest to further explore the mechanisms through which Hsp can be utilized to trigger immunoregulatory pathways, capable of suppressing such a wide and diversified spectrum of inflammatory diseases.

Address for correspondence: Willem van Eden, M.D., Ph.D., Department of Infectious Diseases and Immunology, Faculty of Veterinary Medicine, Yalelaan 1, Utrecht University, 3584CL Utrecht, the Netherlands. Voice: +31-30-2534358; fax: +31-30-2533555.
w.vaneden@uu.nl

KEYWORDS: regulatory T cells; inflammation; Hsp60; arthritis; atherosclerosis

INTRODUCTION

Heat shock proteins (Hsp) received attention by immunologists since their discovery and especially since it was observed that under various pathological conditions immune responses to Hsp were readily developing. Through their evolutionary conservation, their role in maintaining integrity of cellular proteins and their stress inducibility, their potential impact on the organization of immune reactivity in mammals is considered to be broad and multifaceted. Their role as chaperones for intracellular proteins, such as tumor antigens, their role as proteins to deliver protein fragments for cross-presentation to $CD8^+$ T cells, and some other aspects have been the subject of various reviews.[1,2] This review will concentrate on immune responses induced by Hsp as immunogenic antigens, and the potential of such responses to control inflammation.

MICROBIAL HSP AND INDUCTION OF T CELL REGULATION

Immunization with whole (heat-killed) mycobacteria in rats leads to an autoimmune arthritis known as adjuvant arthritis. The analysis of T cell responses in rats with adjuvant arthritis revealed the presence of T cells recognizing a mycobacterial antigen, which had a regulatory activity in the inflammatory process. Later on, these T cells were seen to recognize the so-called 65-kDa antigen of mycobacteria, which happened to be the mycobacterial Hsp60.[3] Hsp60 of *E. coli* (GroEL) was already known as the "common antigen of Gram negatives" indicating the dominant immunogenicity of the Hsp60 family of molecules.

From these early observations in adjuvant arthritis it had become apparent that immune exposure to Hsp60 was inducing inflammation inhibitory regulatory activity at the level of T cells and that such regulatory T cells were triggered even in the context of a disease producing response elicited by immunization with whole mycobacteria.

Subsequent experiments carried out in various experimental models have further substantiated the arthritis inhibitory effect of mycobacterial Hsp60. This is the case for streptococcal cell wall-induced arthritis in rats,[4] adjuvant arthritis in rats,[5] pristane-induced arthritis in mice,[6] and to some extent collagen-induced arthritis in rats.[7] Ragno *et al.* have shown the protective effect of mycobacterial Hsp60 as a naked DNA vaccine in adjuvant arthritis.[8]

Several studies now also have shown the protective effects of mycobacterial Hsp60-derived peptides in adjuvant arthritis.[9–13] Besides nasal administration of peptide, oral administration of mycobacterial Hsp60 has been successful.[14] Studies on the potential of repetitive oral administration of mycobacterial Hsp60 to suppress adjuvant arthritis, have revealed that in the presence of soybean trypsin inhibitor, very low dosages of Hsp60 (30 μg) can suppress disease.

The oral administration, by gastric gavage, was also effectively suppressing disease, when the procedure was started at the time of already clinically overt disease.[15] Also, other Hsp family members were tested and found to inhibit experimental arthritis. In rat adjuvant arthritis this was done with Hsp10, and the protective effect was seen for mycobacterial Hsp10 and not for the *E. coli* Hsp 10 (GroES) or the rat Hsp10.[16] For mycobacterial Hsp70, disease prevention by prior Hsp immunization was seen in rat adjuvant arthritis, avridine arthritis, and to a lower extent collagen II arthritis.[7,17] Recently, we also have seen disease inhibition of aggrecan-induced arthritis in BALB/c mice[18] using preimmunization with mycobacterial Hsp70 in DDA as an adjuvant (Berlo *et al.* in prep.). When analyzing an *E. coli*-derived bacterial extract called OM-89, which is used as an oral slow-acting antirheumatic drug in humans, it appeared that one of its major protein constituents was Hsp70 and some lower amounts of Hsp60. Upon oral administration in adjuvant arthritis rats prior to disease induction, OM-89 induced T cell reactivity to Hsp60 and Hsp70 and had a profound disease suppressive activity,[19] indicating that also GroEL and DnaK of *E. coli* have disease-suppressive activity.

Wendling defined several T cell epitopes in mycobacterial Hsp70 in Lewis rats. One of these peptides, Hsp70 111–125, was found to stimulate production of interleukin (IL)-10 in responding T cells. Upon nasal administration the 111–125 peptide was found to prevent the subsequent induction of adjuvant arthritis.[20]

Given the fact that the various Hsp families are antigenically unrelated and that they have seemingly equal capacities to inhibit arthritis, Prakken *et al.* analyzed whether or not other non-Hsp antigens, sharing immunological characteristics with Hsp, would also have arthritis inhibitory effects. Therefore, other evolutionary conserved bacterial immunogens of a conserved nature were tested for their protective qualities in experimental arthritis. For this we used the antigens superoxide dismutase (SOD) of *E.coli*, glyceraldehyde-3-phosphate dehydrogenase of Bacillus (G3PDH), and aldolase of Staphylococcus. These antigens were found to be immunogenic, as they induced proliferative T cell responses and delayed-type hypersensitivity reactions. All three antigens were having "self-homologues" present in mammalian cells. Clearly enough, upon immunization none of these antigens was seen to affect arthritis to any extent and this was apparent in both adjuvant- and avridine-induced arthritis.[17] Therefore, the evidence collected so far, is strongly indicative of the induction of arthritis-protective mechanisms as a shared characteristic of microbial heat shock proteins, which is absent in other immunogenic and conserved bacterial antigens.

In nonobese diabetes (NOD) the effects of mycobacterial Hsp60 have been found to vary with the form of their administration.[21] In phosphate-buffered saline (PBS), mycobacterial Hsp60 was found to inhibit disease development and in incomplete Freund's adjuvant (IFA) it induced more rapid diabetes. The spontaneous onset of β cell destruction went together with the development of antimycobacterial Hsp60 T cells, the release of self-Hsp60 in the blood and the subsequent production of anti-Hsp60 antibodies. These studies have led to a further analysis of the mouse Hsp60 and especially the p277 peptide of Hsp60 in NOD.[22] Other antigens, such as mycobacterial Hsp70, were not seen to have any measurable effect on the development of diabetes. Thus, in contrast to what has been seen in arthritis, in the NOD the effects of mycobacterial Hsp60 cannot be explained as a mere result of immunization to any bacterial Hsp. From this analysis in NOD mice it seems that Hsp60 may have a unique function in β cells and in the destructive process of type I diabetes.[23]

In agreement with this were findings in NOD and in genetically protected NOD-asp mice. The development of insulitis was characterized by the *in vivo* priming of mycobacterial or human Hsp60 responsive T cells that produced IL-10 upon *in vitro* restimulation with Hsp60. This was not seen in control mice without developing insulitis. Apparently, expression of endogenous Hsp60 in insulitis was associated with the regulation of insulitis.[24] When administered at a high dose, also the nonmicrobial Hsp gp96 was found to suppress diabetes in NOD mice.[25]

Other studies have indicated that, intracellularly, Hsp may play a role in protecting β cells. For Hsp70 a role as a molecule involved in cellular repair in diabetes was postulated, when it was shown that (heat stress) induction of endogenous Hsp70 in islets was impaired in diabetes-prone BioBreeding (BB) rats at a young age but also at the older diabetes-sensitive age.[26]

In the rat model of myelin basic protein (MBP)-induced experimental autoimmune encephalomyelitis (EAE), a 12-kDa purified protein derivative (PPD)-derived protein with sequence homologies to Hsp10 was found to suppress disease.[27] In a mouse EAE study the reduced EAE susceptibility of conventionally reared animals compared to specific pathogen free (SPF) animals, was associated with skewing of Hsp60-specific T cell cytokines toward a Th2 pattern. These findings were indicative of the fact that frequent exposure to infectious agents leads to a Th2 skewing of immune responses to Hsp and that this is associated with milder and less frequent relapses of EAE.[28] Immunization with nonmicrobial Hsp gp96 suppressed MBP-induced EAE in SJL mice.[25]

Rha *et al.*[29] reported a distinctive quality of *M. leprae* Hsp60, not found for other microbial Hsp. *M. leprae* Hsp60 suppressed disease in a murine model of allergic airway inflammation and airway hyperresponsiveness. In the bronchoalveolar lavage fluid, IL-4 and IL-5 production were found to be suppressed, while IL-10 and interferon (IFN)-γ production was increased.

CELL STRESS UPREGULATED HSP ARE TARGETS OF T CELL REGULATION

Steve Anderton[9] was able to generate a series of mycobacterial Hsp60-specific T cell lines from rats after immunization with mycobacterial Hsp60.

These T cells were tested in adoptive transfer experiments and one of them was transferring resistance to arthritis induction. This particular T cell line was found to recognize a highly conserved sequence of Hsp60 and as expected, this T cell did cross-recognize the mammalian homologue of the mycobacterial protein: rat Hsp60. The other T cells that did not transfer resistance were also mapped and none of them cross-recognized mammalian Hsp60.

The same was also found for mycobacterial Hsp70.[20] Immunization with mycobacterial Hsp70 protected against adjuvant arthritis in rats, and also protected in a model induced with a synthetic oily compound called avridine. Also here mapping studies revealed that only T cells recognizing highly conserved sequences were transferring protection and were cross-responding to the mammalian homologous proteins. Various studies using different routes of exposure to peptides representing conserved Hsp60 and Hsp70 T cell epitopes, have shown immunoregulatory effects of these conserved sequences in a multitude of disease models.

From these studies we must conclude that immune exposure to microbial Hsp was capable of inducing a regulatory T cell response and that such regulation depended on a T cell response that included a repertoire of (endogenous) self-Hsp- specific T cells.

The existence of self-Hsp-reactive T cells has been demonstrated in many different studies, even in human umbilical cord lymphocytes[30] and in mice transgenic for Hsp60.[31] Apparently thymus selection allows such a repertoire to develop. Microbial Hsps are dominantly immunogenic and frequencies of Hsp-specific T cells can be very high following microbial exposure. It makes sense to suppose that this T cell stems from a repertoire that has been selected in the thymus by self (endogenous)-Hsp peptides. Interactions with such self-Hsp must have stimulated positive selection and must have allowed the self-Hsp-reactive T cells to survive negative selection. In the peripheral immune system, when the cells have left the central lymphoid organs, microbial Hsp can be the full agonists for these T cells compatible with the immunodominant character of these Hsps. At the same time, when self-Hsp is expressed in the periphery, under conditions of cell stress, the self-Hsp can act as a partial agonist producing a regulated or actively regulatory response in these T cells.[32]

In addition, at the site of inflammation Hsp will be upregulated in all (stressed) cells of which the majority will be tissue cells lacking costimulatory molecules. In the absence of costimulatory molecules T cells will adopt a state of anergy or regulation.[33] Through these mechanisms Hsp-specific T cells can adopt a regulatory phenotype upon antigen recognition in the periphery of the immune system. And this may explain the tolerant state of these cells and that they can be present at high frequencies without causing damage. It is possible that this tendency of these cells to stay in a tolerant or regulatory state is further promoted by mucosal tolerance in the gut-associated lymphoid system (GALT) through contact with abundantly present microbial Hsp from the gut flora. The tolerance for this collection of Hsp in the gut microbiota will be dominated by tolerance for the conserved sequences, as these are shared among the variety of bacterial species that are present. Thus, there may be a natural focus on the conserved sequences of Hsp to drive a repertoire of regulatory T cells.

In various studies Hsp-reactive T cells have been seen to produce the immunoregulatory cytokine IL-10.[17,34,35] These cells are most likely regulatory T cells, induced in the periphery and reflecting mechanisms known to contribute to development of peripheral tolerance.

Some studies have also suggested that Hsps are antigens seen by natural regulatory T cells (Tregs).[36] Along the same line, studies in children with juvenile idiopathic arthritis (JIA) and adults with rheumatoid arthritis (RA) have indicated the potential of Hsp to trigger in T cells the activity of FoxP3, which is a marker for natural Tregs.

Thus, given the fact that Tregs induced by microbial Hsp have been seen to cross-recognize endogenous Hsp through their specificity for conserved Hsp sequences, it seems that upregulated endogenous Hsp in stressed cells is the target and the possible initiator of the local regulatory activity of these T cells.[37,38]

Regulation by T Cell Reactivity to Endogenous Hsp60

Adjuvant arthritis (AA) can be prevented by vaccinating rats with naked DNA of the mycobacterial Hsp65 molecule or by DNA vaccination with DNA-encoding mammalian Hsp60; the self-Hsp60 DNA vaccine was found to be more effective than the foreign Hsp65 vaccine.[39] Indeed, effective vaccination with either construct induced Tregs that responded to mammalian Hsp60 by the secretion of IFN-γ and high amounts of tumor growth factor (TGF)-β and IL-10. The response of the pathogenic T cells to epitopes of mycobacterial Hsp65 manifested downregulation of IFN-γ and upregulation of TGF-β.[39] Thus, populations of T cells in a single rat can show differential cytokine responses to epitopes of Hsp65 (the target of effector T cells) and to epitopes of Hsp60 (the target of Tregs).

The role of Hsp60 in regulation was also evident when we studied the control of AA by self-Hsp molecules other than Hsp60 using DNA vaccines coding for the human 70-kDa (Hsp70) or 90-kDa (Hsp90) Hsps.[40] Both Hsp70 and Hsp90 vaccines downregulated the arthritogenic T cell response and inhibited AA. In addition, both Hsp70 and Hsp90 vaccines induced T cell responses to Hsp60, and vaccination with Hsp70 triggered the release of endogenous Hsp60 to the circulation. Thus, the regulatory activities of Hsp on inflammation might involve two types of cross-reactivity: molecular cross-reactivity, which exists between microbial and self-Hsps, and network cross-reactivity, which exists between different self-Hsps.[40] Hsp60 reactivity is key to regulation.

Using overlapping Hsp60 peptides, it was possible to identify a regulatory peptide (Hu3) that was specifically recognized by the T cells of rats vaccinated with Hsp60.[41] Vaccination with Hu3, or transfer of splenocytes from Hu3-vaccinated rats, inhibited the development of AA. Vaccination with the mycobacterial homologue peptide of Hu3 was not effective. Effective Hsp60 peptide vaccination was associated with the downregulation of IFN-γ secretion and enhanced secretion of IL-10 and/or TGF-β in response to the epitope of Hsp65 recognized by the effector T cells. The regulatory response to Hsp60 or its Hu3 epitope included both IFN-β and IL-10/TGF-β secretors. These results show that regulatory mechanisms can be activated by immunization with relevant self-Hsp60 peptide epitopes.[41]

Major Histocompatibility Complex (MHC)II Presentation of Self-Hsp and Reduced Hsp Upregulation in Aging

In order to act as a target for Tregs, upregulated endogenous Hsp needs to be presented by MHC class II molecules. A mechanism essential for the presentation of endogenous (cytosolic) Hsp peptides in the context of MHC II molecules seems to be autophagy. The so-called chaperone-mediated autophagy (CMA) is known to be the primary mechanism for Hsp70 epitopes to be loaded onto MHC class II molecules, and is expected to be true of all Hsp. Interestingly also, autophagy is a mechanism that cells optimize at times of stress, such as under nutrient deprivation. Altogether the induction of the cellular stress response and the upregulation of autophagy seem to act in synergy to display Hsp in the context of MHCII for T cell recognition.

It makes sense to suppose that such fundamental features of the expression of cell stress need to be perceived by the immune system and that the healthy immune system translates the information obtained into mechanisms of control, such as the induction of regulation to manage the ongoing and potentially damaging immune reactivity. Overt disease may be the expression of a failure of this system and therefore it remains to be seen how easy this system is to be manipulated in order to restore immune homeostasis during disease. Nonetheless, the data collected so far, in experimental models of disease and in diseased humans, are suggestive of the fact that Hsp can be good targets for therapeutic interventions.

Most chronic inflammatory diseases, such RA and atherosclerosis, are diseases of older age. The reasons that cause the aging immune system to loose control and to produce disease seem manifold. One reason may be that Hsp inducibility gradually decreases under aging. Expression of Hsps in cells depends on the activity of so-called heat shock factors (HSF). The function of these transcription elements in cells is compromised in older individuals. Therefore, a major factor in the biology of aging is the reduced stress resistance of cells due to an inadequate Hsp upregulation. This would imply that a reduced Hsp expression under conditions of cell stress would lead to an impaired immune regulation.

In addition, autophagy turns out to become compromised with rising age. Thus, a combination of factors may be jointly responsible for decreased immunoregulation through recognition of Hsp in the elderly. With this in mind it becomes an attractive and challenging task to further explore innovative strategies to manipulate the expression of cell stress and the associated induction of Hsp with the purpose of optimizing immune recognition of Hsp.

Hsp60 Is a Ligand for Innate Toll-Like Receptors (TLRs)

Although Hsp60 is recognized by the antigen receptors (T cell receptor; TCR) of regulatory and effector T cells and by the BCR of B cells, Hsp60 is also a ligand for innate TLR signaling in various immune cells; Hsp60 is pleiotropic in its interactions with the immune system. Various studies have shown that human Hsp60 can induce proinflammatory responses in mouse or human macrophages and dendritic cells.[42–44] Such interactions can upregulate inflammation. In contrast, however, human Hsp60 also acts as a costimulator of human $CD4^+CD25^+$ Tregs.[45] Treatment of Tregs with Hsp60, or its peptide p277, before anti-CD3 activation significantly enhanced the ability of relatively low concentrations of the Tregs to downregulate $CD4^+CD25^-$ or $CD8^+$ effector T cells, detected as inhibition of target T cell proliferation and IFN-γ and tumor necrosis factor (TNF)-α secretion. The enhancing effects of Hsp60 costimulation on Tregs involved innate signaling via TLR-2. Hsp60-treated Tregs suppressed target T cells both by cell-to-cell contact and by secretion of TGF-β and IL-10. Thus, Hsp60 can downregulate adaptive immune responses by upregulating Tregs innately through TLR2 signaling.[45]

Hsp60 also activates B cells innately.[46] Human Hsp60 (but not the *E. coli* GroEL or the *Mycobacterial* Hsp65 molecules) induced naïve mouse B cells to proliferate and to secrete IL-10 and IL-6. In addition, the Hsp60-treated B cells upregulated their expression of MHCII and accessory molecules CD69, CD40, and B7–2. We tested the functional ability of Hsp60-treated B cells to activate an allogeneic T cell response, and found enhanced secretion of both IL-10 and IFN-γ by the responding T cells. The effects of Hsp60 were found to be largely dependent on TLR4 and MyD88 signaling.

In summary, we now know that Hsp60 is recognized by the immune system by several different cell types and by both innate and adaptive immune receptors: The cells that respond to Hsp60 using adaptive receptors are T effector cells (Th1), T regulatory cells of various types (Th2, Tregs), and B cells; the cells that respond innately to Hsp60 are dendritic cells, macrophages, Tregs, and B cells. Some of these responses upregulate inflammation and some downregulate inflammation.[47] It would appear that the unprecedented focus of the immune system on Hsp60 stems from the fact that Hsp60 has been discovered by the immune system during its evolution to function as a "biomarker" that indicates the state of the body.[48] Not only is Hsp60 a marker, it can be used as a therapeutic agent to manage inflammation.[49]

ATHEROSCLEROSIS AND IMMUNE RESPONSES TO HSP

Hsp60-Specific T Cells

After the group of plain infectious diseases, cardiovascular diseases, notably those due to atherosclerosis, are the second most frequent killer worldwide, constantly increasing their impact even in less developed countries. As a matter of fact, a close correlation of lifelong infectious load with a significantly increased risk to develop atherosclerosis has also been demonstrated.[50] This was one of the reasons for intensified endeavors to identify a possible common denominator between the activity of classical risk factors and inflammatory conditions in the development of atherosclerosis.[51,52] The innate and adaptive immune response against Hsps, above all Hsp60, may represent this common denominator. This hypothesis has received increased attention since the original description of the induction of atherosclerosis by immunization of normocholesterolemic rabbits with complete Freund's adjuvant (CFA) or recombinant mHsp65, respectively.[53] In these experiments, rabbits were first immunized with delipidated atherosclerotic plaque proteins and CFA using ovalbumin (OVA) plus CFA for control purposes. Interestingly, both plaque protein- and OVA-immunized animals developed the early inflammatory stage of atherosclerosis, which is characterized by mononuclear intimal cell infiltration without concomitant foam cell formation at the known atherosclerosis predilection sites. In subsequent experiments, mHsp65 was identified as the culprit disease-inducing antigen and Hsp65-reactive T cells were found to be enriched in early atherosclerotic lesions compared to T cells derived from the peripheral blood of the same animals.[54] When mHsp65 immunized rabbits also received a cholesterol-rich diet, atherosclerotic lesions became more prominent including the appearance of abundant foam cells. In contrast to the lesions in the former, those in the cholesterol-rich group turned out not to be reversible during a prolonged observation period.[55] Importantly, mHsp65 immunized rabbits only developed atherosclerosis but no arthritis. In contrast, as mentioned earlier in this review, rats develop adjuvant arthritis but no atherosclerosis after immunization with mHsp65.[3] These results already pointed to the existence of different arthritogenic and atherogenic Hsp60 epitopes, respectively. In contrast to rat arthritis, where arthritogenic and arthritoprotective Hsp60 epitopes have already been identified,[38] the situation is less advanced for atherosclerosis.

Wild-type mice are notoriously resistant against the induction of atherosclerosis, for example, by feeding a cholesterol-rich diet, but also by immunization with Hsp60. However, immunizing hypercholesterolemic C57 BL/6 mice with mHsp65 significantly aggravates the extent and severity of atherosclerotic lesions.[56] Interestingly, the immunization of ApoE KO mice with biochemically modified low-density lipoprotein (LDL) has a protective effect.[57] Adoptive transfer of the disease can be achieved by spleen and lymphnode-derived T cells from mHsp65-immunized donors.[58]

In humans, T cells reacting with bacterial and/or human Hsp60 can be found in the circulation of all healthy people due to previous vaccinations and infections as well as *bona fide* autoimmunity destined to remove cellular debris and biochemically altered Hsp60. In the BRUNECK-Study, an ongoing prospective atherosclerosis prevention study that started in 1990, the frequency of Hsp60-reactive T cells was not increased in 50–60-year-old individuals with sonographically demonstrable atherosclerotic lesions.[59] However, a significant association of peripheral blood hHsp60-reactive T cells, and less pronounced mHsp65-reactive T cells, with an increased intima-media thickness was demonstrated in the young healthy male (17–18 years) cohort that participated in the so-called Atherosclerosis Risk Factors in Male Youngsters (ARMY)-Study.[60] Similar data recently emerged in the analogous Atherosclerosis Risk Factors in Female Youngster (ARFY)-Study (submitted). From these and the earlier mouse and rabbit data, it was concluded that Hsp60-reactive T cells play an important role in the initial, clinically still inapparent stage of atherogenesis, and that both bacterial–human cross-reactive Hsp60 epitopes as well as specific human Hsp60 epitopes are involved in this reaction. So far, proof for this concept by phenotypic and functional analyses of T cells derived from early human lesions, is still lacking. In this respect, it is noteworthy that immunohistological investigations have shown that the majority of T cells in early lesions carry the α/β TCR but a considerable proportion (10–15%) are γ/δ TCR$^+$ compared to only 1–2% γ/δTCR$^+$ cells in the peripheral blood.[61] γ/δ TCR$^+$ cells are known to react with Hsps in a non-MHC-restricted fashion, and this issue should therefore also be addressed in the future functional studies.[62] Benagiano *et al.*[63] reported that advanced lesions, that is, atherosclerotic plaques, harbor *in vivo*-activated CD4$^+$ T cells that recognize hHsp60. However, these authors were not able to detect hHsp60-specific T cells in the peripheral blood. In patients with positive serology and polymerase chain reaction (PCR) for *Chlamydia pneumoniae* DNA, but not in patients negative for both, most of the plaque-derived T cells specific for hHsp60 also recognized chlamydial Hsp60. The T cells derived from these advanced lesions identified both self- and bacterial–human cross-reactive epitopes. Upon challenge with human hHsp60 the majority of these T cells expressed Th-type I functions. In similar experiments, Rossmann *et al.* (submitted) were able to demonstrate hHsp60-reactive T cells both in the peripheral blood and in advanced atherosclerotic lesions with a significant accumulation in the latter, similar to the situation described in the rabbit model. In addition, these authors also provided evidence for the presence of oligo- and monoclonal T cell populations in the lesions compared to the peripheral blood of the same patients, as determined by PCR spectratyping pointing to an antigen-driven process. It will be interesting to assess the potential role of Hsp60-specific Tregs in addition to the non-antigen-specific Tregs during atherogenesis.

Furthermore, Ramage et al.[64] demonstrated that human T-lymphocytes proliferate *in vitro* upon exposure to highly purified Hsp60, a response that was confined to the CD54 RA$^+$ RO$^-$ subset with only minimal responses by adult CD45 RA$^-$ RO$^+$ T cells, that is, cells that are designated as naïve and memory cells, respectively. In contrast, both CD45 RA$^+$ RO$^-$ and CD45 RA$^-$ RO$^+$ T cells proliferated to bacterial Hsp60 from various sources and only CD45 RA$^-$ RO$^+$ (memory) T cells responded to a mHsp65-derived peptide previously defined as a major bacteria-specific epitope. These authors speculated that *in vivo* challenge with bacterial Hsp60 will activate T cells capable of seeing conserved and nonconserved epitopes, but only that seeing nonconserved epitopes finally maintain the CD45 RA$^-$ RO$^+$ memory phenotype.

An interesting aspect of this line of research is the possible role of bacterial–human Hsp60 cross-reactivity as a common denominator for the well-established association of periodontal disease and atherosclerosis.[65] As a matter of fact, Yamazaki et al.[44] have shown an accumulation of hHsp60-reactive T cells in the gingival tissues of periodontitis patients and Schett et al.[66] found a significant increase of salivary IgA-antibodies against defined Hsp60 epitopes in patients with gingivitis.

Hsp60-Specific Antibodies

Information on the presence and specificity of humoral antibodies to Hsp60 in association with atherosclerosis has proceeded much further than that dealing with T cells. As a matter of fact, the role of bacterial–human cross-reactive antibodies to Hsp60 as a new, independent biomarker has first been shown in the course of the BRUNECK-Study, where the notion emerged that everybody has such antibodies based on previous vaccinations and infections, as expected. However, the antibody titers were significantly increased in the individuals with sonographically demonstrable atherosclerotic lesions.[67] In the meantime, this finding has been confirmed by many laboratories[68–72] including the role of this new biomarker not only for diagnostic but also prognostic purposes.[73] Due to the fact that Hsp60, an originally mitochondrial protein, is also transported to the cytosol and the cell surface[74,67] where it acts as a danger signal, the cytotoxic potential of these antibodies deserves special mentioning. Endothelial cells subjected to various forms of stress can be lysed by anti-Hsp60 antibodies in a complement-dependent fashion or via antibody-dependent cellular cytotoxicity (ADCC).[75] As a matter of fact, it has recently been shown that the transfer of either purified human anti-Hsp60 antibodies derived from atherosclerotic patients or murine monoclonal antibodies specific for eukaryontic Hsp60 leads to vascular damage upon passive transfer into ApoE KO mice.[76] Atherogenic Hsp60 B cell epitopes recognized by human anti-Hsp60 antibodies have already been identified.[77,78] Based on the results of immunohistological studies, the analyses of T cell reactivity and the cytotoxic potential of humoral antibodies, it can thus be hypothesized that atherosclerosis is initiated by Hsp60-reactive T cells and aggravated by the concomitant presence of cytotoxic antibodies.

Interestingly, the titer of Hsp60 antibodies decreases in the peripheral blood of patients that have undergone a myocardial infarction.[79] Judging from analogous animal experiments, this drop in titer is due to immune complex formation between preexisting antibodies and Hsp60 derived from necrotic myocardial tissue.[80]

Hsp60 and Innate Immunity

In addition to adaptive immunity, innate immune processes certainly also play a major role in the development of atherosclerosis. Circulating Hsp60 (soluble Hsp60–sHsp60) can be found in the peripheral blood of all human beings, again with a significant increase in individuals with sonographically demonstrable atherosclerotic lesions.[81] Due to biochemical alterations, sHsp60 can also be considered to be the target for *bona fide* adaptive autoimmunity. However, it has also been shown that Hsp60 can bind to aggregates of TLR-4 and CD14 as well as TLR-2/TLR-6,[82,83] respectively, entailing the transcription of genes coding for proinflammatory cytokines, such as TNF-α and IFN-γ. So far, it has not yet been studied if various stress factors induce the expression of different TLRs, for example, at the known predilections sites for the development of atherosclerotic lesions.

Autoimmune Hypothesis of Atherogenesis

Based on results discussed above, the autoimmune hypothesis of atherogenesis was developed.[84] In essence, this hypothesis postulates that classical atherosclerosis risk–factors, such as high blood pressure, high blood cholesterol levels, and biochemically altered LDL (e.g., oxidized LDL–oxLDL), diabetes, cigarette smoking, etc., the well-established proatherogenic role of which is not disputed, first act as endothelial stressors in the earliest stages of the disease. Such Hsp60-expressing endothelial cells become the target of preexisting anti-bacterial–human cross-reactive adaptive and innate immunity and/or *bona fide* anti-hHsp60 autoimmunity. This leads to the incipient inflammatory stage characterized by infiltration of the intima with T cells, followed by blood-derived monocytes/macrophages and media-derived smooth muscle cells. When the presence of these risk factors persists, these early lesions progress to more severe, complicated alterations and finally atherosclerotic plaques that present with a conundrum of immunologic–inflammatory hallmarks. It has been shown for all of the risk factors investigated so far that they lead to a concomitant expression of Hsp60 and adhesion molecules, thus providing the prerequisites for an interaction of Hsp60-reactive T cells with stressed endothelial cells.

Hsp60-Based Therapeutic Approaches

Maron et al.[85] showed in atherosclerosis-prone LDL-R KO mice, which were fed a cholesterol-rich diet, that nasal administration of mHsp65 caused a decrease of atherosclerotic plaques in the aortic arch. The reduction of the size of the plaques was accompanied by reduction of macrophage-positive areas and increased expression of IL-10 in the plaques. In another experimental atherosclerosis model, Harats et al.[86] studied the effect of oral tolerance induced with mHsp65. In this case, early atherosclerosis was attenuated and the effects seem to be mediated by IL-4. Very similar results were obtained by van Puyvelde et al.[87] in LDL-R KO mice on a high-cholesterol diet, where a cuff is placed around the carotid artery. In this model, low-dose oral mHsp65 caused a dramatic drop in the size of atherosclerotic plaques. A very similar effect was seen after the oral administration of mHsp65-derived peptide that comprised a previously mapped T cell epitope, indicating that the effect on plaque size was T cell mediated. The dream of developing a vaccine against atherosclerosis thus seems to become a reality.

Other Hsps and Atherosclerosis

From over 20 known families of Hsps, only very few in addition to Hsp60 have been studied with respect to their possible role in the development of atherosclerosis. These included Hsp27 and Hsp70.

The possible role of Hsp27 in atherogenesis has only been addressed recently. Park et al.[88] studied the expression of Hsp27 in human atherosclerotic plaques as well as the increased plasma levels of sHsp27 in patients with acute coronary syndrome. The expression of Hsp27 decreased from still normal appearing arterial areas toward the core of the atherosclerotic plaque as did the degree of Hsp27 phosphorylation that is necessary for the fulfillment of its chaperone function.

As compared to the abundant body of data relating to the role of Hsp60 in the pathogenesis of atherosclerosis, much less work has been done with respect to Hsp70.

Since Hsp70 is a very sensitive indicator of cellular stress, it is not surprising that Hsp70 family members have been found to be expressed in rabbit and human atherosclerotic lesions.[89–91] Surgical stress leads to an upregulation of inducible Hsp70 and Hsp27 in rats on both RNA- and protein levels. The greatest abundance of Hsp70 expression was observed in the adrenal system, followed by the vascular system, notably the aorta.[92] Also, serum levels of sHsp70 have been determined and high concentrations were found to be associated with low-coronary artery disease risk—an effect that was contributed to its cell protection effects.[93] Both increased and decreased Hsp70 antibody titers have been described to correlate with atherosclerosis.[94,95]

Data on the effect of the immunization of experimental animals with Hsp70 are very scarce. However, in at least one instance an accelerated intimal thickening in balloon-injured carotid arteries of rats after immunization against Hsp70 was found.[96]

FIRST CLINICAL EXPERIENCE WITH HSP-BASED IMMUNOTHERAPY

Immune responses to Hsp are ubiquitously found in inflammatory conditions. Teleologically, immune recognition of Hsp evokes a potent and ancestral proinflammatory response that has as objective the clearance of a perceived microbial infection. This mechanism has evolved into a sophisticated tool to modulate inflammation independently of its original trigger and it is often found in many conditions in which amplification of inflammation may be physiologically useful. Indeed, inflammation in itself is a stressful stimulus, which leads to local overexpression of Hsp, leading to a potential amplification and perpetuation of the inflammatory process, even if the original trigger has become irrelevant to the process. It is therefore not surprising that immune responses to Hsp are found ubiquitously in both physiologic and pathologic inflammatory conditions, including infection, autoimmunity, and atherosclerosis. As outlined in the previous section of this review, regulatory mechanisms have evolved to modulate intensity and duration of this trigger-independent inflammatory loop, in order to prevent damage. These regulatory mechanisms are often impaired in autoimmunity and their restoration may be one of the objectives of a trigger-independent approach to antigen-specific immunotherapy. The task is not, however, an easy one. Hsp are large proteins which, when processed, give raise to a multitude of potential epitopes of which only a few are dominant. In addition, as we have recently shown, different epitopes from the same Hsp in the same disease may have very different functional effects on the immune response, some being pro-inflammatory, some others tolerogenic.[97] Hence, the first hurdle is to identify epitopes, rather than proteins, which are recognized by a majority of patients, and to characterize appropriately the quality of such immune recognition in order to identify the most "proinflammatory" epitope if induction of tolerance is the desired goal. In our experience, we have predesigned *in silico* the peptides based on the computerized prediction of agretopic motifs that may enable the putative antigens to be a strong pan-HLA class II binder.[98] Theoretical binding scores are then matched with measurement of actual binding to isolated HLA molecules *in vitro*. This approach generates a pool of candidate peptides whose immunogenicity is then screened *in vitro* against biological samples from patients. The quality of immune response toward the individual peptides, and in particular the proinflammatory or tolerogenic connotation of such response is the parameter that identifies the lead target for further development. This strategy has proven to be very effective in rapidly identifying epitopes in various disease settings, such as RA, MS, IBD, and JIA. In our most advanced program, we have identified a 15mer derived from the *E. coli* Hsp dnaJ (dnaJP1) as a strong proinflammatory epitope in patients with active RA.[99] With a primary objective to restore mechanisms of impaired regulation of the Hsp proinflammatory system, we embarked on a clinical program that has recently completed pilot phase II trial. The peptide was given orally once a day for 6 months. Results from the first two phases could be summarized as follows: (i) the approach is certainly safe and well tolerated; (ii) an immune deviation from proinflammatory to tolerogenic T cell responses was induced in treated patients. This immune deviation, in particular the decline in TNF-α production and a corresponding increase in IL-10, may have a value as biomarker predictor or surrogate of clinical efficacy; (iii) an improvement of signs and symptoms of RA was observed.[100]

An analysis of the mechanisms of molecular immunology, which underlie the immunological and clinical effects of the treatment, is still under way at the time of this writing. Available data point to an overlapping of several different mechanisms in determining both clinical and immunological effects. These mechanisms include immune deviation of dnajp1-specific T cells as well as restoration of Treg function, as evidenced by the significant increase in IL-10 and Foxp3 expression by dnaJP1, CD4-CD25^{++} T cells.

As this aspect will heavily influence the next steps of development, it may be worth discussing briefly about the potential positioning for this therapeutic approach. The main advantages of trigger-independent, epitope-specific Hsp immunotherapy are its safety profile, route of administration, and its versatility as a "work with" drug in combination with currently used disease modifying antirheumatic drugs (DMARDs) and biologics. This concept has been clearly evidenced by both our animal studies and, still preliminary, by the clinical data collected to date. Hence, beyond the intuitive use as first-line agents together with DMARDs, Hsp immunotherapy can be used as a complement to biologics in regimens in which dosing or scheduling of biologic is adjusted once clinical control has been achieved.[101]

REFERENCES

1. MORIMOTO, R.I. 1998. Regulation of the heat shock transcriptional response: cross talk between a family of heat shock factors, molecular chaperones, and negative regulators. Genes Dev. **12:** 3788–3796.

2. NICCHITTA, C.V. 2003. Re-evaluating the role of heat-shock protein-peptide interactions in tumour immunity. Nat. Rev. Immunol. **3:** 427–432.
3. VAN EDEN, W. *et al.* 1988. Cloning of the mycobacterial epitope recognized by T lymphocytes in adjuvant arthritis. Nature **331:** 171–173.
4. VAN DEN BROEK, M.F. *et al.* 1989. Protection against streptococcal cell wall-induced arthritis by pretreatment with the 65-kD mycobacterial heat shock protein. J. Exp. Med. **170:** 449–466.
5. BILLINGHAM, M.E. *et al.* 1990. A mycobacterial 65-kD heat shock protein induces antigen-specific suppression of adjuvant arthritis, but is not itself arthritogenic. J. Exp. Med. **171:** 339–344.
6. THOMPSON, S.J. *et al.* 1998. An immunodominant epitope from mycobacterial 65-kDa heat shock protein protects against pristane-induced arthritis. J. Immunol. **160:** 4628–4634.
7. KINGSTON, A.E. *et al.* 1996. A 71-kD heat shock protein (hsp) from Mycobacterium tuberculosis has modulatory effects on experimental rat arthritis. Clin. Exp. Immunol. **103:** 77–82.
8. RAGNO, S. *et al.* 1997. Protection of rats from adjuvant arthritis by immunization with naked DNA encoding for mycobacterial heat shock protein 65. Arthritis Rheum. **40:** 277–283.
9. ANDERTON, S.M. *et al.* 1995. Activation of T cells recognizing self 60-kD heat shock protein can protect against experimental arthritis. J. Exp. Med. **181:** 943–952.
10. MOUDGIL, K.D. *et al.* 1997. Diversification of T cell responses to carboxy-terminal determinants within the 65-kD heat-shock protein is involved in regulation of autoimmune arthritis. J. Exp. Med. **185:** 1307–1316.
11. ULMANSKY, R. *et al.* 2002. Resistance to adjuvant arthritis is due to protective antibodies against heat shock protein surface epitopes and the induction of IL-10 secretion. J. Immunol. **168:** 6463–6469.
12. FEIGE, U. & W. VAN EDEN. 1996. Infection, autoimmunity and autoimmune disease. Experientia **77:** 359–373.
13. PRAKKEN, B.J. *et al.* 1997. Peptide-induced nasal tolerance for a mycobacterial heat shock protein 60 T cell epitope in rats suppresses both adjuvant arthritis and nonmicrobially induced experimental arthritis. Proc. Natl. Acad. Sci. USA **94:** 3284–3289.
14. HAQUE, M.A. *et al.* 1996. Suppression of adjuvant arthritis in rats by induction of oral tolerance to mycobacterial 65-kDa heat shock protein. Eur. J. Immunol. **26:** 2650–2656.
15. COBELENS, P.M. *et al.* 2000. Treatment of adjuvant arthritis by oral administration of mycobacterial HSP65 during disease. Arthritis Rheum. **43:** 2964–2702.
16. RAGNO, S. *et al.* 1996. A synthetic 10-kD heat shock protein (hsp10) from Mycobacterium tuberculosis modulates adjuvant arthritis. Clin. Exp. Immunol. **103:** 384–390.
17. PRAKKEN, B.J. *et al.* 2001. Induction of IL-10 and inhibition of experimental arthritis are specific features of microbial heat shock proteins that are absent for other evolutionarily conserved immunodominant proteins. J. Immunol. **167:** 4147–4153.
18. BERLO, S.E. *et al.* 2005. Naive transgenic T cells expressing cartilage proteoglycan-specific TCR induce arthritis upon *in vivo* activation. J. Autoimmun. **25:** 172–180.

19. BLOEMENDAL, A. *et al.* 1997. Experimental immunization with anti-rheumatic bacterial extract OM-89 induces T cell responses to heat shock protein (hsp)60 and hsp70; modulation of peripheral immunological tolerance as its possible mode of action in the treatment of rheumatoid arthritis (RA). Clin. Exp. Immunol. **110:** 72–78.
20. WENDLING, U. *et al.* 2000. A conserved mycobacterial heat shock protein (hsp) 70 sequence prevents adjuvant arthritis upon nasal administration and induces IL-10-producing T cells that cross-react with the mammalian self-hsp70 homologue. [In Process Citation]. J. Immunol. **164:** 2711–2717.
21. ELIAS, D. *et al.* 1991. Vaccination against autoimmune mouse diabetes with a T-cell epitope of the human 65-kDa heat shock protein. Proc. Natl. Acad. Sci. USA **88:** 3088–3091.
22. BOCKOVA, J. *et al.* 1997. Treatment of NOD diabetes with a novel peptide of the hsp60 molecule induces Th2-type antibodies. J. Autoimmun. **10:** 323–329.
23. BRUDZYNSKI, K. *et al.* 1992. Immunocytochemical localization of heat-shock protein 60-related protein in beta-cell secretory granules and its altered distribution in non-obese diabetic mice. Diabetologia **35:** 316–324.
24. VAN HALTEREN, A.G. *et al.* 2000. T cell reactivity to heat shock protein 60 in diabetes-susceptible and genetically protected nonobese diabetic mice is associated with a protective cytokine profile. J. Immunol. **165:** 5544–5551.
25. CHANDAWARKAR, R.Y. *et al.* 2004. Immune modulation with high-dose heat-shock protein gp96: therapy of murine autoimmune diabetes and encephalomyelitis. Int. Immunol. **16:** 615–624.
26. WACHLIN, G. *et al.* 2002. Stress response of pancreatic islets from diabetes prone BB rats of different age. Autoimmunity **35:** 389–395.
27. BEN-NUN, A. *et al.* 1995. A 12-kDa protein of Mycobacterium tuberculosis protects mice against experimental autoimmune encephalomyelitis. Protection in the absence of shared T cell epitopes with encephalitogenic proteins. J. Immunol. **154:** 2939–2948.
28. BIRNBAUM, G. *et al.* 1998. Heat shock proteins and experimental autoimmune encephalomyelitis. II: environmental infection and extra-neuraxial inflammation alter the course of chronic relapsing encephalomyelitis. J. Neuroimmunol. **90:** 149–161.
29. RHA, Y.H. *et al.* 2002. Effect of microbial heat shock proteins on airway inflammation and hyperresponsiveness. J. Immunol. **169:** 5300–5307.
30. FISCHER, H.P. *et al.* 1992. High frequency of cord blood lymphocytes against mycobacterial 65-kDa heat-shock protein. Eur. J. Immunol. **22:** 1667–1669.
31. BIRK, O.S. *et al.* 1996. A role of Hsp60 in autoimmune diabetes: analysis in a transgenic model. Proc. Natl. Acad. Sci. USA **93:** 1032–1037.
32. PAUL, A.G. *et al.* 2000. Highly autoproliferative T cells specific for 60-kDa heat shock protein produce IL-4/IL-10 and IFN-gamma and are protective in adjuvant arthritis. J. Immunol. **165:** 7270–7277.
33. TAAMS, L.S. *et al.* 1998. Anergic T cells actively suppress T cell responses via the antigen-presenting cell. Eur. J. Immunol. **28:** 2902–2912.
34. TANAKA, S. *et al.* 1999. Activation of T cells recognizing an epitope of heat-shock protein 70 can protect against rat adjuvant arthritis. J. Immunol. **163:** 5560–5565.
35. KAMPHUIS, S. *et al.* 2005. Tolerogenic immune responses to novel T-cell epitopes from heat-shock protein 60 in juvenile idiopathic arthritis. Lancet **366:** 50–56.

36. NISHIKAWA, H. *et al*. 2005. Definition of target antigens for naturally occurring CD4(+) CD25(+) regulatory T cells. J. Exp. Med. **201:** 681–686.
37. WIETEN, L., F. HAUET-BROERE, R. VAN DER ZEE *et al*. Cell stress induced HSP are targets of regulatory T cells: a role for HSP inducing compounds as anti-inflammatory immuno-modulators? FEBS Lett. In press.
38. VAN EDEN, W. *et al*. 2005. Heat-shock proteins induce T-cell regulation of chronic inflammation. Nat. Rev. Immunol. **5:** 318–330.
39. QUINTANA, F.J. *et al*. 2002. Inhibition of adjuvant arthritis by a DNA vaccine encoding human heat shock protein 60. J. Immunol. **169:** 3422–3428.
40. QUINTANA, F.J. *et al*. 2004. Inhibition of adjuvant-induced arthritis by DNA vaccination with the 70-kd or the 90-kd human heat-shock protein: immune cross-regulation with the 60-kd heat-shock protein. Arthritis Rheum. **50:** 3712–3720.
41. COHEN, I.R. *et al*. 2003. HSP60 and the regulation of inflammation: physiological and pathological. *In* Heat Shock Proteins and Inflammation. W.V. Eden, Ed.: 1–13, Birkhauser Verlag, Basle.
42. FLOHE, S.B. *et al*. 2003. Human heat shock protein 60 induces maturation of dendritic cells versus a Th1-promoting phenotype. J. Immunol. **170:** 2340–2348.
43. CHEN, W. *et al*. 1999. Human 60-kDa heat-shock protein: a danger signal to the innate immune system. J. Immunol. **162:** 3212–3219.
44. YAMAZAKI, K. *et al*. 2002. Accumulation of human heat shock protein 60-reactive T cells in the gingival tissues of periodontitis patients. Infect. Immun. **70:** 2492–2501.
45. ZANIN-ZHOROV, A. *et al*. 2006. Heat shock protein 60 enhances CD4+ CD25+ regulatory T cell function via innate TLR2 signaling. J. Clin. Invest. **116:** 2022–2032.
46. COHEN-SFADY, M. *et al*. 2005. Heat shock protein 60 activates B cells via the TLR4-MyD88 pathway. J. Immunol. **175:** 3594–3602.
47. QUINTANA, F.J. & I.R. COHEN. 2005. Heat shock proteins as endogenous adjuvants in sterile and septic inflammation. J. Immunol. **175:** 2777–2782I.R.
48. COHEN, I.R. 2007. Real and artificial immune systems: computing the state of the body. Nat. Rev. Immunol. **7:** 569–574.
49. RAZ, I. *et al*. 2001. Beta-cell function in new-onset type 1 diabetes and immunomodulation with a heat-shock protein peptide (DiaPep277): a randomised, double-blind, phase II trial. Lancet **358:** 1749–1753.
50. EPSTEIN, S.E. *et al*. 2000. Infection and atherosclerosis: potential roles of pathogen burden and molecular mimicry. Arterioscler. Thromb. Vasc. Biol. **20:** 1417–1420.
51. LIBBY, P. *et al*. 2002. Inflammation and atherosclerosis. Circulation **105:** 1135–1143.
52. HANSSON, G.K. 2005. Inflammation, atherosclerosis, and coronary artery disease. N. Engl. J. Med. **352:** 1685–1695.
53. XU, Q. *et al*. 1992. Induction of arteriosclerosis in normocholesterolemic rabbits by immunization with heat shock protein 65. Arterioscler. Thromb. **12:** 789–799.
54. XU, Q. *et al*. 1993. Increased expression of heat shock protein 65 coincides with a population of infiltrating T lymphocytes in atherosclerotic lesions of rabbits specifically responding to heat shock protein 65. J. Clin. Invest. **91:** 2693–2702.
55. XU, Q. *et al*. 1996. Regression of arteriosclerotic lesions induced by immunization with heat shock protein 65-containing material in normocholesterolemic, but not hypercholesterolemic, rabbits. Atherosclerosis **123:** 145–155.

56. GEORGE, J. et al. 1999. Enhanced fatty streak formation in C57BL/6J mice by immunization with heat shock protein-65. Arterioscler. Thromb. Vasc. Biol. **19:** 505–510.
57. GEORGE, J. et al. 1998. Hyperimmunization of apo-E-deficient mice with homologous malondialdehyde low-density lipoprotein suppresses early atherogenesis. Atherosclerosis **138:** 147–152.
58. GEORGE, J. et al. 2001. Cellular and humoral immune responses to heat shock protein 65 are both involved in promoting fatty-streak formation in LDL-receptor deficient mice. J. Am. Coll. Cardiol. **38:** 900–905.
59. KNOFLACH, M. et al. 2007. T-cell reactivity against HSP60 relates to early but not advanced atherosclerosis. Atherosclerosis pubmed ahead of print.
60. KNOFLACH, M. et al. 2003. Cardiovascular risk factors and atherosclerosis in young males: ARMY study (Atherosclerosis Risk-Factors in Male Youngsters). Circulation **108:** 1064–1069.
61. KLEINDIENST, R. et al. 1993. Immunology of atherosclerosis. Demonstration of heat shock protein 60 expression and T lymphocytes bearing alpha/beta or gamma/delta receptor in human atherosclerotic lesions. Am. J. Pathol. **142:** 1927–1937.
62. YOUNG, D.B. 1992. Heat-shock proteins: immunity and autoimmunity. Curr. Opin. Immunol. **4:** 396–400.
63. BENAGIANO, M. et al. 2005. Human 60-kDa heat shock protein is a target autoantigen of T cells derived from atherosclerotic plaques. J. Immunol. **174:** 6509–6517.
64. RAMAGE, J. et al. 1999. T cell responses to heat shock protein 60: Differential responses by CD4+ T cell subsets according to their expression of CD45 isotypes. J. Immunol. **162:** 704–710.
65. FORD, P. et al. 2005. Characterization of heat shock protein-specific T cells in atherosclerosis. Clin. Diagn. Lab. Immunol. **12:** 259–267.
66. SCHETT, G. et al. 1997. Salivary anti-hsp65 antibodies as a diagnostic marker for gingivitis and a possible link to atherosclerosis. Int. Arch. Allergy Immunol. **114:** 246–250.
67. XU, Q. et al. 1993. Association of serum antibodies to heat-shock protein 65 with carotid atherosclerosis. Lancet **341:** 255–259.
68. PROHASZKA, Z. et al. 1999. Antibodies against human heat-shock protein (hsp) 60 and mycobacterial hsp65 differ in their antigen specificity and complement-activating ability. Int. Immunol. **11:** 1363–1370.
69. BURIAN, K. et al. 2001. Independent and joint effects of antibodies to human heat-shock protein 60 and Chlamydia pneumoniae infection in the development of coronary atherosclerosis. Circulation **103:** 1503–1508.
70. HUITTINEN, T. et al. 2002. Autoimmunity to human heat shock protein 60, *Chlamydia pneumoniae* infection, and inflammation in predicting coronary risk. Arterioscler. Thromb. Vasc. Biol. **22:** 431–437.
71. POCKLEY, A.G. & J. FROSTEGARD. 2005. Heat shock proteins in cardiovascular disease and the prognostic value of heat shock protein related measurements. Heart **91:** 1124–1126.
72. BIRNIE, D.H. et al. 2005. Increased titres of anti-human heat shock protein 60 predict an adverse one year prognosis in patients with acute cardiac chest pain. Heart **91:** 1148–1153.
73. XU, Q. et al. 1999. Association of serum antibodies to heat-shock protein 65 with carotid atherosclerosis : clinical significance determined in a follow-up study. Circulation **100:** 1169–1174.

74. SOLTYS, B.J. & R.S. GUPTA. 1997. Cell surface localization of the 60 kDa heat shock chaperonin protein (hsp60) in mammalian cells. Cell Biol. Int. **21:** 315–320.
75. SCHETT, G. et al. 1995. Autoantibodies against heat shock protein 60 mediate endothelial cytotoxicity. J. Clin. Invest. **96:** 2569–2577.
76. FOTEINOS, G. et al. 2005. Anti-heat shock protein 60 autoantibodies induce atherosclerosis in apolipoprotein E-deficient mice via endothelial damage. Circulation **112:** 1206–1213.
77. METZLER, B. et al. 1997. Epitope specificity of anti-heat shock protein 65/60 serum antibodies in atherosclerosis. Arterioscler. Thromb. Vasc. Biol. **17:** 536–541.
78. PERSCHINKA, H. et al. 2006. Identification of atherosclerosis-associated conformational heat shock protein 60 epitopes by phage display and structural alignment. Atherosclerosis
79. HOPPICHLER, F. et al. 1996. Changes of serum antibodies to heat-shock protein 65 in coronary heart disease and acute myocardial infarction. Atherosclerosis **126:** 333–338.
80. SCHETT, G. et al. 1999. Myocardial injury leads to a release of heat shock protein (hsp) 60 and a suppression of the anti-hsp65 immune response. Cardiovasc. Res. **42:** 685–695.
81. XU, Q. et al. 2000. Serum soluble heat shock protein 60 is elevated in subjects with atherosclerosis in a general population. Circulation **102:** 14–20.
82. BULUT, Y. et al. 2002. Chlamydial heat shock protein 60 activates macrophages and endothelial cells through Toll-like receptor 4 and MD2 in a MyD88-dependent pathway. J. Immunol. **168:** 1435–1440.
83. HABICH, C. et al. 2002. The receptor for heat shock protein 60 on macrophages is saturable, specific, and distinct from receptors for other heat shock proteins. J. Immunol. **168:** 569–576.
84. WICK, G. et al. 2004. Heat shock proteins and stress in atherosclerosis. Autoimmun. Rev. 3(Suppl 1): S30–S31.
85. MARON, R. et al. 2002. Mucosal administration of heat shock protein-65 decreases atherosclerosis and inflammation in aortic arch of low-density lipoprotein receptor-deficient mice. Circulation **106:** 1708–1715.
86. HARATS, D. et al. 2002. Oral tolerance with heat shock protein 65 attenuates Mycobacterium tuberculosis-induced and high-fat-diet-driven atherosclerotic lesions. J. Am. Coll. Cardiol. **40:** 1333–1338.
87. VAN PUYVELDE, G.H. 2007. Regulation of T Cell Responses in Atherosclerosis. Thesis, University of Leiden, Leiden, Germany.
88. PARK, J.E. 2006. Expression of heat shock protein 27 in human atherosclerotic plaques and increased plasma level of heat shock protein 27 in patients with acute coronary syndrome. Circulation **114:** 886–893.
89. XU, Q. et al. 2000. Mechanical stress-induced heat shock protein 70 expression in vascular smooth muscle cells is regulated by Rac and Ras small G proteins but not mitogen-activated protein kinases. Circ. Res. **86:** 1122–1128.
90. GEETANJALI, B. et al. 2002. Changes in heat shock protein 70 localization and its content in rabbit aorta at various stages of experimental atherosclerosis. Cardiovasc. Pathol. **11:** 97–103.
91. ZHU, W.M. et al. 1995. Oxidized LDL induce hsp70 expression in human smooth muscle cells. FEBS Lett. **372:** 1–5.
92. UDELSMAN, R. et al. 1993. Vascular heat shock protein expression in response to stress. Endocrine and autonomic regulation of this age-dependent response. J. Clin. Invest. **91:** 465–473.

93. Zhu, J. *et al*. 2003. Increased serum levels of heat shock protein 70 are associated with low risk of coronary artery disease. Arterioscler. Thromb. Vasc. Biol. **23:** 1055–1059.
94. Chan, Y.C. *et al*. 1999. Anti-heat-shock protein 70 kDa antibodies in vascular patients. Eur. J. Vasc. Endovasc. Surg. **18:** 381–385.
95. Kocsis, J. *et al*. 2002. Antibodies against the human heat shock protein hsp70 in patients with severe coronary artery disease. Immunol. Invest. **31:** 219–231.
96. George, J. *et al*. 2001. Accelerated intimal thickening in carotid arteries of balloon-injured rats after immunization against heat shock protein 70. J. Am. Coll. Cardiol. **38:** 1564–1569.
97. Massa, M., M. Passalia, S. Mansoni, *et al*. 2007. Differential recognition of heat-shock protein dnaj-derived epitopes by effector and treg cells leads to modulation of inflammation in juvenile idiopathic arthritis. Arthritis Rheum. **56:**1648–1657.
98. Tremoulet, A.H. & S. Albani. 2006. Novel therapies for rheumatoid arthritis. Exp. Opin. Investig. Drugs **15:** 1427–1441.
99. Prakken, B.J. *et al*. 2004. Epitope-specific immunotherapy induces immune deviation of proinflammatory T cells in rheumatoid arthritis. Proc. Natl. Acad. Sci. USA **101:** 4228–4233.
100. van de Ven, A. *et al*. 2007. Immunological tolerance in the therapy of rheumatoid arthritis. Discov. Med. **7:** 46–50.
101. Roord, S.T. *et al*. 2006. Modulation of T cell function by combination of epitope specific and low dose anticytokine therapy controls autoimmune arthritis. PLoS ONE 1, 87.

New Molecular Mechanisms of Duodenal Ulceration

SANDOR SZABO, XIAOMING DENG, TETYANA KHOMENKO,
LONGCHUAN CHEN, GANNA TOLSTANOVA, KLARA OSAPAY,
ZSUZSANNA SANDOR, AND XIMING XIONG

VA Medical Center and University of California-Irvine, School of Medicine, Long Beach, California, USA

ABSTRACT: Stress is a major etiologic factor in the pathogenesis of gastric and duodenal ulceration, as first described in rats by Hans Selye. In patients with "peptic ulcers" duodenal ulcers are more frequent than gastric ulcers (except in Japan). Thus, our research during the last three decades focused on the molecular mechanisms of duodenal ulcer in rodent models of chemically induced duodenal ulceration, and here we review our three recent findings: Endothelins (ET-1), the immediate early gene egr-1 and imbalance of angiogenic/antiangiogenic molecules. Namely, we found an enhanced expression and release of ET-1 within 15–30 min after the administration of duodenal ulcerogen cysteamine, resulting in local ischemia that triggers the expression of hypoxia-inducible factors (HIF-1α). Our gene expression studies also revealed an early (0.5–2 h) increase in the expression of egr-1 that is followed (12–24 h) by upregulation of angiogenic growth factors (e.g., VEGF, bFGF, PDGF). Surprisingly, this event is also associated with an enhanced production of angiostatin and endostatin that probably counteract the beneficial effect of angiogenic molecules. Thus, the initial injury to endothelial and epithelial cells in duodenal ulceration seems to be aggravated (and not initiated) by HCl and proteolytic enzymes. The resulting mucosal necrosis does not rapidly heal because of the imbalance of VEGF and angiostatin/endostatin, hence duodenal ulcers develop. The experimental ulcers Selye described morphologically are now characterized at the molecular and genome level, involving unexpected mediators like ET-1, egr-1 and angiogenesis-related molecules.

KEYWORDS: endothelin-1; early growth response 1; angiogenesis; VEGF; endostatin/angiostatin; duodenal ulceration

INTRODUCTION

Ulcer disease, that is, gastric and duodenal ulcers, is more prevalent than the idiopathic ulcerative colitis and Crohn's disease, especially since some of the

Author for correspondence: Sandor Szabo, M.D., Ph.D., M.P.H., D.Sc. (h.c.), VA Medical Center (05/113), 5901 East 7th Street, Long Beach, CA, USA. Voice: 562-826-5403; fax: 562-826-5623. sandor.szabo@med.va.gov gastroduodenal ulcers are related to stress, environmental exposure, excessive alcohol, and drug use. In most countries of the world, duodenal ulcer is the most prevalent form of "peptic ulcers."[1] At one time during their lives, 8–10% of the U.S. population suffers from duodenal ulcer disease and another 1% from gastric ulcer.[1,2] Despite recent advances and better understanding of the etiology and the pathogenesis of gastric ulceration, including the role of *Helicobacter pylori* (*H. pylori*), the molecular events leading to ulcer development and healing remain poorly understood.[3,4] For example, despite the progress regarding the roles of *H. pylori* and other endogenous and exogenous pathogenic factors in gastritis and gastric ulcers, it remains essentially unknown how duodenal ulceration is triggered, and what are the cellular and molecular target sites.

The initial investigations in our laboratory were related to the development of new animal models of gastrointestinal ulcers. Namely, the preulcerogenic biochemical and molecular mechanisms of ulceration and healing can be best studied in animal models. Selye[5] was the first to recognize that the morphologic triad of stress response includes gastric erosions/ulcers, in addition to adrenal enlargement and thymolymphatic atrophy.

Gastric erosions and ulcers were always relatively easy to induce in small rodents (e.g., rats, mice) by severe stress or large doses of nonsteroidal anti-inflammatory drugs (NSAID), such as aspirin, indomethacin.[5–7] In contrast, until the early 1970s, there were no easily reproducible models of the most frequent form of "peptic ulcers," that is, duodenal ulceration but we found that a few chemicals (e.g., propionitrile, cysteamine) selectively induce perforating duodenal ulcers in the first 5–7 mm of proximal duodenum in 24–48 h in rats.[8,9] Furthermore, we established a structure–activity correlation between duodenal ulcerogens[9,10] and also identified nonulcerogenic structural analogs, such as ethanolamine, which often served as an excellent control in numerous studies for cysteamine.

These animal models of acute and chronic duodenal ulcers were similar to human duodenal ulcer by several functional and morphologic criteria.[11,12] Therefore, most of the pathogenesis research has been performed with the duodenal ulcerogens cysteamine or propionitrile in rats.

In our early pathogenic studies, we demonstrated increased secretion of gastric acid and decreased duodenal neutralization of acid after the duodenal ulcerogens cysteamine and propionitrile but not after ethanolamine in rats. This suggested that the effects were not due to the toxic properties of the duodenal ulcerogens. The decrease in acid handling may contribute to duodenal susceptibility to acid after treatment with ulcerogens and possibly reflects early pathophysiologic changes in duodenal ulceration.[13] We also demonstrated that although acid in the duodenum is required for ulcer formation, the hypersecretion of acid induced by cysteamine is not the only factor responsible for the development of duodenal ulcer.[14]

We, therefore, hypothesized that specific cellular targets, for example, endothelial cells and molecular targets, such as ET-1, transcription factors like

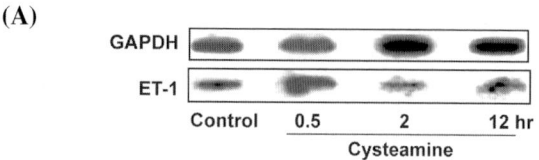

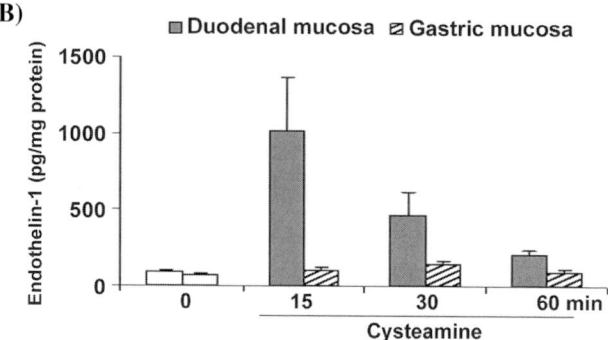

FIGURE 1. Gene expression of endothelin-1 (ET-1) by customer blot (**A**) and concentration of ET-1 measured by RIA (**B**).

early growth response 1 (egr-1) and Sp1, and growth factors VEGF, PDGF, and bFGF may be involved in the early and/or late stages of duodenal ulceration. Thus, our investigations were focused on these targets in last 10 years, for example, on vascular approach to gastroduodenal ulceration,[15] transcription factors, and growth factors in ulcer development and healing demonstrating that intragastric administration of bFGF, PDGF, VEGF, or gene therapy with naked DNA or adenoviral vectors of VEGF and PDGF enhance the local synthesis of PDGF and VEGF in injured duodenal mucosa and significantly accelerate chronic duodenal ulcer healing in rats.[16–19]

In this article, we provide an overview focusing on our new studies on ET-1, egr-1, and the imbalance of angiogenic and antiangiogenic factors.

UPREGULATION OF ET-1 AND ET RECEPTOR B: MECHANISMS OF ACUTE MICROVASCULAR INJURY IN DUODENAL ULCERATION

ET-1 is a potent vasoconstrictor[20] that plays an important role in the pathophysiology of hypertension, myocardial ischemia, and other diseases. ET-1 is produced primarily in endothelial cells and is also expressed in vascular smooth muscle cells, myocytes, bronchial epithelial cells, mesangial cells, neurons, astrocytes, and macrophages.[21] In our initial studies, we tested the hypothesis that local release of endogenous ET-1 might play a role in the pathogenesis of gastric and duodenal ulcers.[22] We subsequently found a very early increased gene expression of ET-1 (FIG. 1A) and rapid local release of ET-1 protein

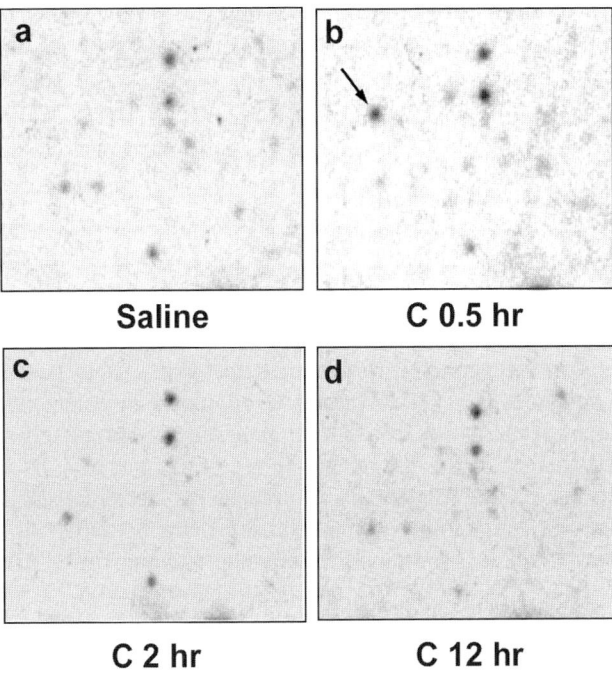

FIGURE 2. Expression of endothelin receptor B (arrow) by microarray in duodenal mucosa of control, or 0.5, 2, and 12 h after cysteamine (C).

with an approximate 10-fold increase in duodenal mucosa in 15 min after cysteamine administration (FIG. 1B), much preceding the development of duodenal ulcer.[17] We also demonstrated a simultaneous increase of ET receptor B expression by both Affymetrix and custom gene microarrays (FIG. 2), which was confirmed by real-time PCR,[23] while pretreatment of rats with the ET-A, -B receptor antagonist bosentan or neutralizing anti-ET-1 antibody decreased the severity and incidence of cysteamine-induced duodenal ulcers (TABLE 1).

Rapid local release of ET-1 was also demonstrated prior to the development of gastric hemorrhagic erosions after intragastric administration of ethanol or HCl in rats.[22] Plasma and gastric mucosal ET-1 concentrations were elevated in the gastric mucosa during indomethacin- and stress-induced ulceration.[24,25] Local gastric ischemia-reperfusion resulted in gastric mucosal damage accompanied by increased ET-1 tissue concentration. ET-1 levels were also elevated in the gastric mucosa of patients with peptic ulcer.[26] Thus, release of ET-1 seems to be a common mechanism of gastric and duodenal ulceration.

Recently, it was shown[27] that intestinal epithelial cells released ET-1 under unstimulated condition and significantly enhanced ET-1 release during stimulation. The mechanism of regulation of ET-1 mRNA expression in response to hypoxia is mediated by HIF-1.[28,29] Increased expression of ETRB-mRNA

TABLE 1. Effect of pretreatment with ET-1 antibodies or antagonist bosentan on cysteamine-induced duodenal ulcers in rats

Group	Pretreatment	Duodenal ulcer (mm^2)
1	Controls	20.9 ± 2.7
2	Bosentan, 3 mg/100 g × 1	18.6 ± 3.2
3	Bosentan, 3 mg/100 g × 2	11.4 ± 1.7 (P< 0.05)
4	Bosentan, 10 mg/100 g × 1	14.1 ± 2.0 (P< 0.05)
5	Bosentan, 10 mg/100 g ×2	8.7 ± 2.0 (P< 0.01)
6	Anti-ET-1 antibodies × 1	5.9 ± 0.8 (P = 0.23)
7	Anti-ET-1 antibodies × 3	3.0 ± 1.4 (P = 0.05)

and synthesis of ET-1 (up to 10-fold) were detected in various pathologic conditions.[30] Exogenous ET-1 (10 nM) induced apoptosis via enhanced caspase-3 activity, which appeared to be ETRB-mediated, as it was completely suppressed by the ETRB antagonist BQ 788 but not by the ETRA antagonist.[30] ET-1 binding to ETRB also enhanced nitric oxide-induced cell death in vascular smooth muscle cells.[31] These findings indicate that ET-1 may participate in the mechanisms of cell death (e.g., apoptosis) through the activation of ETRB-mediated pathways.

UPREGULATION OF EGR-1: AN ORGAN- AND/OR DISEASE-DEPENDENT GENE IN DUODENAL ULCERATION

In our gene expression studies initially by the Clontech DNA microarray (for about 1,200 genes) and subsequently with Affymetrix gene chip (8,000–15,000 genes), we identified about 40 genes with markedly changed expression in proximal duodenal mucosa during the preulcerogenic stage of duodenal ulceration, for example, often causing a 5- to 20-fold increase or decrease.[23] Of these markedly changed genes, we confirmed not only the ET-1 with very early expression but also other very early expressed genes, such as egr-1. Egr-1 (Krox-24, NGFI-A, zif268), a protein with three tandem Cys2-His2 zinc finger motifs, is a transcription factor that interacts with consensus GC-rich region, GCG(T/G)GGGCG, to influence the transcription of genes in response to a wide variety of mitogenic and nonmitogenic stimuli, including growth factors, shear stress, mechanical injury, hypoxia.[32]

Thus, targeting this transcription factor may be useful in efforts to increase proliferation or repair after mucosal injury.

Endothelins and Egr-1

It has been demonstrated that early released ET-1 induces a rapid expression of egr-1.[33,34] The stimulation of protein synthesis in cardiac myocytes in the presence of ET-1 was induced by the activation of the egr-1 gene.[35] An antisense oligonucleotide of egr-1 abolished the stimulation of protein synthesis induced by ET-1.[36] The expression of the Egr-1 was closely correlated with the proliferation of mesangial cells.[37] The mitogenic neuropeptide, ET-3 can modulate the proliferation of astrocytes through the production of Egr-1 protein,[38] which bound specifically to several early growth-related proteins (e.g., Egr-1-binding sites on the bFGF promoter and activated bFGF transcription).[39] Thus, ET-1 seems to be one of the chemical mediators between cell injury and the Egr-1 expression resulting in elevated growth factors production.

Corepressors of Egr-1 Activity, NAB1 and NAB2

Egr-1 activity, like many transcription factors is regulated by direct interactions with other proteins that modulate the level of transcriptional activation. Egr-1 associates with corepressor proteins that can modulate the expression of Egr-dependent genes. Two corepressors of Egr-1, NGFI-A-binding proteins, NAB1 and NAB2, have been identified.[40] These factors bind to Egr-1 R1 region and repress the transactivating potential of Egr-1. Whereas NAB1 is constitutively expressed in most tissues and appears to be a general transcriptional regulator, NAB2 may function as an important inducible regulator of gene expression.[41] NAB2 induces inhibition of Egr-1-dependent synthesis of growth factors (PDGF-AB, HGF, TGFβ, VEGF, and VEGF receptor 1).[42,43] Gene regulation mediated by the interplay of Egr-1 and NAB2 might be a unifying principle in growth processes, such as neuron outgrowth, wound healing, angiogenesis, and tumor invasion.

Expression and Distribution of Egr-1 in Duodenal Ulceration

Since gene expression is most effectively controlled by transcription factors, proteins that bind to *cis*-acting elements of DNA and guide the binding of polymerase II to start the transcription of specific mRNA, we tested the hypothesis that the expression of immediate early genes and their transcription factor products, such as Egr-1 and Sp1, might precede the increased synthesis of bFGF, PDGF, and VEGF in duodenal ulcer healing. We hypothesized that ET-1 induced a rapid expression of egr-1, while increased egr-1 transcription activity led to increased bFGF, PDGF, and possibly VEGF synthesis in duodenal mucosa during cysteamine-induced ulceration. Indeed, the duodenal ulcerogen cysteamine, but not its nonulcerogen and toxic analogue ethanolamine, rapidly increased duodenal (but not gastric) mucosal levels of ET-1, which were followed by enhanced expression of egr-1 gene that was confirmed by RT-PCR and real-time PCR (FIG. 3) in the preulcerogenic stage of duodenal ulceration.

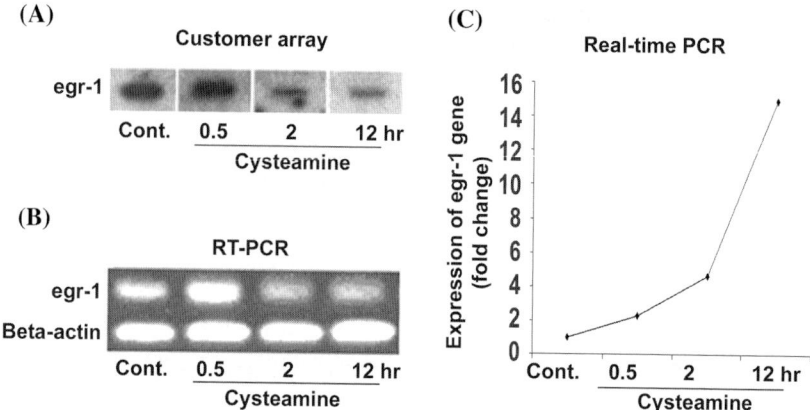

FIGURE 3. Gene expression of egr-1 detected by customer array (**A**), RT-PCR (**B**), and real-time RCR (**C**).

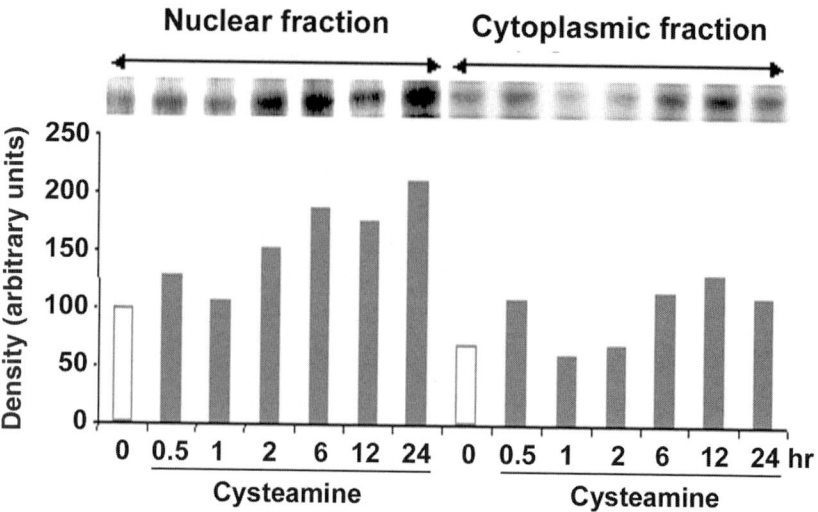

FIGURE 4. Western blots for detection of Egr-1 in nuclear and cytoplasmic fractions of rat duodenal mucosa after the administration of saline or cysteamine at different time points.

Expression of Egr-1 protein was detected by Western blotting. Rats exposed to the duodenal ulcerogen cysteamine displayed an early increase followed by a later time-dependent induction of Egr-1 in proximal duodenal mucosa. Western blot analysis of nuclear and cytoplasmic extracts showed an increase in Egr-1 expression. At 2 h after cysteamine the expression of Egr-1 in nuclear fraction was enhanced 1.5–2.0-fold, with further elevation (2–3-fold) at 12 and 24 h. (FIG. 4). In order to identify whether the prominently changed egr-1 genes were the organ- or disease-dependent, we further examined egr-1 expression

TABLE 2. Expression of the potential ulcer-specific gene egr-1 in distal organs other than duodenum in rats after administration of cysteamine (C) by real-time PCR (fold changes)

Organ	Saline	C 0.5 h	C 2 h	C 12 h
Duodenal mucosa	1.00	2.30	6.62	14.93
Ileal mucosa	1.00	0.50	0.30	2.10
Colonic mucosa	1.00	1.10	1.00	0.90

All data were normalized with GAPDH expression.

in other GI tract tissues, for example, ileum and colon. We found that egr-1 gene expression was not markedly changed in these tissues after cysteamine (TABLE 2).

We concluded that early-enhanced expression of egr-1 gene in the duodenal mucosa at 0.5 h after cysteamine was followed by a time-dependent increase in Egr-1 protein expression in nuclear fraction up to 24 h. These changes in the expression and distribution of Egr-1 in duodenal mucosa may play an organ- and/or disease-dependent role in experimental duodenal ulceration.

Role of Egr-1 in Duodenal Ulceration and Healing

We tested the hypothesis that inhibition of egr-1 RNA synthesis by egr-1 antisense may suppress the expression of Egr-1 protein, decrease growth factors synthesis and aggravate the cysteamine-induced duodenal ulcers in rats. Rats were given antisense egr-1 at 0.02, 0.2 mg/rat 30 min before every dose of cysteamine. The changes in egr-1 gene were detected by RT-PCR. Expression of Egr-1 and VEGF, PDGF and bFGF were determined by Western blot. The egr-1 antisense at 0.2 mg but not at 0.02 mg/rat aggravated duodenal ulcers 48 h after cysteamine and increased the size of ulcers from 8.1 ± 1.8 to 20.7 ± 4.0 mm^2 ($P < 0.01$) (FIG. 5A). Antisense egr-1 inhibited the transcription factor RNA and protein expressions (FIG. 5B). The expression of PDGF and VEGF in duodenal mucosa of rats exposed to cysteamine-induced ulceration was decreased after the administration of egr-1 antisense (FIG. 6). We conclude that depletion of egr-1 RNA expression in duodenal mucosa by egr-1 antisense oligonucleotides aggravated cysteamine-induced duodenal ulcer. The suppression of transcription factor Egr-1 expression was followed by inhibition of bFGF, PDGF, and VEGF synthesis in duodenal mucosa of rats treated with egr-1 antisense. This implies that Egr-1 may play a beneficial role in pathogenesis of duodenal ulceration, for example, by regulation of growth factors synthesis.

Thus, ET-1 seems to be one of the chemical mediators between cell injury and the Egr-1 expression resulting in elevated production of growth factors, such as bFGF, PDGF, VEGF, and VEGF receptor Flt-1.[39,44–46]

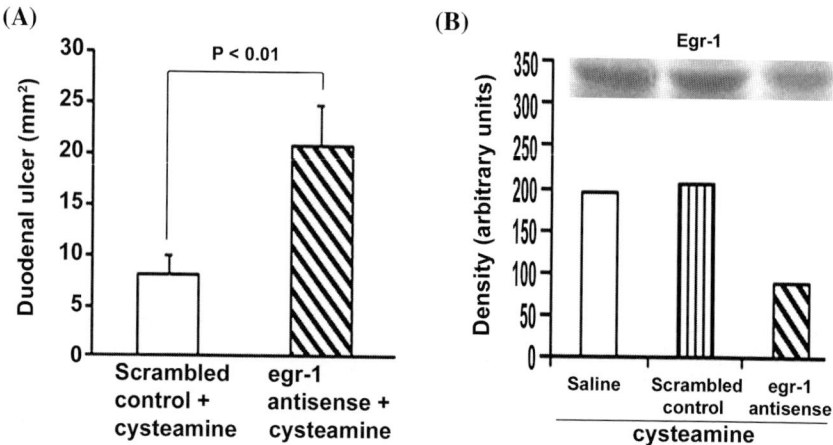

FIGURE 5. (A) Effect of egr-1 inhibition on cysteamine-induced duodenal ulcer formation and (B) Egr-1 expression in duodenal mucosa of rats pretreated with saline, control oligo, or antisense of egr-1 before cysteamine.

IMBALANCE OF ANGIOGENIC FACTOR VEGF AND ANTIANGIOGENIC ANGIOSTATIN/ENDOSTATIN: A NOVEL EXPLANATION OF DELAYED AND IMPAIRED ULCER HEALING

Based on our gene expression and pharmacologic studies on angiogenic growth factors (e.g., bFGF, PDGF, and VEGF), we demonstrated elevated levels of bFGF, PDGF, and VEGF in chemically induced acute duodenal ulceration[17,47] and accelerated healing of chronic duodenal ulcer in rats treated with peptides or genes of bFGF, PDGF, or VEGF.[19,48,49] These results were surprising in the light of perceived protective roles of these growth factors. However, we could not understand and explain why the healing of duodenal ulcers would be impaired or delayed in spite of elevated local tissue expression and concentration of angiogenic peptides.[50] This puzzle became more complicated after our most recent experiments demonstrating enhanced expression of angiostatin and endostatin in the pathogenesis of both cysteamine- or propionitrile-induced duodenal ulceration[51] and iodoacetamide-caused ulcerative colitis.[52] New biochemical, molecular, biological, and immunohistochemical studies indicate that bFGF, PDGF, and VEGF play a pathophysiologic role in the natural history of ulcer healing.[18]

We have thus developed a new hypothesis of why ulcers develop and, especially, why they do not heal. Namely, it appears that the apparent "response to injury" resulting in the rapidly enhanced expression of angiogenic growth factors cannot achieve its goal of rapid healing, since their effect is counteracted by the equally fast and robust expression of antiangiogenic angiostatin

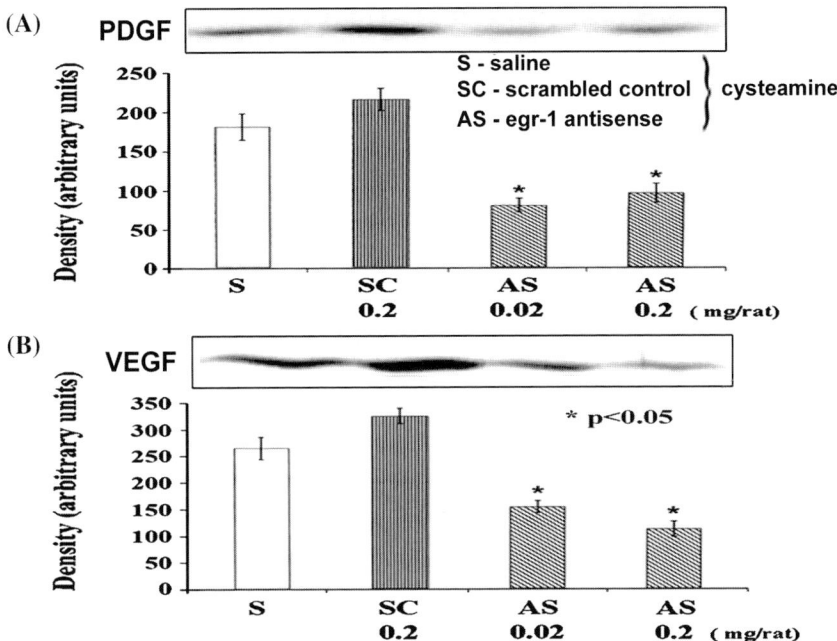

FIGURE 6. Expression of PDGF (**A**) and VEGF (**B**) by Western blotting after cysteamine-induced duodenal ulcer pretreated with control oligo or egr-1 antisense.

and endostatin, which have been demonstrated to be potent inhibitors of angiogenesis.[53]

Angiogenesis, that is, endothelial cell proliferation and tube formation in postembryonic tissue, is a crucial element in external (e.g., skin) and internal (e.g., gastrointestinal) wound/ulcer healing that needs granulation tissue, which forms the basis of proliferating and migrating epithelial cells to complete the healing process.[54] The process is governed by the balance between angiogenic factors, such as VEGF, bFGF, and PDGF, and antiangiogenic factors, such as endostatin and angiostatin.[55] The critical switch to angiogenesis involves a change in the local equilibrium between these positive and negative regulators of microvessels.[55]

Angiostatin was originally isolated as a 38 kDa fragment of plasminogen from urine of tumor-bearing mice.[56] Endostatin is a 20 kDa fragment of collagen XVIII originally isolated from mouse hemangioendothelioma (EOMA) conditioned media.[57] *In vitro*, both endostatin and angiostatin inhibit endothelial cell proliferation and migration.[58,59] However, the relative antiangiogenic activity of these fragments *in vivo* is unclear.

Recently, we demonstrated a markedly increased expression of collagen XVIII gene (endostatin precursor), which had an up to fivefold increase in duodenal mucosa in the early stages of duodenal ulceration induced by

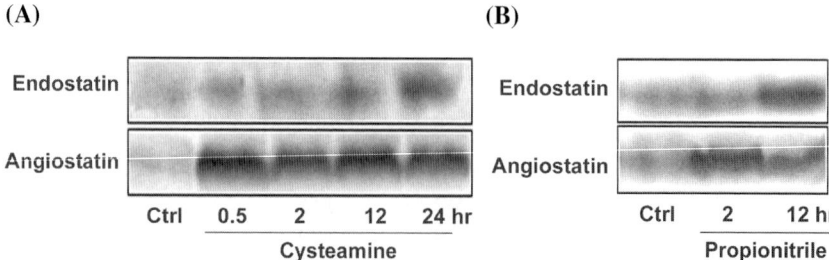

FIGURE 7. Expression of endostatin (19 kDa) and angiostatin (50 kDa) in duodenal mucosa after cysteamine (**A**) or propionitrile (**B**).

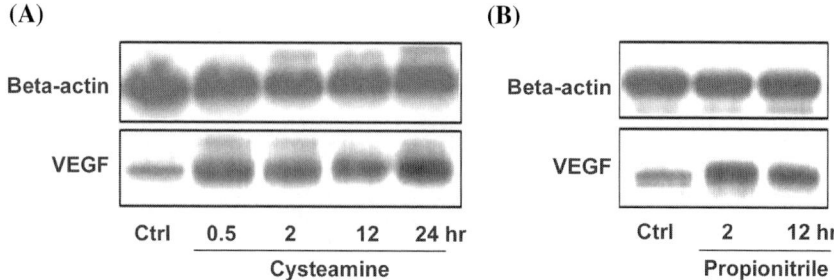

FIGURE 8. Expression of VEGF (23 kDa) in duodenal mucosa after cysteamine (**A**) or propionitrile (**B**).

cysteamine. The increased gene expression of collagen XVIII was followed by elevated levels of endostatin in duodenal mucosa in the late stage after the ulcerogenic cysteamine or propionitrile (FIG. 7). There was an increased angiostatin level within the duodenal mucosa after the administration of both duodenal ulcerogens, although the gene expression did not show increased expression of plasminogen (angiostatin precursor). When we further investigated the enzymatic activity of MMP, which cleaves both collagen XVIII and plasminogen to generate endostatin and angiostatin, respectively, we found an early increased activity of MMP2 in the duodenal mucosa prior to the generation of endostatin and angiostatin. This implies that MMP2 may play an important role in generation of endostatin and angiostatin during duodenal ulceration. Recent studies have highlighted the function of MMPs as negative regulators of angiogenesis by their release of antiangiogenic fragments, such as angiostatin and endostatin.[60–64]

Another interesting finding was that the angiogenic factor VEGF was also markedly increased in the duodenal mucosa at a similar time course (FIG. 8), while this study showed that endostatin and angiostatin were increased during the ulceration induced by cysteamine or propionitrile. Thus, seem to face an altered balance between pro- and antiangiogenic factors in duodenal ulceration. This was also found by our other very recent study on the expression of angiogenic and antiangiogenic factors in rats with ulcerative colitis[52] and the levels of elevated endostatin was parallel to the severity of the ulcerative colitis.[65] These unexpected findings may actually explain for the first time the initially surprising results demonstrating increased levels of angiogenic growth factors (e.g., bFGF, PDGF, VEGF) in the early stages of experimental duodenal ulceration. Namely, increased local concentrations of angiogenesis stimulators should lead to rapid healing, but we actually see an acute ulcer development. We speculate that the potentially beneficial effect of VEGF is apparently antagonized by the simultaneously elevated levels of angiostatin and endostatin. Thus, these results indicating an "angiogenic imbalance" suggest a novel mechanism of gastroduodenal ulceration.

Simultaneous increase of angiogenic and antiangiogenic factors in ulceration is a novel finding[66] but the precise mechanisms are unclear. Several studies have shown that fluid from venous leg ulcers, particularly those that heal slowly, contains endostatin and angiostatin, which inhibit *in vitro* angiogenesis.[67–69] Drinkwater *et al.*[70] demonstrated that the increased levels of antiangiogenic factors, for example, angiostatin and endostatin, were not associated with VEGF downregulation and, on the contrary, leg ulcers express elevated levels of VEGF relative to levels found in normal skin, while in tumors the increase of the antiangiogenic activity of endostatin and angiostatin was through a downregulation of VEGF expression both at mRNA and protein levels. A potential mechanism of increased endostatin and angiostatin in duodenal ulcer might be that the ulcer environment alters the expression of various components that could potentially modify antiangiogenic activity compared to normally healing wounds. For example, chronic wounds have altered expression of the proteoglycan glypican that binds and antagonizes bFGF in the wound environment.[71] In addition, the elevated levels of several proteinases may degrade the angiogenic factors, such as VEGF.[61,72–74]

The biologic confirmation of our molecular biochemical results came from our pharmacologic experiments when rats receiving daily injections of endostatin peptide showed a markedly aggravated duodenal ulcer induced by cysteamine. Most rats with endostatin administration had a very extensive necrotic duodenal mucosa with large and deep ulcer craters that often perforated or penetrated into the adjacent organs, such as liver or pancreas. Other *in vitro* studies demonstrated that endostatin induced endothelial cell apoptosis and inhibited the proliferation and migration of some types of endothelial cells.[75] It was suggested that interactions of endostatin with tropomyosin result in disruption of the integrity of microfilaments and might thereby contribute to the angiogenic effect of endostatin by inhibiting cell motility.[76,77] An *in vivo* study showed that the healing of full-thickness skin wounds in mice was delayed by systemic administration of endostatin.[78]

SUMMARY

In this review we summarized our findings in recent studies on duodenal ulceration with upregulated vascular factors, such as ET-1 and its receptor B, and transcription factor, such as Egr-1 as well as a novel finding on imbalance of angiogenic factor VEGF and antiangiogenic factors endostatin/angiostatin in duodenal mucosa. Our results have now demonstrated for the first time that the molecular regulator of ulcer development and healing is the transcription factor Egr-1-product of immediate early gene egr-1. Furthermore, we have also learned that the initial event leading to egr-1 activation is duodenal mucosal vascular injury and hypoxia, followed by the early release of potent vasoconstrictor ET-1 along with an early increased expression of its receptor ETRB. These are specific to duodenal mucosa since we did not see these changes in other GI tissue in rats given cysteamine. It is associated with enhanced Egr-1 gastrointestinal to DNA, leading to increased synthesis of bFGF, PDGF, and VEGF. One of the most convincing results was the marked aggravation of cysteamine-induced duodenal ulcer and decreased expression of growth factors, such as bFGF, PDGF, and VEGF by egr-1 antisense.

Our most recent studies demonstrated that the duodenal ulcerogens cysteamine and propionitrile induce both angiogenic factors, such as VEGF and antiangiogenic factors, such as endostatin/angiostatin at similar time course during the duodenal ulceration, while neither angiostatin nor endostatin showed changes in the duodenal mucosa after nonulcerogenic analog ethanolamine. The demonstration that daily administration of endostatin significantly aggravated the cysteamine-induced duodenal ulcers and delayed ulcer healing confirmed the biological effects of endostatin. We, therefore, conclude that rapid released ET-1 and early increased expression of ETRB may play a critical role in the initial injury of endothelial and epithelial cells in duodenal ulceration and that elevated ET-1 also induced a very early increase of Egr-1expression, which leads to sequential production of angiogenic growth factors, such as bFGF, PDGF, and VEGF. Egr-1 is an organ- and/or disease-specific transcription factor in duodenal ulceration, while an imbalance of angiogenic and antiangiogenic factors in duodenal ulceration seems to be a novel explanation for delayed and impaired healing of duodenal ulcers. Thus, the experimental duodenal ulcers Hans Selye described morphologically are now characterized at molecular and genome levels involving unexpected mediators, such as ET-1, egr-1, and angiogenesis-related molecules.

REFERENCES

1. HUNT, R.H. 1990. Peptic ulcer disease. Gastroenterol. Clin. N. Am. **19:** 1–225.
2. SONNENBERG, A. & J.E EVERHART. 1997. Health impact of peptic ulcer in the United States. Am. J. Gastroenterol. **92:** 614–620.

3. LEVIN, T.R., J.A. SCHMITTDIEL, J.M. HENNING, et al. 1998. A cost analysis of a *Helicobacter pylori* eradication strategy in a large health maintenance organization. Am. J. Gastroenterol. **93:** 743–747.
4. MARSHALL, B.J. &, H.M. WINDSOR. 2005. The relation of *Helicobacter pylori* to gastric adenocarcinoma and lymphoma: pathophysiology, epidemiology, screening, clinical presentation, treatment, and prevention. Med. Clin. North Am. **89:** 313–344.
5. SELYE, H. 1950. Stress. The press, Acta Medica Publishers, Montreal.
6. ROBERT, A. & S. SZABO. 1983. Stress ulcers. *In* Selye's Guide to Stress Research. H. Selye, Ed.: Volume 3. The press, Van Nostrand Reinhold Publishing, New York, NY.
7. SZABO, S., W.F. SPILL & K.D. RAINSFORD. 1989. Non-steroidal anti-inflammatory drug-induced gastropathy: mechanisms and management. Med. Toxicol. Adv. Drug Exp. **4:** 77–94.
8. SZABO, S. & H. SELYE. 1972. Duodenal ulcers produced by propionitrile in rats. Arch. Pathol. **93:** 390–391.
9. SELYE, H. & S. SZABO. 1973. Experimental model for production of perforating duodenal ulcers by cysteamine in the rat. Nature **244:** 458–459.
10. POULSEN, S.S. & S. SZABO. 1977. Mucosal surface morphology and histologic changes in the duodenum of the rat following administration of cysteamine. Br. J. Exp. Pathol. **58:** 1–8.
11. SZABO, S., E.S. REYNOLDS & S.H. UNGER. 1982. Structure-activity relations between alkyl nucleophilic chemicals causing duodenal ulcer and adrenocortical necrosis. J. Pharmacol. Exp. Ther. **223:** 68–76.
12. GIAMPAOLO, C., A.T. GRAY, R.A. OLSHEN & S. SZABO. 1991. Predicting chemically induced duodenal ulcer and adrenal necrosis with classification trees. Proc. Natl. Acad. Sci. USA **88:** 6298–6302.
13. ADLER, R.S., G.T. GALLAGHER & S. SZABO. 1983. Duodenal ulcerogens cysteamine and propionitrile decrease duodenal neutralization of acid in the rat. Dig. Dis. Sci. **28:** 716–723.
14. KIRKEGAARD, P., S.S. POULSEN, F.B. LOUD, et al. 1980. Cysteamine-induced duodenal ulcer and acid secretion in the rat. Scand. J. Gastroenterol. **15:** 621–624.
15. SZABO, S., T. KHOMENKO, Z. GOMBOS, et al. 2000. Transcription factors and growth factors in ulcer healing. Aliment. Pharmacol. Ther. Suppl. **1:** 33–43.
16. SZABO, S. & Z. SANDOR. 1996. Basic fibroblast growth factor and PDGF and GI diseases. Bailliere's Clin. Gastroenterol. **11:** 922–927.
17. SZABO, S., A. VINCZE, Z. SANDOR, et al. 1998. Vascular approach to gastroduodenal ulceration: new studies with endothelins and VEGF. Dig. Dis. Sci. **43**(Suppl): 40S–45S.
18. SZABO, S., X. DENG, T. KHOMENKO, et al. 2001. Gene expression and gene therapy in experimental duodenal ulceration. J. Physiol. Paris **95:** 325–335.
19. DENG, X., S. SZABO, T. KHOMENKO, et al. 2004. Gene therapy with adenoviral plasmids or naked DNA of vascular endothelial growth factor and platelet-derived growth factor accelerates healing of duodenal ulcer in rats. J. Pharmacol. Exp. Ther. **311:** 982–988.
20. YANAGISAWA, M., H. KURIHARA, S. KIMURA, et al. 1988. A novel potent vasoconstrictor peptide produced by vascular endothelial cells. Nature **322:** 411–415.
21. MACCUMBER, M.W., C.A. ROSS & S.H. SNYDER. 1990. Endothelin in brain: receptors, mitogenesis, and biosynthesis in glial cells. Proc. Natl. Acad. Sci. USA **87:** 2359–2363.

22. MORALES, R.E., B.R. JOHNSON & S. SZABO. 1992. Endothelin induces vascular and mucosal lesions, enhances the injury by HCl/ethanol, and the antibody exerts gastroprotection. FASEB J. **6:** 2354–2360.
23. DENG, X.M., S. SZABO, T. KHOMENKO, et al. 2007. Detection of duodenal ulcer-associated genes in rats. Dig. Dis. Sci. Accepted.
24. MATSUMARU, K., H. KASHIMURA, M. HASSAN, et al. 1997. A novel synthetic mixed-type endothelin receptor antagonist, attenuates acute gastric mucosal lesions induced by indomethacin and HCl in rat: role of endothelin-1. J. Gastroenterol. **32:** 164–170.
25. MICHIDA, T., S. KAWANO, E. MASUDA, et al. 1994. Role of endothelin 1 in hemorrhagic shock-induced gastric mucosal injury in rats. Gastroenterology **106:** 988–993.
26. MASUDA, E., S. KAWANO, T. MICHIDA, et al. 1997. Plasma and gastric mucosal endothelin-1 concentration in patients with peptic ulcer. Dig. Dis. Sci. **42:** 314–318.
27. SHIGEMATSU, T., S. MIURA, M. HIROKAWA, et al. 1998. Induction of endothelin-1 synthesis by IL-1 and its modulation of rat intestinal cell growth. Am. J. Physiol. **38:** G556–G563.
28. SEMENZA, G.L. 2000. HIF-1: mediator of physiological and pathophysiological responses to hypoxia. J Appl. Physiol. **88:** 1474–1480.
29. KOTCH, L.E., N.V. IYER, E. LAUGHNER & G.L. SEMENZA. 1999. Defective vascularization of HIF-1a-null embryos is not associated with VEGF deficiency but with mesenchymal cell death. Dev. Biol. **209:** 254–267.
30. CATTARUZZA, M., C. DIMIGEN, H. EHRENREICH & M. HECKER. 2000. Stretch-induced endothelin B receptor-mediated apoptosis in vascular smooth muscle cells. FASEB J. **14:** 991–998.
31. NAKAHASHI, T., K. FUKUO, H. NISHIMAKI, et al. 1998. Endothelin-1 enhances nitric oxide-induced cell death in cultured vascular smooth-muscle cells. J. Cardiovasc. Phamacol. **31:** S351–S353.
32. YAN, S.F., J.L. LU, Y.S. ZOU, et al. 1999. Hypoxia-associated induction of early response-1 gene expression. J. Biol. Chem. **274:** 15030–15040.
33. SHUBEITA, H.E., P.M. MCDONOUGH, A.N. HARRIS, et al. 1990. Endothelin induction of inositol phospholipid hydrolysis, sarcomere assembly and cardiac gene expression in ventricular myocytes. J. Biol. Chem. **265:** 20555–20562.
34. NEYSES, L., J. NOUSKAS & H. VETTER. 1991. Inhibition of endothelin-1 induced myocardial protein synthesis by an antisense oligonucleotide against the early growth response gene1. Biochem. Biophys. Res. Comm. **181:** 22–27.
35. MAAS, A., C. GROHE, C. KUBISCH, et al. 1995. Hormonal induction of an immediate-early gene response in myogenic cell lines—a paradigm for heart growth. Eur. Heart J. **16**(Suppl.C): 12–14.
36. BRUNEAU, B.G., L.A. PIAZZA & A.J. DE BOLD. 1997. BVP gene expression is specifically modulated by stretch and ET-1 in a new model of isolated rat atria. Am. J. Physiol. **273:** H2678–H2686.
37. HOFER, G., C. GRIMMER, V.P. SUKHATME, et al. 1996. Transcription factor Egr-1 regulates glomerular mesangial cell proliferation. J. Biol. Chem. **271:** 28306–28310.
38. BIESIADA, E., M. RAZANDI & E.R. LEVIN. 1996. Egr-1 activates basic fibroblast growth factor transcription. Mechanistic implications for astrocyte proliferation. J. Biol. Chem. **271:** 18576–18581.

39. SANTIAGO, F.S., H.C. LOWE, F.L. DAY, et al. 1999. Early growth response factor-1 induction by injury is triggered by release and paracrine activation by fibroblast growth factor-2. Am. J. Pathol. **154:** 937–944.
40. QU, Z., L.A. WOLFRAIM, J. SVAREN, et al. 1998. The transcriptional corepressor NAB2 inhibits NGF-induced differentiation of PC12 cells. J. Cell Biol. **142:** 1075–1082.
41. SILVERMAN, E.S., L.M. KHACHIGIAN, F.S SANTIAGO, et al. 1999. Vascular smooth muscle cells express the transcriptional corepressor NAB2 in response to injury. Am. J. Pathol. **155:** 1311–1317.
42. QU, Z., L.A. WOLFRAIM, J. SVAREN, et al. 1998. The transcriptional corepressor NAB2 inhibits NGF-induced differentiation of PC12 cells. J. Cell Biol. **142:** 1075–1082.
43. LUCERNA, M., D. MECHTCHERIAKOVA, A. KADL, et al. 2003. NAB2, a corepressor of EGR-1, inhibits vascular endothelial growth factor-mediated gene induction and angiogenic responses of endothelial cells. J. Biol. Chem. **278:** 11433–11440.
44. KHACHIGIAN, L.M., A.J. WILLIAMS & T. COLLINS. 1995. Interplay of Sp1 and Egr-1 on the proximal platelet-derived growth factor A-chain promoter in cultured vascular endothelial cells. J. Biol. Chem. **270:** 27679–27686.
45. KHACHIGIAN, L.M., V. LINDNER, A.J. WILLIAMS & T. COLLINS. 1996. Egr-1-induced endothelial gene expression: a common theme in vascular injury. Science **271:** 1427–1431.
46. VIDAL, F., J. ARAGONES, A. ARANTZAZU & M.O. DE LANDAZURI. 2000. Up-regulation of vascular endothelial growth factor receptor Flt-1 after endothelial denudation: role of transcription factor Egr-1. Blood **95:** 3387–3394.
47. VINCZE, A., M. NAGATA, Z. SANDOR & S. SZABO. 1996. ELISA and Western blot studies with basic fibroblast growth factor (bFGF) and platelet-derived growth factor (PDGF) in experimental duodenal ulcer development and healing. Inflammopharmacology **4:** 261–265.
48. FOLKMAN, J., S. SZABO, M. STOVROFF, et al. 1991. Duodenal ulcer. Discovery of a new mechanism and development of angiogenic therapy that accelerates healing. Ann. Surg. **241:** 414–425.
49. SZABO, S. et al. 1995. Growth factors: "New endogenous drugs" for ulcer healing. Scand. J. Gastroenterol. **30**(Suppl. 210): 15–18.
50. SZABO, S. et al. 2001. Angiogenesis and growth factors in ulcer healing. In New Angiogenesis. D. Fan Tai-Ping and E.C. Kohn, Eds.: 119–211. Totowa. Humana Press, NJ.
51. DENG, X.M. et al. 2006. Increased expression of both angiogenic VEGF, and anti-angiogenic endostatin and angiostatin in ulcerative colitis in rats, mice and humans: an explanation for impaired and delayed healing. Gastroenterology **128:** A998.
52. SANDOR, ZS., X.M. DENG, T. KHOMENKO, et al. 2006. Altered angiogenic balance in ulcerative colitis: A key to impaired healing? Biochem. Biophys. Res. Commun. **350:** 147–150.
53. HANAHAN, D. & J. FOLKMAN. 1996. Patterns and emerging mechanisms of the angiogenic switch during tumorigenesis. Cell **86:** 353–364.
54. FOLKMAN, J. 2006. Angiogenesis. Annu. Rev. Med. **57:** 1–18.
55. PANDYA, N.M., N.S. DHALLA & D.D. SANTANI. 2006. Angiogenesis–a new target for future therapy. Vascul. Pharmacol. **44:** 265–274.

56. O'REILLY, M.S. *et al.* 1994. Angiostatin: a novel angiogenesis inhibitor that mediates the suppression of metastases by a Lewis lung carcinoma. Cell **79:** 315–328.
57. O'REILLY, M.S. *et al.* 1997. Endostatin: an endogenous inhibitor of angiogenesis and tumor growth. Cell **88:** 277–285.
58. CLAESSON-WELSH, L *et al.* 1998. Angiostatin induces endothelial cell apoptosis and activation of focal adhesion kinase independently of the integrin-binding motif RGD. Proc. Natl. Acad. Sci. USA **95:** 5579–5583.
59. TADDEI, L. *et al.* 1999. Inhibitory effect of full-length human endostatin on in vitro angiogenesis. Biochem. Biophys. Res. Commun. **263:** 340–345.
60. MOTT, J.D. & Z. WERB. 2004. Regulation of matrix biology by matrix metalloproteinases. Curr. Opin. Cell Biol. **16:** 558–564.
61. SOTTILE, J. 2004. Regulation of angiogenesis by extracellular matrix. Biochem. Biophys. Acta **1654:** 13–22.
62. CORNELIUS, L.A. *et al.* 1998. Matrix metalloproteinases generate angiostatin: effects on neovascularization. J. Immunol. **161:** 6845–6852.
63. FERRERAS, M., U. FELBOR, T. LENHARD, *et al.* 2000. Generation and degradation of human endostatin proteins by various proteinases. FEBS Lett. **86:** 247–251.
64. NILSSON, U.W. & C. DABROSIN. 2006. Estradiol and tamoxifen regulate endostatin generation via matrix metalloproteinase activity in breast cancer *in vivo*. Cancer Res. **66:** 4789–4704.
65. SZABO, S., TOLSTANOVA, G., DENG, X.M., *et al.* 2007. New molecular mechanisms of GI inflammation and neoplasia [abstract]. The 12th Taishotoyama International Symposium on Gastroenterology, Shimoda, Japan, 12.
66. SZABO, S. 2007. New molecular mechanisms of duodenal ulcer—The most frequent "peptic ulcer" [abstract]. Hans Selye Centennial Symposium, Quebec, Canada, 28.
67. SMITH, E. & R. HOFFMAN. 2005. Multiple fragments related to angiostatin and endostatin in fluid from venous leg ulcers. Wound Rep. Reg. **13:** 148–157.
68. BUCALO, B., W.H. EAGLSTEIN & V. FALANGA. 1993. Inhibition of cell proliferation by chronic wound fluid. Wound Rep. Reg. **1:** 181–186.
69. DRINKWATER, S.L., S. SMITH, B.M. SAWYER & K.G. BURNAND. 2002. Effect of venous ulcer exudates on angiogenesis *in vitro*. Br. J. Surg. **89:** 709–713.
70. DRINKWATER, S.L., K.G BURNAND, R. DING & A. SMITH. 2003. Increased but ineffectual angiogenic drive in nonhealing venous leg ulcers. J. Vasc. Surg. **38:** 1106–1112.
71. KATO, M. *et al.* 1998. Physiological degradation converts the soluble syndecan-1 ectodomain from an inhibitor to a potent activator of FGF-2. Nat. Med. **4:** 691–697.
72. GRINNELL, F. & M. ZHU. 1996. Fibronectin degradation in chronic wounds depends on the relative levels of elastase, alpha1-proteinase inhibitor, and alpha2-macroglobulin. J. Invest. Dermatol. **106:** 335–341.
73. WECKROTH, M., A. VAHERI, J. LAUHARANTA, *et al.* 1996. Matrix metalloproteinases, gelatinase and collagenase, in chronic leg ulcers. J. Invest. Dermatol. **106:** 1119–1124.
74. LAUER, G. *et al.* 2000. Expression and proteolysis of vascular endothelial growth factor is increased in chronic wounds. J. Invest. Dermatol. **115:** 12–18.
75. BOEHM, T., J. FOLKMAN, T. BROWDER & M.S. O'REILLY. 1997. Antiangiogenic therapy of experimental cancer does not induce acquired drug resistance. Nature **390:** 404–407.

76. YAMAGUCHI, N. *et al.* 1999. Endostatin inhibits VEGF-induced endothelial cell migration and tumor growth independently of zinc binding. EMBO J. **18:** 4414–4423.
77. MACDONALD, N.J. *et al.* 2001. Endostatin binds tropomyosin. A potential modulator of the antitumor activity of endostatin. J. Biol. Chem. **276:** 25190–25196.
78. BLOCH, W. *et al.* 2000. The angiogenesis inhibitor endostatin impairs blood vessel maturation during wound healing. FASEB J. **14:** 2373–2376.

Metabolic Syndrome

Psychosocial, Neuroendocrine, and Classical Risk Factors in Type 2 Diabetes

N.G. ABRAHAM,[a] E.J. BRUNNER,[b] J.W. ERIKSSON,[c] AND R.P. ROBERTSON[d]

[a] *New York Medical College, Valhalla, New York, USA*
[b] *University College London, United Kingdom*
[c] *Sahlgrenska University Hospital, Gothenburg, Sweden*
[d] *University of Washington, Seattle, Washington, USA*

> ABSTRACT: This article summarizes some aspects of stress in the metabolic syndrome at the psychosocial, tissue, and cellular levels. The metabolic syndrome is a valuable research concept for studying population health and social–biological translation. The cluster of cardiovascular risk factors labeled the metabolic syndrome is linked with low socioeconomic status. Systematic differences in diet and physical activity contribute to social patterning of the syndrome. In addition, psychosocial factors including chronic work stress are linked with its development. Psychosocial factors could lead to metabolic perturbations and increase cardiovascular risk via activation of neuroendocrine responses, for example, in the autonomic nervous system and in several hormonal pathways. High glucocorticoid levels will promote lipid storage in visceral rather than subcutaneous adipose tissue. Adipocytes secrete several proinflammatory cytokines, which considered major contributors to increase in oxidants and cell injury. Upregulation of heme oxygenase 1 (HO-1) and peroxidase in the early development of diabetes produces a decrease in oxidative-mediated injury. Increased HO activity is associated with a significant decrease in superoxide, endothelial cell shedding and blood pressure. Finally, it is proposed that overexpression of glutathione peroxidase in β cells may protect β cell deterioration from oxidative stress during development of diabetes and hyperglycemia and this may result in attenuation of β cell failure. If this proves to be the case, then the scene will be set to develop glutathione peroxidase mimetics for use in preclinical and clinical trials.

Address for correspondence: N.G. Abraham, New York Medical College, Department of Pharmacology, Basis Science Building/Room 527, Valhalla, NY 10595. Voice: 914-594-4132; fax: 914-591-4273.
nader_abraham@nymc.edu

KEYWORDS: work stress; social epidemiology; neuroendocrine; adipokines; insulin resistance; antioxidants; adipocyte; glutathione; cell; bilirubin

INTRODUCTION

Accumulating epidemiological, biological, and mechanistic evidence indicates that psychosocial, neuroendocrine, immunogenic, and oxidative stress play critical roles in the development of metabolic syndrome, cardiovascular disease, and type 2 diabetes. Each of these three "stress-related" conditions are linked with low social status in developed economies.[1,2] This fact highlights the importance of a holistic approaches to preventing chronic disease as well as to understanding its etiology.[3,4] Brunner points out that prior to the industrial revolution the metabolic syndrome and type 2 diabetes were rare and that the use of the Whitehall model in present day populations can predict "at risk" individuals. Eriksson describes the role of neuroendocrine pathways in the development of insulin resistance in the progression of metabolic syndromes and type 2 diabetes.[5] Insulin resistance occurs mainly in muscle, fat, and liver and the underlying mechanism(s) appears different in individual tissues.[6,7] The role of neuroendocrine pathways as possible mediators of the stress response in the brain is examined.[8] Stress leads to neuroendocrine responses that, over time, may be of significance in the development of insulin resistance and type 2 diabetes. This has led to the conclusion that both genetic and environmental factors are important and that humoral or neural mechanisms rather than intrinsic genetic defects on the target cells of insulin are the primary source of perturbations in stress-related disease. Antioxidant gene, heme oxygenase-1 (HO-1) expression, is crucial in micro- and macrovascular disease in animal models of type 1 and 2 diabetes and that induction of HO-1, plays a critical role in preventing vascular endothelial cells death, and attenuating cardiovascular complications.[9,10] HO-1 induction results in increased levels of the heme degradation products carbon monoxide (CO) and bilirubin. These compounds are known to counteract the detrimental effects of oxidative stress in type 1 and 2 diabetes rendering endothelial cells resistant to diabetes-induced apoptosis by increasing the levels of antioxidant genes including extracellular superoxide dismutase (EC-SOD) and catalase. The experimental basis for chronic oxidative stress as an underlying mechanism for glucose toxicity in β cells is examined.[11,12] A role for the overexpression of glutathione peroxidase in β cells is proposed by Robertson's group as a cell protection and to the normalization of a potential approach to limit blood glucose levels in diabetic patients. This concise report details the interactions of psychosocial and neuroendocrine factors, HO-1, and the underlying mechanisms involved in oxidative-mediated injury in cardiovascular disease and diabetes.

Psychosocial Factors in the Metabolic Syndrome

Life expectancy is at an all-time high in developed economies but health is not evenly distributed across social strata. Persistent health inequalities are observed in many causes of morbidity and mortality, and among the causes with major public health importance, coronary heart disease (CHD) and type 2 diabetes mellitus stand out. Consistent with an underlying role in these health inequalities, there are inverse social gradients in prevalence of the metabolic syndrome in many populations, the lower the social position, the greater the risk of the syndrome.

Why are we interested in the metabolic syndrome? A holistic approach to evaluating chronic disease risk is useful. While it is appropriate to consider serum cholesterol, blood pressure, and other risk factor levels one at a time, whether in an individual or among populations, summary indexes of risk, such as the Framingham score offer a global assessment of disease risk that single factors do not capture. The metabolic syndrome is, thus, a biological indicator of disease risk. Per Bjorntorp and others have proposed a specific causal link between chronic stress and the metabolic syndrome, but the evidence for such specificity has been elusive.

THE PSYCHOSOCIAL HYPOTHESIS

As an internal response, stress is probably most often regarded as psychological, but the implication is that there may be (patho-) physiological consequences. Stress means different things in different situations: it usually refers to individuals, with secondary consideration of the wider context. In contrast, the psychosocial hypothesis builds on the observation that the social environment has strong psychological elements. The social hierarchy influences access to material resources. It also shapes the availability of psychological resources. Beliefs and emotions are socially patterned, and the regularity of such social differences across place and time suggests they may contribute to social differences in mental and physical health. The idea is intrinsic to a variety of psychosocial concepts including power, control (or autonomy), demands and reciprocity (effort and reward), justice and fairness as well as gender and racial discrimination.

METABOLIC SYNDROME DEFINITIONS

Various definitions of the metabolic syndrome exist. A scoring system identifies co-occurrence of several risk factors, typically three or more. Epidemiological studies have used cut-points based on observed distributions, for example, 3 or more of 5 variables above the top quintile: serum triglycerides, HDL-cholesterol (bottom quintile), 2-h postload glucose, systolic blood pressure

(SBP), and waist hip ratio, taking account of medication. Using this definition, prevalence of the syndrome in 1991–1993 was 12% among healthy office workers aged 50 years in the Whitehall II study.[1] NHANES III (1988–1994) found that prevalence of the metabolic syndrome was 24% in U.S. adults (32% among Mexican Americans).[13] Prevalence rose from 7% among 20–29-year olds to more than 40% over 60 years of age.

RESEARCH MODELS

Selye's view was that repeated stress perceptions have cumulative physiological impact (the general adaptation syndrome). Risk of the metabolic syndrome and its component risk factors supplies a research model. Krieger's concept of embodiment provides an account of ways in which social position and racial discrimination gets under the skin to influence blood pressure levels. Building on Cannon's work showing that physiological systems maintain stability (homeostasis) via sensitive feedback and control mechanisms,[14] McEwen and Seeman's allostasis concept proposes that altered stable states are produced in response to repeated activation of neuroendocrine mechanisms, particularly the autonomic nervous system and hypothalamic-pituitary-adrenal axis.[15] Transition to some alternative state may be temporary or permanent.

The metabolic syndrome is a particularly common "alternative state" today, but this was not the case in the past. In preindustrial societies, metabolic syndrome and type 2 diabetes were rare if not unknown. Clues about its psychosocial origins come from the observation that it is linked to low social status.[1]

Our research model connects psychosocial factors to disease development via two pathways, direct and indirect (FIG. 1). There may be interaction between adverse psychosocial circumstances and behavior-related factors, such as obesity. In addition, pathophysiological states, such as glucose intolerance are potentially reversible.

EVIDENCE FROM WHITEHALL II

Findings from a nested case-control study within the Whitehall II cohort of office-based Civil Servants are consistent with a psychosocial component in the origins of the metabolic syndrome, showing that it is linked simultaneously with several indicators of adverse autonomic and neuroendocrine function.[2] A randomly chosen group of healthy men in civil service employment was studied. Nurses blind to the risk factor status of participants organized collection of a 24-h urine sample over a working day. Compared with the control group, metabolic syndrome cases had higher urinary cortisol and normetanephrine (norepinephrine metabolites) outputs, higher heart rates, and lower heart rate variability (HRV).

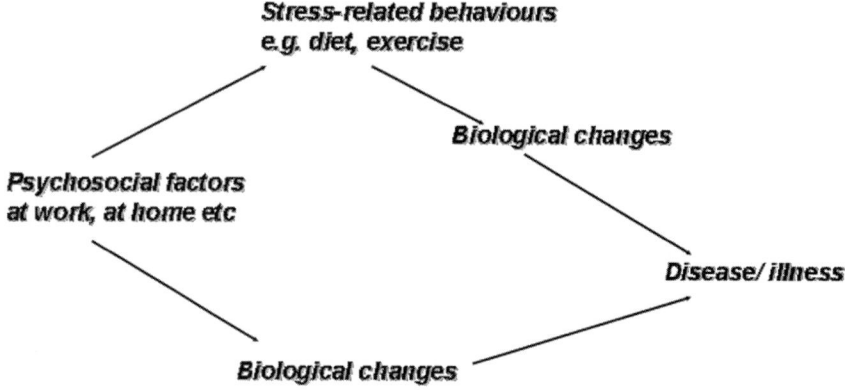

FIGURE 1. Direct and indirect pathways link psychosocial factors to disease.

Having obtained such fascinating results, we conducted a larger study of the Whitehall II cohort. We examined the interrelationships of HRV and metabolic syndrome with employment grade and psychosocial factors. HRV measurement was made by means of a 5-min electrocardiogram recording. Lower social position in the sample of 2,197 men was associated with higher heart rate and lower HRV, indicating low vagal tone and sympathetic predominance. All five components and metabolic syndrome factor itself were associated with low HRV. In addition, the relationship between lower social position and higher risk of metabolic syndrome was mediated by HRV, behavioral factors and low job control. This study provided population-based evidence that disturbances of autonomic function are involved in mediating the excess coronary risk associated with low social position.

These two cross-sectional studies support other research suggesting psychosocial pathways are plausible.[4] Behavioral as well as psychosocial influences may account for the differences in autonomic and neuroendocrine activity. We show, for example, that level of physical activity is related both to the likelihood of having the metabolic syndrome and to heart rate variability in the Whitehall II study.[16]

Other cross-sectional studies have linked work stress with components of the metabolic syndrome in Whitehall II and other large samples.[3,17] Reviews show these associations are not consistent,[18,19] adding to the need for prospective studies. Accumulated data over 14 years of Whitehall II follow-up made a robust study possible. Repeated measurements of work characteristics over this period gave good characterization of psychosocial exposure. The association between four phases of measurement and ATPIII metabolic syndrome was analyzed to test the hypothesis that there was a dose–response association.

The study proved to show such a dose–response relationship (FIG. 2). Measurement of work "stress" was based on the iso-strain model using self-report questionnaires, which capture perceptions of the psychosocial work environment. Iso-strain is defined as high job demands, low job control, and low social support at work from co-workers and/or supervisors. The social gradient in metabolic syndrome is explained to a small degree by taking account of iso-strain (approximately 6%) compared to 50% when iso-strain, health behaviors and obesity were entered into the model. Nevertheless, a causal inference can be drawn from this study. The prospective stress–metabolic syndrome relation was robust to adjustment for social position and adverse health behaviors.

A related 19-year prospective study showed that work stress was linked to development of obesity, including central obesity, during mid-life.[20] The study shows that in addition to the known effects of positive energy balance, there may be a psychosocial aspect to the secular trend in obesity prevalence.

BEYOND WORK?

There are parallels between work hierarchies and wider social structure. Political philosophers such as John Rawls have argued that the distribution of autonomy (or control), reciprocity and justice are important qualities of the social order, rooted in human preference. These qualities are partly psychosocial in nature, and provide a novel basis to consider why health and disease are unequally distributed across social strata. In the work context, Whitehall II shows that both perceived lack of reciprocity (from by the effort reward imbalance questionnaire) and a sense of injustice in the workplace predict CHD.[21] Further follow-up of the cohort as participants move into retirement will enable us to examine health inequalities and the roles of psychosocial factors and the metabolic syndrome during the third age.

Role of Neuroendocrine Pathways in Insulin Resistance and Type 2 Diabetes

There is now much support for psychosocial stress as a risk factor for metabolic syndrome and type 2 diabetes, and development of insulin resistance can be a common pathway in this context.[22,23] Insulin resistance can be defined as an attenuated effect of insulin in target tissues, mainly muscle, fat, and liver.[24] Type 2 diabetes is in most cases caused by a combination of β cell dysfunction and insulin resistance. Physical inactivity, adiposity due to overeating, stress, and smoking are risk factors that interact with susceptibility genes

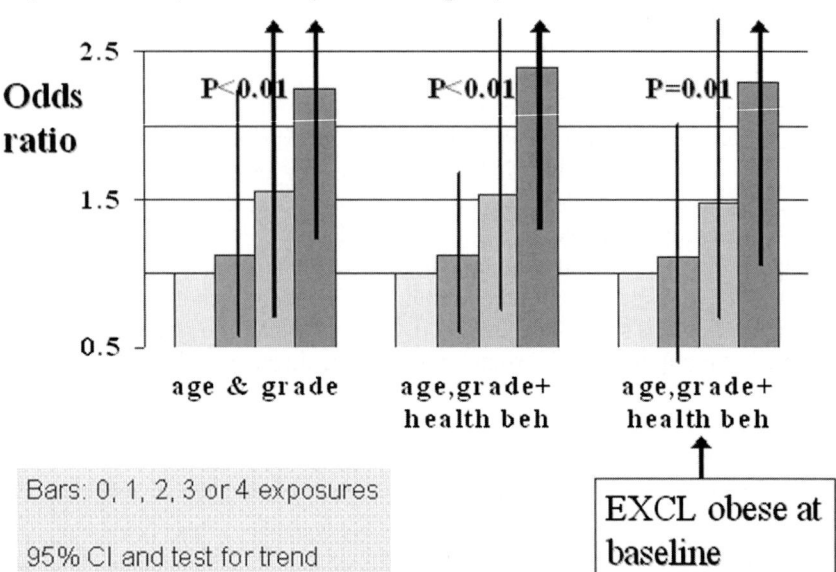

FIGURE 2. Dose–response effect of job stress (iso-strain) on incident metabolic syndrome. Whitehall II study. X-axis labels show adjustment variables. Health beh = health behaviors over four exposure phases (current smoking, no daily fruit and vegetable consumption, heavy alcohol consumption, no exercise) Source: Chandola et al. 2006.

in the development of the disease. The metabolic syndrome is often used to define a cluster of risk markers that predict cardiovascular disease but also type 2 diabetes. Accumulating evidence suggests that abdominal obesity is a central component of this syndrome.[5] Obviously, there is much evidence supporting the importance of genetic factors in human insulin resistance,[25] and there are common polymorphisms that are associated with type 2 diabetes and insulin resistance. However, there is evidence suggesting that insulin resistance is not a primary cellular defect and that there are factors in the surrounding "tissue environment" that can have a causative role. These factors include metabolic, neural, and hormonal signals, and for example, it is well recognized that high levels of glucose and free fatty acids, which are hallmarks of type 2 diabetes, will have detrimental effects in some tissues, for example, muscle and liver.

Neuroendocrine Mechanisms in Metabolic Regulation

Neuroendocrine pathways have received much attention as possible mediators of stress response of the brain. In general, stress can be seen as a threat of the organism's homeostasis, and reactions are elicited at the central nervous, neurohormonal, cellular, and molecular levels that aim to disarm the stressor and restore the equilibrium.[22] However, when overactive or prolonged these defense mechanisms will have detrimental effects on brain function, the cardiovascular system as well as regulation of nutrient metabolism. For rapid signaling from the CNS to peripheral tissues, the autonomic nervous system is used. Essentially all metabolically active tissues, for example, skeletal muscle, heart, adipose, and liver, have autonomic innervations with both sympathetic and parasympathetic nerves. Catecholamine release from the adrenal medulla is also regulated mainly by autonomic nerve activity. For more long-term communication the brain can use the hypothalamopituitary hormonal systems and other neuroendocrine pathways. In this context, prolonged elevation of insulin-antagonistic hormones like cortisol [26] and growth hormone can contribute to insulin resistance in various tissues. Moreover, dysregulation of the autonomic nervous system might be a potential mechanism for early insulin resistance in the development of type 2 diabetes.[27]

Clinical conditions with endogenous or exogenous glucocorticoid excess are associated with insulin resistance, glucose intolerance, central obesity, and hypertension, that is, features of the metabolic syndrome. In clinical obesity, there are alterations in cortisol metabolism, and an enhanced local conversion of cortisone to cortisol via 11β-hydroxysteriod dehydrogenase type 1 in the adipose tissue may contribute to the development of the metabolic syndrome. The insulin-antagonistic effects of cortisol include impairment of insulin signaling and glucose uptake as well as enhanced hepatic glucose output and an increased adipose tissue lipolysis.[22] Moreover, there are data suggesting that there are detrimental effects on insulin secretion and β cell survival.[28] Besides glucocorticoids, gonadal steroid hormones, mainly estrogen and testosterone, are of interest. Altered levels of these hormones have been implicated in abdominal obesity and insulin resistance, obviously with different patterns in men and women.[29]

The autonomic nervous system can also be involved in insulin resistance and development of type 2 diabetes.[27] In healthy subjects, insulin resistance appears to be associated with an altered balance in the autonomic nervous system with a relative increase in sympathetic versus parasympathetic activity following standardized stress [8] or following hyperinsulinemia.[30]

Adipokines in Inflammation and Insulin Resistance

The role of hormones and other molecules secreted by the adipocyte and other cell types, for example, macrophages, in adipose tissue has been extensively investigated during recent years. Cytokine-like factors secreted by adipose tissue are often designated "adipokines." Some of these, for example, TNF-α and interleukin (IL)-6, have been indicated as culprits in the development of insulin resistance. IL-6 levels display a strong association with insulin resistance and type 2 diabetes,[31] and this cytokine can directly inhibit insulin receptor signal transduction in hepatocytes as well as adipocytes.[32] However, the metabolic effects of IL-6 are complex, and in the brain, there is both an activation of the HPA axis and also effects that prevent obesity, as demonstrated in animal experiments. Leptin is a crucial signal from the adipose tissue to the brain that induces satiety and promotes CNS-mediated increase in energy expenditure.[33] These effects of leptin are exerted in the hypothalamus via inhibition of NPY neurons and activation of POMC neurons in the arcuate nucleus, hence leading to MC4 receptor activation in the paraventricular nucleus. Leptin levels are associated with insulin resistance and obesity, whereas there is an inverse relationship for adiponectin.[34] There are probably also several other known and unknown factors secreted by adipose tissue, such as retinol-binding protein 4, MCP-1 etc., which are linked to insulin resistance. High FFA levels directly and indirectly will contribute to insulin resistance. FFAs in the portal vein can lead to activation of the HPA axis, but in addition may elicit a sympathoadrenergic response.[35] In type 2 diabetic patients, we observed that hyperglycemia is associated with both elevated cytokine and cortisol levels in the circulation.[36] Taken together, there are data that support a vicious circle connecting visceral adiposity, neuroendocrine overactivity, insulin resistance/hyperglycemia, and proinflammatory cytokines (FIG. 3).

Stress, Visceral Adiposity, and Diabetes

Stressful situations will lead to neuroendocrine responses, which in the long-term perspective might be important in the development of insulin resistance and type 2 diabetes. The HPA axis together with the sympathetic nervous system can mediate effects of perceived stress in different organs. The downstream hormones of the HPA axis, and the sympathoadrenergic system, that is, cortisol and adrenaline and noradrenaline, respectively, are known to oppose the effects of insulin. Some studies suggest elevated cortisol levels in situations, such as work stress and unemployment.[37,38] A chronically stressed HPA axis appears to display a decreased diurnal variability.[22] Bjorntorp and co-workers proposed that a hypothalamic arousal may contribute to insulin resistance via excess cortisol production,[23] but then there might be a shift to a "burn-out" phenomenon with low secretion of cortisol, growth hormone, and sex steroids.

There is a clear relationship between central fat storage, that is, visceral obesity, and features of the metabolic syndrome. The causal relationship is not established, but the association of visceral fat accumulation in the development of insulin resistance and type 2 diabetes has been generally accepted.[39] A link between central obesity and HPA axis dysregulation has also been suggested.[40] Moreover, an altered activity in the sympathetic and parasympathetic nervous

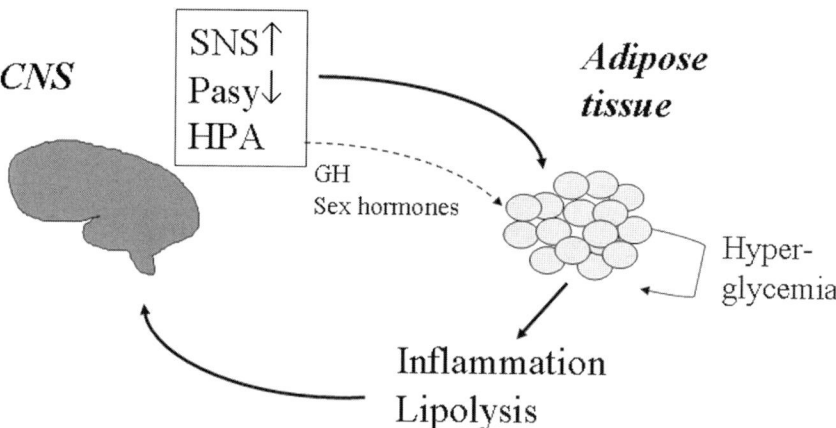

FIGURE 3. A hypothetical vicious circle. Neuroendocrine stress response leads to adipose tissue insulin resistance and lipolysis. This promotes hyperglycemia, via interactions with pancreas, muscle and liver, which in turn can elicit release of proinflammatory cytokines from adipose tissue. Such cytokines can together with elevated FFA levels further activate the neuroendocrine pathways in the CNS. GH = growth hormone; HPA = hypothalamopituitary-adrenal system; Pasy = parasympathetic nervous system; SNS = sympathetic nervous system.

systems may be associated with visceral obesity.[41] One attractive hypothesis is that in a situation of calorie overload, subcutaneous adipose tissue eventually reaches its upper limit for further triglyceride storage, and that this triggers adipose inflammation and lipid "spill-over" that is diverted to visceral fat and with time also to ectopic locations, that is, in liver and muscle.

As mentioned before in this article, there are now several studies that link psychosocial factors to the metabolic syndrome. Stressful life events, low educational level, low sense of coherence, work stress, low emotional support as well sleeping disorders have all been associated with the development of type 2 diabetes.[42,43] Taken together, much evidence support that psychosocial stress that leads to long-term neuroendocrine dysregulation is likely to increase the risk for type 2 diabetes.

Antioxidants Gene, HO-1 in Type 2 Diabetes

Heme oxygenase-1 (HO-1) is the rate-limiting enzymatic step that catalyzes the breakdown of heme into equimolar amounts of biliverdin, an antioxidant rapidly converted to bilirubin, and carbon monoxide (CO), an antiapoptotic vasodilator, with the release of its iron moiety.[9] Oxidant stress strongly induces heme oxygenase-1 (the inducible form of HO), which guards against oxidative insult. Upregulation of HO-1 decreases oxidative stress, attenuates endothelial

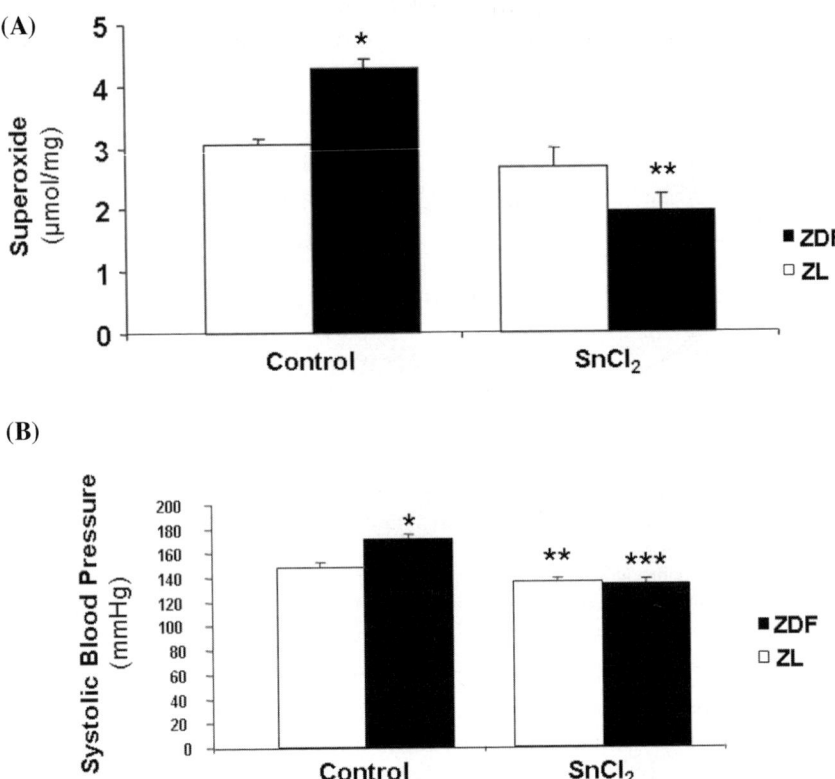

FIGURE 4. (**A**) Superoxide levels in aortic tissue from 22-week-old Zucker lean (ZL) and Zucker diabetic fatty (ZDF) (mean ± SE; $n = 4$), $*P < 0.005$ versus ZL control; $**P < 0.05$ versus ZDF control. (**B**) Systolic blood pressure (SBP) measurements from 22-week-old ZL and ZDF, $*P < 0.001$ versus ZL control, $**P < 0.05$ versus ZL control; $***P < 0.000001$ versus ZDF control and $P < 0.05$ versus ZL control.

cell sloughing and fragmentation, and restores endothelial cell function in experimental diabetes.

Effect of HO-1 on Vascular O_2^- Levels and Blood Pressure in Diabetes

Levels of both O_2^- (FIG. 4A) and blood pressure (FIG. 4B) 22-week-old Zucker diabetic fatty (ZDF) were significantly elevated compared to Zucker lean (ZL) and were reduced by the induction of HO. O_2^- levels were 4.26 ± 0.16 μmol/mg in ZDF compared with 3.05 ± 0.09 μmol/mg in ZL ($P < 0.005$). HO induction significantly reduced levels of O_2^- in ZDF to 1.96 ± 0.27 μmol/mg ($P < 0.05$) while an observed decrease in ZL (2.676 ± 0.31 μmol/mg) was not statistically significant. ZDF treated with $SnCl_2$ demonstrated O_2^- levels equivalent to 8-week-old control ZL while heme levels were reduced to levels significantly below 8-week-old ZL ($P < 0.05$) and ZDF ($P < 0.005$) controls.

As an assessment of the potential clinical benefits of HO-1 induction, we measured SBP and circulating endothelial cells (CEC) following treatment with $SnCl_2$. SBP was significantly lower in control ZL than in ZDF, demonstrating the link between T2DM and hypertension (FIG. 4B). Following HO induction with $SnCl_2$, SBP was decreased in ZL and in ZDF.

Type 2 Diabetes, Hyperglycemia, and Chronic Oxidative Stress

The pathogenesis of type 2 diabetes, also known as adult-onset diabetes, is usually attributed to a combination of pancreatic islet β cell dysfunction and resistance to the action of insulin in important targets, such as liver, muscle, and fat tissue. The cause of β cell failure is polygenic in nature, whereas the cause of insulin resistance is at least partially explained by associated obesity. It is important to note, however, that many type 2 diabetic people are not obese, and that the majority of obese individuals do not develop type 2 diabetes. Thus, it appears that type 2 diabetes is primarily a genetic disease that can be made worse, but is not caused, by excessive body fat.

A major issue in the field of type 2 diabetes research is why a continual and inexorable decline in glucose control occurs despite optimal drug treatment. This decline in β cell function is associated with chronically elevated blood glucose levels leading to the notion of β cell exhaustion because of either continual stimulation by glucose, or that high concentrations of glucose are chemically toxic to the β cell. There is an intrinsic paradox at play since glucose is considered to be a physiologic compound and supportive of β cell function at many levels from insulin gene transcription through insulin secretion.

One concept that has emerged is that of glucose toxicity, that is, chronically high glucose levels form metabolites that can be harmful. This has led to the idea that glucose toxicity of the β cell might be attributable to formation of excess levels of reactive oxygen species (ROS). One can envision that the normal route of glycolysis and oxidative phosphorylation might become oversaturated with glucose. This in turn might lead to shunting excess traffic of glucose molecules along any of several alternative routes, including methylglyoxal formation and glycation; enediol and α-ketoaldehyde formation (glucoxidation); diacylglycerol formation and protein kinase C activation; glucosamine formation and hexosamine metabolism; and sorbitol metabolism.[11] One needs only to imagine a flooding of all these pathways by glucose as a mechanism for excessively high concentrations of ROS in many tissues, including pancreatic β cells.

This pathophysiologic construction suggests that ROS, which like glucose function as positive chemical mediators in physiologic processes, become negative forces in excess concentrations. Both ROS and glucose have good and evil sides, depending on whether their levels are normal or excessive leading to the term glucose toxicity with a major mechanism of action being chronic oxidative stress.

Experimental Evidence for Chronic Oxidative Stress as a Mechanism for Glucose Toxicity of the β Cell

The concept of glucose toxicity was first proposed in 1985[44] and the first biochemical and molecular evidence to support this hypothesis at the level of the β cell was reported in the early 1990s. In studies using the β cell line HIT-T15, serial observations were made over many passages that demonstrated chronic exposure of β cells led to decreased insulin gene expression, insulin stores, and glucose-induced insulin secretion.[12] Protein levels of two critically important transcription factors, PDX-1 and MafA, were low to nondetectable after prolonged culturing of HIT-T15 cells in media containing suprphysiologic concentrations of glucose.[45,46] It was determined that the molecular mechanism for decreased PDX-1 levels was posttranscriptional while the mechanism for decreased MafA was posttranslational.[47] The transcriptional machinery needed for insulin gene expression was not abnormal,[45] so attention was focused on the insulin promoter region through a series of studies involving mutation of the PDX-1 and MafA DNA-binding sites and reconstitution of glucotoxic β cells with these two proteins.[45–47] Mutation of the insulin promoter-binding sites for either protein in nonglucose toxic β cells led to marked decreased in promoter activity. Reconstitution of glucotoxic β cells by transient transfection of either PDX-1 or MafA lead to improved promoter activity. It was recently shown that adenoviral reconstitution of HIT-T15 cells with both PDX-1 and MafA fully restored insulin promoter reporter activity and partially normalized levels of insulin mRNA[47] as shown in FIGURE 5.

It became evident from the concepts put forward by Wolff and colleagues[48] that the mechanism of glucose toxicity might involve generation of ROS, specifically from glucose autoxidation. This led to early work assessing whether the defects in insulin gene expression and abnormal insulin secretion associated with exposure to high glucose concentrations could be ameliorated by antioxidants. Treatment of db/db mice and ZDF rats with NAC preserved insulin gene expression and β cell function.[49,50] Inclusion of NAC or aminoguanidine in media containing suprphysiologic concentrations of glucose protected HIT-T15 cells against the loss of PDX-1 and insulin gene expression[51] as shown in FIGURE 6.

A Central Role for Glutathione Peroxidase in the β Cell

The status of antioxidant enzyme expression in pancreatic β cells is of major importance to the thesis that chronic oxidative stress may cause β cell

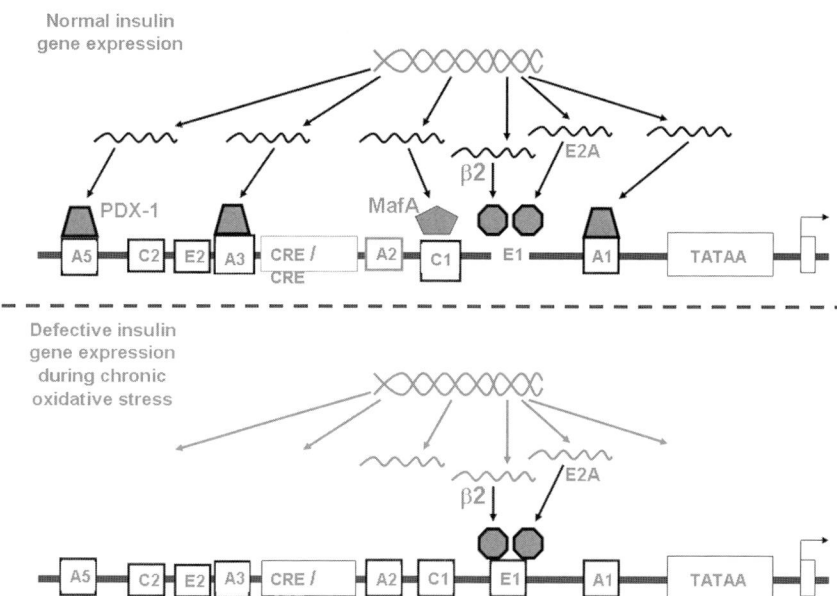

FIGURE 5. Molecular mechanisms of actions leading to defective insulin gene expression in glucotoxic β cells. In this model, insulin gene transcription is intrinsically normal, but gene expression of two critical transcription factors, PDX-1 and MafA, is not. Normally, PDX-1 binds to the insulin promoter at three sites, A1, A3, and A5. MafA binds to C1 only. During the development of glucose toxicity, PDX-1 fails to become expressed because of a posttranscriptional defect, and MafA fails to become expressed because of a posttranslational defect. Consequently, insulin gene expression at the mRNA level decreases, as does insulin stores and glucose-induced insulin secretion. Taken from Ref. 11.

dysfunction. The pancreatic islet is unusual in that it contains the lowest complement of antioxidant enzymes of any other tissue. Isolated islets from rodents contain very low levels of SOD-1, SOD-2, catalase, and glutathione peroxidase mRNA, protein, and activity levels.[52] Similar observations have been made using human isolated islets raising the question whether the islet is designed so that low levels of ROS are purposely encouraged to facilitate processes, such as gene transcription and exocytosis. These low levels of antioxidant enzymes become, however, a handicap when excess levels of ROS are formed in the β cell or surrounding tissue.

This paradoxical situation wherein low levels of ROS might be good for the β cell, but higher levels might not be, led to a series of experiments in which rodent and human islets exposed to high concentrations of glucose formed greater ROS concentrations than islets exposed to normal glucose concentrations.[51] Therefore, the potential protective effect of glutathione peroxidase overexpression was tested in islets exposed to high concentrations of glucose. Glutathione peroxidase was chosen because it metabolizes both hydrogen

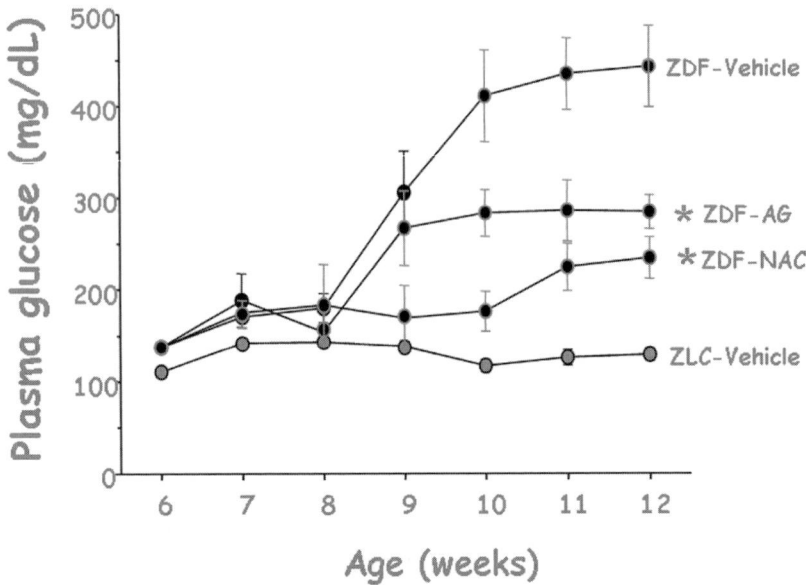

FIGURE 6. Plasma glucose levels in ZDF rats, a genetic model of type 2 diabetes, who were treated with placebo, n-acetylcysteine, or aminoguanidine beginning at 6 weeks of age. Both drugs are antioxidants and both drugs ameliorated the degree of hyperglycemia developed by the animals. ZL controls that do not develop hyperglycemia are shown for comparison. Taken from Ref. 49.

peroxide and lipid peroxides. SOD-1 and SOD-2 were excluded because their catabolization of superoxide forms hydrogen peroxide, an ROS. Catalase was not chosen because it catabolizes hydrogen peroxide, but not lipid peroxides. Glutathione peroxide overexpression increased activity of the enzyme sixfold, roughly equivalent to the activity present in liver, thus preventing the deleterious effects of ribose on insulin gene expression, insulin content, and glucose-induced insulin secretion.[51]

This led the Robertson lab to assess the potential benefits of transgenic experiments in which glutathione peroxidase is overexpressed in the β cells of animal models of type 2 diabetes. Overexpression of this enzyme was examined in β cells in db/db mice to determine if it would protect them from β cell deterioration as they begin to develop hyperglycemia and, if so, whether this would result in attenuation of β cell failure (FIG. 7), as seen in the case of NAC and aminoguanidine treatment of ZDF rats. If this proves to be the case, it will encourage the development of glutathione peroxidase mimetics for use in preclinical and clinical trials. This approach will determine whether such drugs will provide a novel, ancillary layer of protection to patients being treated with conventional antidiabetic drugs, which usually do not completely normalize blood glucose concentrations in patients with diabetes.

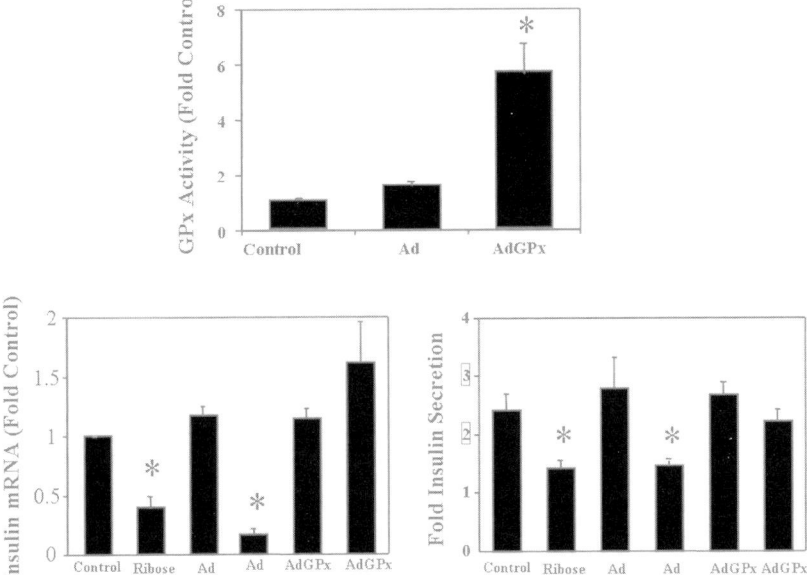

FIGURE 7. Preventive effects of adenoviral infection of GPx cDNA in isolated islets against the oxidative effects of ribose. Ribose in noninfected cells decreases insulin gene expression and glucose-induced insulin secretion from rat islets *in vitro* (two lower panels). Adenoviral infection of islets with GPx cDNA increases intrinsic GPx activity in islets sixfold (top panel) and prevents the adverse effects of ribose on islets (bottom panels). Infection with virus not containing GPx (Ad) has no effects on islets under control conditions or on the adverse effects of ribose on islet function (bottom panels). Taken from Ref 51.

CONCLUSION

There are now several studies that link psychosocial factors to the metabolic syndrome and type 2 diabetes. Much effort has been made to find the mechanisms of insulin resistance in type 2 diabetes and the metabolic syndrome, there is, however, still a need for better understanding. Obviously, both genetic and acquired factors are of importance. It is likely that either humoral or neural mechanisms rather than intrinsic defects in insulin's target cells are primary perturbations. An enhanced neuroendocrine "vulnerability" upon stressful environmental stimuli could be of importance, and this could be defined by inherited as well as acquired factors. The intrinsically low level of antioxidant gene expression in β cells puts them at particular risk for oxidative damage from environmental oxidants. Furthermore, uncontrolled hyperglycemia generates ROS, which further damage already compromised β cells in type 2 diabetes. Therapy directed toward increasing antioxidant protection may facilitate the management of type 2 diabetes.

ACKNOWLEDGMENTS

This work was supported by NIH grants DK068134, HL55601, and HL34300 (N.G.A), financial support was given by the Swedish Research Council (Medicine 14287) and the Swedish Diabetes Association, and the scientific contributions by Jonas Burén, Stina Lindmark, Magdalena Lundgren, Frida Renström, and Maria Svensson are gratefully acknowledged (J.W. E) and NIH grant DK-38325 (RPR).

REFERENCES

1. BRUNNER, E.J., M.G. MARMOT, K. NANCHAHAL, et al. 1997. Social inequality in coronary risk: central obesity and the metabolic syndrome. Evidence from the WII study. Diabetologia **40:** 1341–1349.
2. BRUNNER, E.J., H. HEMINGWAY, B.R. WALKER, et al. 2002. Adrenocortical, autonomic, and inflammatory causes of the metabolic syndrome. Circulation **106:** 2659–2665.
3. BRUNNER, E.J., G. Davey SMITH, M.G. MARMOT, et al. 1996. Childhood social circumstances and psychosocial and behavioural factors as determinants of plasma fibrinogen. Lancet **347:** 1008–1013.
4. BRUNNER E & M. MARMOT. 2006. Social organization, stress, and health. *In Social Determinants of Health*. M. Marmot & R.G Wilkinson, Eds.: p. 6–30. Second edition. Oxford University Press. Oxford.
5. NORBERG, M., H. STENLUND, C. ANDERSSON, et al. 2007. Metabolic syndrome in the prediction of type 2 diabetes- no independent role of inflammation or dyslipidemia. Obesity. In press.
6. ZIERATH, J.R., D. GALUSKA, L.A. NOLTE, et al. 1994. Effects of glycaemia on glucose transport in isolated skeletal muscle from patients with NIDDM: *in vitro* reversal of muscular insulin resistance. Diabetologia **37:** 270–277.
7. BUREN, J., S. LINDMARK, F. RENSTROM & J.W. ERIKSSON. 2003. *In vitro* reversal of hyperglycemia normalizes insulin action in fat cells from type 2 diabetes patients: is cellular insulin resistance caused by glucotoxicity *in vivo*? Metabolism **52:** 239–245.
8. LINDMARK, S., U. WIKLUND, P. BJERLE & J.W. ERIKSSON. 2003. Does the autonomic nervous system play a role in the development of insulin resistance? A study on heart rate variability in first-degree relatives of Type 2 diabetes patients and control subjects. Diabet. Med. **20:** 399–405.
9. ABRAHAM, N.G. & A. KAPPAS. 2005. Heme oxygenase and the cardiovascular-renal system. Free Radic. Biol. Med. **39:** 1–25.
10. KRUGER, A.L., S. PETERSON, S. TURKSEVEN, et al. 2005. D-4F induces heme oxygenase-1 and extracellular superoxide dismutase, decreases endothelial cell sloughing and improves vascular reactivity in rat model of diabetes. Circulation **23:** 3126–3134.
11. ROBERTSON, R.P. 2004. Chronic oxidative stress as a central mechanism for glucose toxicity in pancreatic islet beta cells in diabetes. J. Biol. Chem. **279:** 42351–42354.

12. ROBERTSON, R.P., H.J. ZHANG, K.L. PYZDROWSKI & T.F. WALSETH. 1992. Preservation of insulin mRNA levels and insulin secretion in HIT cells by avoidance of chronic exposure to high glucose concentrations. J. Clin. Invest. **90:** 320–325.
13. FORD, E.S., W.H. GILES & W.H. DIETZ. 2002. Prevalence of the metabolic syndrome among US adults. Findings from the Third National Health and Nutrition Examination Survey. JAMA **287:** 356–359.
14. CANNON, W.B. 1929. Organization for physiological homeostasis. Physiol. Rev. **9:** 399–431.
15. MCEWEN, B.S. 1998. Protective and damaging effects of stress mediators. N. Engl. J. Med. **338:** 171–179.
16. RENNIE, K.L., N. MCCARTHY, S. YAZDGERDI, et al. 2003. Association of the metabolic syndrome with both vigorous and moderate physical activity. Int. J. Epidemiol. **53:** 600–606.
17. SIEGRIST, J. & R. PETER. 1997. Chronic work stress is associated with atherogenic lipids and elevated fibrinogen in middle-aged men. J. Int. Med. **242:** 149–156.
18. SCHNALL, P.L. & P.A. LANDSBERGIS. 1994. Job strain and cardiovascular disease. Ann. Rev. Public Health **15:** 381–411.
19. VRIJKOTTE, T.G.M., L.J.P. VAN DOORNEN & E.J.C. DE GEUS. 1999. Work stress and metabolic and hemostatic risk factors. Psychosom. Med. **61:** 796–805.
20. BRUNNER, E.J., T. CHANDOLA & M.G. MARMOT. 2007. Prospective effect of job strain on general and central obesity in the Whitehall II study. Am. J. Epidemiol. **165:** 828–837.
21. KIVIMAKI, M., J.E. FERRIE, E.J. BRUNNER, et al. 2005. Justice at work and reduced risk of coronary heart disease among employees: the Whitehall II study. Arch. Intern. Med. **165:** 2245–2251.
22. KYROU, I., G.P. CHROUSOS & C. TSIGOS. 2006. Stress, visceral obesity, and metabolic complications. Ann. N. Y. Acad. Sci. **1083:** 77–110.
23. BJORNTORP, P., G. HOLM & R. ROSMOND. 1999. Hypothalamic arousal, insulin resistance and type 2 diabetes mellitus. Diabet. Med. **16:** 373–383.
24. BIDDINGER, S.B. & C.R. KAHN. 2006. From mice to men: insights into the insulin resistance syndromes. Annu. Rev. Physiol. **68:** 123–158.
25. SLADEK, R., G. ROCHELEAU, J. RUNG, et al. 2007. A genome-wide association study identifies novel risk loci for type 2 diabetes. Nature **445:** 881–885.
26. RIZZA, R.A., L.J. MANDARINO & J.E. GERICH. 1982. Cortisol-induced insulin resistance in man: impaired suppression of glucose production and stimulation of glucose utilization due to a postreceptor detect of insulin action. J. Clin. Endocrinol. Metab. **54:** 131–138.
27. BRUCE, D.G., D.J. CHISHOLM, L.H. STORLIEN, et al. 1992. The effects of sympathetic nervous system activation and psychological stress on glucose metabolism and blood pressure in subjects with type 2 (non-insulin-dependent) diabetes mellitus. Diabetologia **35:** 835–843.
28. DELAUNAY, F., A. KHAN, A. CINTRA, et al. 1997. Pancreatic beta cells are important targets for the diabetogenic effects of glucocorticoids. J. Clin. Invest. **100:** 2094–2098.
29. BJORNTORP, P. & R. ROSMOND. 2000. Neuroendocrine abnormalities in visceral obesity. Int. J. Obes. Relat. Metab. Disord. **24**(Suppl 2): S80–S85.
30. LAITINEN, T., I.K. VAUHKONEN, L.K. NISKANEN, et al. 1999. Power spectral analysis of heart rate variability during hyperinsulinemia in nondiabetic offspring of type 2 diabetic patients: evidence for possible early autonomic dysfunction in insulin-resistant subjects. Diabetes **48:** 1295–1299.

31. PRADHAN, A.D., J.E. MANSON, N. RIFAI, et al. 2001. C-reactive protein, interleukin 6, and risk of developing type 2 diabetes mellitus. JAMA **286:** 327–334.
32. ROTTER, V., I. NAGAEV & U. SMITH. 2003. Interleukin-6 (IL-6) induces insulin resistance in 3T3-L1 adipocytes and is, like IL-8 and tumor necrosis factor-alpha, overexpressed in human fat cells from insulin-resistant subjects. J. Biol. Chem. **278:** 45777–45784.
33. JEQUIER, E. 2002. Leptin signaling, adiposity, and energy balance. Ann. N. Y. Acad. Sci. **967:** 379–388.
34. WHITEHEAD, J.P., A.A. RICHARDS, I.J. HICKMAN, et al. 2006. Adiponectin—a key adipokine in the metabolic syndrome. Diabetes Obes. Metab. **8:** 264–280.
35. BENTHEM, L., K. KEIZER, C.H. WIEGMAN, et al. 2000. Excess portal venous long-chain fatty acids induce syndrome X via HPA axis and sympathetic activation. Am. J. Physiol. Endocrinol. Metab. **279:** E1286–E1293.
36. LINDMARK, S., J. BUREN & J.W. ERIKSSON. 2006. Insulin resistance, endocrine function and adipokines in type 2 diabetes patients at different glycaemic levels: potential impact for glucotoxicity *in vivo*. Clin. Endocrinol. (Oxf.) **65:** 301–309.
37. ELLER, N.H., B. NETTERSTROM & A.M. HANSEN. 2006. Psychosocial factors at home and at work and levels of salivary cortisol. Biol. Psychol. **73:** 280–287.
38. MAIER, R., A. EGGER, A. BARTH, et al. 2006. Effects of short- and long-term unemployment on physical work capacity and on serum cortisol. Int. Arch. Occup. Environ. Health **79:** 193–198.
39. KAHN, B.B. & J.S. FLIER. 2000. Obesity and insulin resistance. J. Clin. Invest. **106:** 473–481.
40. MARIN, P., N. DARIN, T. AMEMIYA, et al. 1992. Cortisol secretion in relation to body fat distribution in obese premenopausal women. Metabolism **41:** 882–886.
41. LINDMARK, S., L. LONN, U. WIKLUND, et al. 2005. Dysregulation of the autonomic nervous system can be a link between visceral adiposity and insulin resistance. Obes. Res. **13:** 717–728.
42. AGARDH, E.E., A. AHLBOM, T. ANDERSSON, et al. 2003. Work stress and low sense of coherence is associated with type 2 diabetes in middle-aged Swedish women. Diabetes Care **26:** 719–724.
43. NORBERG, M., H. STENLUND, B. LINDAHL, et al. 2007. Work stress and low emotional support is associated with increased risk of future type 2 diabetes in women. Diabetes Res. Clin. Pract. **76:** 368–377.
44. UNGER, R.H. & S. GRUNDY. 1985. Hyperglycaemia as an inducer as well as a consequence of impaired islet cell function and insulin resistance: implications for the management of diabetes. Diabetologia **28:** 119–121.
45. OLSON, L.K., J.B. REDMON, H.C. TOWLE & R.P. ROBERTSON. 1993. Chronic exposure of HIT cells to high glucose concentrations paradoxically decreases insulin gene transcription and alters binding of insulin gene regulatory protein. J. Clin. Invest. **92:** 514–519.
46. SHARMA, A., L.K. OLSON, R.P. ROBERTSON & R. STEIN. 1995. The reduction of insulin gene transcription in HIT-T15 beta cells chronically exposed to high glucose concentration is associated with the loss of RIPE3b1 and STF-1 transcription factor expression. Mol. Endocrinol. **9:** 1127–1134.
47. HARMON, J.S., R. STEIN & R.P. ROBERTSON. 2005. Oxidative stress-mediated, post-translational loss of MafA protein as a contributing mechanism to loss of insulin gene expression in glucotoxic beta cells. J. Biol. Chem. **280:** 11107–11113.

48. WOLFF, S.P. & R.T. DEAN. 1987. Glucose autoxidation and protein modification. The potential role of 'autoxidative glycosylation' in diabetes. Biochem. J. **245:** 243–250.
49. TANAKA, Y., C.E. GLEASON, P.O. TRAN, *et al.* 1999. Prevention of glucose toxicity in HIT-T15 cells and Zucker diabetic fatty rats by antioxidants. Proc. Natl. Acad. Sci. USA **96:** 10857–10862.
50. KANETO, H., Y. KAJIMOTO, J. MIYAGAWA, *et al.* 1999. Beneficial effects of antioxidants in diabetes: possible protection of pancreatic beta-cells against glucose toxicity. Diabetes **48:** 2398–2406.
51. TANAKA, Y., P.O. TRAN, J. HARMON & R.P. ROBERTSON. 2002. A role for glutathione peroxidase in protecting pancreatic beta cells against oxidative stress in a model of glucose toxicity. Proc. Natl. Acad. Sci. USA **99:** 12363–12368.
52. TIEDGE, M., S. LORTZ, J. DRINKGERN & S. LENZEN. 1997. Relation between antioxidant enzyme gene expression and antioxidative defense status of insulin-producing cells. Diabetes **46:** 1733–1742.

Stress Sensitization in Schizophrenia

KUNIO YUII,[a] MICHIO SUZUKI,[b] AND MASAYOSHI KURACHI[b]

[a]*Research Institute of Asperger Disorder, Ashiya University, Graduate School of Education, Hyogo, Japan*

[b]*Department of Neuropsychiatry, University of Toyama Graduate School of Medicine and Pharmaceutical Sciences, Toyama, Japan*

> ABSTRACT: It is well known that environmental factors, such as early life events, perinatal damage, and urbanicity, which interact with multiple genes, induces persistent sensitization to stress possibly through an imbalance in interactions between dopaminergic and glutamatergic systems. This stress sensitization may be critical in the development or relapse of schizophrenia. The neural correlates of a negative mood might be impaired, resulting in stress sensitization and difficulties in social adjustment (Dr. Habel). Urbanicity is associated with later schizophrenia. Metabolic stress induces stress sensitization via dysregulation of dopaminergic and/or noradrenergic systems in activated HVA and cortical response (Dr. Marcelis). The glutamatergic regulation activates HPA axis in stress response (Dr. Zelena). Ameloblast activity in human molar's enamel slowed by exposure to stress, and the segment of enamel rods is smaller, making a particular dark line. Stress sensitization may be induced at the age of 10.5 to 11.5 years resulting from severe emotional stress at the age of 10.5 to 11.5 years (Dr. Yui). It has been reported that volume reductions in the amygdala, hippocampus, superior temporal gyrus, and anterior parietal cortex common to both patient groups may represent the vulnerability to schizophrenia, while volume loss of the prefrontal cortex, posterior parietal cortex, cingulate, insula, and fusiform cortex preferentially observed in schizophrenia may be critical for overt manifestation of psychosis (Dr. Suzuki).
>
> KEYWORDS: schizophrenia; stress sensitization; negative affect; glutamate; ameloblast stress line; volume loss of cortex

INTRODUCTION

The vulnerability of stress model has been widely accepted as a heuristically useful framework for the study of the etiology and clinical course of

Address for correspondence: Kunio Yui, M.D., Ph.D., Research Institute of Asperger Disorder, Ahiya University Graduate School of Education. Rokurokuso-Machi 13–22, Ashiya, 659-8511 Hyogo, Japan. Voice: +81-797-23-0661; fax: +81-797-23-1901.

kyui@r7.dion.ne.jp

schizophrenia.[1] There is accumulating evidence indicating that early environmental adversity, such as prenatal infections and nutrition, early life stressors, and urbanicity, are more common in people with schizophrenia than in general populations. It is well known that these environmental adversities induce enhanced reactivity to a mild and secondary adversity, called "stress sensitization," and environmental factors cause multiple genes to interact, leading to the relapse and/or development of schizophrenia via stress sensitization. Moreover, the effects of the childhood environment, favorable or unfavorable, interact with all process of neurodevelopment. If interconnections with peers are lacking, the brain development of cognition is damaged in a lasting fashion. A cognitive disability of this kind may induce deviated reaction to environments, resulting in sensitization to stress.[2] We will argue below the impact of environmental factors on reactivity to stress, and neurobiological mechanisms underlying stress sensitization.

ENVIRONMENTAL FACTORS

Heightened reactivity to stressors is a cardinal feature of schizophrenia and is considered the core of the constitutional vulnerability that forms the diathesis in the stress vulnerability model. It was well documented that early traumatic events may be important in the cause of stress sensitization via a long-term dysregulation of dopaminergic systems, resulting in the consequent emotional reactivity to daily stress in adult life[3,4] as reflected, in an increase in negative emotion.[5]

Response to Daily Stress: Findings of Habel and Others

Habel and colleagues[6] reported that adult as well as adolescent subjects with schizophrenia experienced more negative affects during happy as well as sad mood induction. They suggested a trend toward reduced specificity for sadness and a higher negative bias.[7] In the light of another study[7] indicating that negative and happy moods reveal distinct cortical activation foci within a common neural network, the neural correlates of a negative mood might be somewhat impaired. Such impairment may lead to a misunderstanding of social communication and may underlie difficulties in social adjustment in schizophrenics, leading to increased reactivity to stress.[6] While, another study suggests that emotional reaction to stress in the daily lives of schizophrenics may not be a consequence of cognitive impairments and in fact depend on clinical outcome or psychotic states (e.g., periodic type, predominance of positive symptoms) of illness.[8] They have further reported that small stressors in daily life associated with emotional responses induced stress sensitization, independent of cognitive impairments, and argued to constitute of an emotional pathway to psychosis.[9,10] The distinct findings between the two studies may be due to methodological difference, that is, an emotion discrimination test[7] versus an assessment of emotional reaction to stress.[8]

Critical Period of Exposure to Stress

Previous studies reported that increased reactivity to daily life stress was significantly prominent in subjects who had experienced trauma either before the age of 10 years[5] or before puberty.[2] More recently, it has been reported that psychological trauma (e.g., physical threat, serious accident, terrible even, sexual abuse, and rape) experienced before the age of 13 years was associated prospectively and in a dose-response fashion with an onset of psychotic symptoms (e.g., delusions, auditory hallucinations, and though interference) at follow-up for average 42 months.[11] Collectively, the consequences of child trauma appear more detrimental when it occurs at a younger age, namely, when brain development is in a vulnerable stage in brain development.[5] Regarding the vulnerability age, 25% of children who had delusions or abnormal body sensitization at the age of 11 years met the criteria for a diagnosis of schizophreniform disorder by 26 years.[12] Collectively, it is plausible that the age before 13 or 11 years appears to be a crucial time related to the development of schizophrenia in adulthood.

Environmental Risk Factors: Findings of Marcelis and Others

Marcelis et al.[13] followed up all live births recorded between 1942 and 1978 in any of the 646 Dutch municipalities through the National Psychiatric Case Register for first admission for psychosis between 1970 and 1992 ($N = 42{,}115$). Urban birth was associated with later schizophrenia. Effect of urban birth was greatest for schizophrenia in individuals from more recent birth cohorts, for men and for early age of onset of illness.[13] Genetically predisposed individuals run a higher risk of developing illness in adverse environments (e.g., divorce, noise pollution, crime, cannabis use, virus infection, nutritional deficiency, and exposure to pollutants) operating either around the time of birth or at the time of illness onset).[13] Urban birth may be related to developmental vulnerability for adult mental illness.[13] The possible etiological mechanisms remain unclear and warrant further investigation. According to recent study including a large sample (1,020,063) of people, there were significant effects of urbanicity and family history of schizophrenia, indicating interactions between a proxy genetic risk factor and a proxy environmental risk factor,[14] around 30% of all schizophrenia incidence, making urbanicity the most important of all environmental factors.[14]

Genetic–Environmental Association

Individuals with a history of childhood trauma who develop schizophrenia may predispose genetic risk.[4] It is recognized that the underlying genetic vulnerability is a function of multiple genes, indication that the risk factor therefore is ubiquitous and variable. Genetic risk factor may show synergism with environmental factors, so that genetic and environmental factors interact with each other, and the combination of their separate weak effects become a joint strong effect. The strong environmental factors may be reduced by genetic factors, or environmental effects may compete with genetic risk factors to cause schizophrenia.[4] It may be concluded that there is no unequivocal association between genetic predisposition and environmental factors.

NEUROBIOLOGICAL BASES OF STRESS SENSITIZATION IN SCHIZOPHRENIA

Traumagenetic Neurodevelopment

Of crucial importance are findings that childhood trauma may induce brain abnormalities that may persist till adulthood. In this regard, the traumagenic neurodevelopmental model is example of a genuine integration of the interactions between social, psychological, and biological factors. A portion of preschizophrenic children shows slight developmental delays, minor cognitive difficulties, and social anxiety, which supports the hypothesis that schizophrenia is a part of neurodevelopmental abnormalities.[15] For example, children who later fulfilled DSM-IV criteria for schizophrenia at the age of 26 years showed significant impairment across a range of developmental demand in neuromotor, language, cognitive, emotional, and interpersonal development from as young as 3 years of age, for example, delay in learning to walk during infancy, an excess of neurologic signs at 3 years of age, and significant impairment on repeated motor testing between 3 and 9 years.[16] They concluded that these impairments may be specific to schizophrenia.[16] These impairments were not merely mediators of the effect of the obstetric complication involving hypoxia fetal growth retardation, and more likely to reflect the expression of schizophrenia-susceptibility gene.[16] Cannon et al.[17] reported that odds of schizophrenia increase linearly with increasing number of hypoxia-associated obstetric complications in specific case with early age at onset or first treatment contact in schizophrenic patients.[17] However, the majority of individuals exposed to hypoxia-associated obstetric complications did not develop schizophrenia; such factors may be incapable of causing schizophrenia on their own. Cannon et al. therefore suggested that hypoxia acts or interactively with genetic predisposition in influencing liability to the illness.[17] As candidate genes that mediate the brain's vulnerability to hypoxic–ischemic neuron injury, Cannon et al.[17] suspected genes participating in N-methyl-D-Aspartate (NMDA) receptor formation or membrane dynamics of glutamatergic neurons since glutamate appears to be a remarkably potent and rapid acting neuronxin that leads from hypoxia to neuron death.[18]

Childhood Trauma and Stress Sensitization

A number of previous studies suggest that early life stress induces long-lived hyperactivity of corticotropin-releasing factor systems and associated smaller hippocampal volume, and consequently induces vulnerability to depression.[10,19] It was reported that left hippocampal volume reduction was only associated with familial liability to schizophrenia, but showed a significant association with a history of obstetric complications.[20] Evidence from neuroimaging studies suggests morphorogicic and functional changes in brain structures involved in the control of stress response in depression. Thus, a reduction in left hippocampal volume has been documented in several studies in subjects with treatment-resistant depression.[19,21] Increased levels of CRF and cortisol during repeated childhood abuse together with persistent hyperactivity and sensitization of the hypothalamopituitary axis associated with depression in adulthood could damage hippocampal neurons.[19]

Perinatal Damage and Dopaminergic Dysregulation

In animal studies, perinatal damage may be related to a labile dopaminergic system vulnerability to sensitization. For example, developmental disruption of temporal cortex may induce dysregulation of the dopaminergic imput to the striatum, increasing the response to mild stress.[22] Kapur[23] hypothesized that the role of dopamine is to mediate the salience of environmental events and internal representations. It has been proposed that a dysregulated, hyperdopaminergic state at a brain level leads to an aberrant assignment of salience to the element of one's experience at a mind levels. Spauwen et al.[11] speculated that exposure to stressors induces permanent dysregulation of the HPA axis, which in turn might underlie the "dopamine sensitization" to stress. Activation of the HPA axis is thus one of the primary manifestations of the stress response. Moreover, dopamine has important role in producing increased sensitivity following prolonged or repetitive exposure to stress, indicating an important ontogenetic mechanism in the formation, even stability, of individual differences in dopamine system reactivity.

Hypothalamopituitary–Adrenal(HPA) Axis: (Findings of Zelena and Others)

The HPA axis is activated in response to acute and chronic stress. Arginine vasopressin (AVP) has been proposed to be an important neuropeptide hormone of the paraventricular neurons during chronic stress in the regulation of the HPA axis.[24] The role of AVP in maintaining adrenocortical responsiveness during chronic stress was studied by Zelena *et al*.[24] They compared plasma levels of adrenocortical hormone (ACTH) and anterior pituitary poopiomelanocortin (POMC) mRNA and ACTH content in the vasopressin-deficient mutant Brattleboro rats, compared to the heterozygous ones (di/+controls). There is no significant difference in the content of POMC mRNA in both mutant and di/+controls by repeated restrain stress. They suggest that AVP is not necessary for activating the HPA axis.[24] They further investigated the role of glutamate in activating the HPA axis via ACTH elevation. They reported that glutamate activates the HPA axis at the level of paraventricular nucleus and that the glutametergic regulation related to AVP had no significant effect.[25]

Homovanillic Acid (HVA) Response: Findings of Marcelis and Others

Marcelis *et al*.[26] studied the response of plasma HVA and cortisol to metabolic stress by using the administration of glucose analogue 2-deoxy-D-glucose (2DG), which induces a glucoprivationstate, in patients with psychosis and nonpsychotic first-degree relatives of patients with psychosis and controls without psychosis. They reported that plasma levels of HVA and cortisol were increased during metabolic stress in patients but not relatives, suggesting that this increased response reflects ill-related effect, possibly the acquired sensitization of neuroendocrine systems by repeated environmental stressors, rather than a genetic vulnerability. They suggested abnormal regulation of dopaminergic and/or noradrenergic systems in activated HVA and cortisol response.[26]

Dopaminergic and Glutamatergic Modulation in Stress Sensitization

Sheitman and Lieberman[27] have proposed "neurochemical sensitization." Genetic and/or epigenetic factors that occur during fetal gestation and early perinatal development induce the failure of neuronal developmental processes and synaptogenesis, which result in a deficiency of the inhibitory capacity of the cortex upon subcortical structures. In adolescence and early adulthood, the redundancy in neural synaptic connections is eliminated through neural pruning and the circuits are refined to the point that the threshold of modulatory capacity is more easily exceeded. In the course of stressful but normative human experiences (e.g., going to college and entering military service), perturbations in neuronal activity that would otherwise be compensated for and equilibrium reestablished, will progressively result in neurochemical sensitization, leading to an onset of schizophrenia. The synaptic regulation of DA or its regulatory neurotransmitters glutamate and GABA may be important in mediating the behavioral pathology.[28] Moreover, Lieberman[28] and Yui et al.[29] suggested endogenous neurochemical sensitization involving mesolimbic–cortical–striatal circuits mediated by DA, and glutamate is important to the onset and in the deteriorative stage. According to another study,[30] in the mesolimbic dopaminergic sensitization, a primary temporolimbic defect involving glutamatergic receptors result in a defect of prefrontal cortex including the widespread cortical glutametergic cells and excitatory amino acid (EAA). A decrease in glutamatergic function results in a reduced tonic DA release in mesolimbic DA cells and thereby reduced DA release upon environmental stimulations in the second trimester of pregnancy. The temporolimbic defect results in a lowering of tonic DA release in NAC and striatum (the third trimester). In early childhood, cortical glutamatergic dysfunction results in decreased striatal GABAergic activity, causing abnormal response rendering the individual's greater sensitization of dopaminergic provocation by stress. When demands upon the system increase in puberty, dopaminergic activation may lead to permanent stress sensitization of the mesolimbic DA system.[30]

The Effects of Cytokine on Stress Sensitization

Infections and other environmental insults during prenatal and perinatal periods may increase reactivity to stress,[31] which may diminish the capacity to compensate for perturbation in neural activity in subsequent stressful experiences. Increased levels of maternal sera cytokine, such as tumor necrosis factor alpha (TNF-α) in relation to infections has been reported as a potential risk factor for schizophrenia among offspring because TNF-α and virus may exert neurotoxic effects, causing altered brain structures and function[31] or induce white matter damage.[32] In regard to *in utero* infection, serum levels of maternal IL-8 (a critically important cytokine in immune response) may be a risk for schizophrenia in offspring,[33] because of a significant association of IL-8 levels during the second trimester and the risk of schizophrenia spectrum disorders in the adult offspring.[34]

Molecular Involvement in Stress Sensitization

There are a few studies on the molecular basis of stress sensitization in psychiatric disorders. Transcriptional regulation of adrenomedullary catecholamine biosynthetic enzymes, such as tyrosine hydroxylase and DA-β hydoxylase, was shown to be a prominent mechanism modulating the response of cold or immobilization stress.[35] Tyrosine hydroxylase transcription is stimulated by the phosphorylation of cAMP response element-binding (CREB) protein.[36] Sensitization to heterotypic stress induced by the preexposure of cold stress exhibited an exaggerated response of phosphorylation/induction of several c-*fos* transcription factors, such as CREB, Fra-2, and Egrl in rat adrenal medulla.[37]

ESTIMATES OF EARLY STRESS USING PERMANENT CHANGES IN HUMAN MOLAR

Objective Marker of Early Stress

Most patients are unlikely to spontaneously disclose sexual or physical abuse to mental staff personnel. We therefore have an obligation to ask. However, assessment of events during development remains highly complex, and early life events are hard to confirm. New approaches to the problem of estimating stress during early brain development are required. Early life events are hard to confirm. New approaches to the problem of estimating stress during early brain development are required. In this regard, human enamel promises accessible repositories of indelible information on stress between gestation and the age of 13 years. Stressful experiences induce long-term activation of the sympathoadrenal system, slowing of tropic parasympathetic functions, and they then induce disrupted secretion of the enamel matrix. During the brain development (in infancy, childhood, and preadolescence), ameloblast activity in human enamel is slowed during 1 to 2 days of extreme stress, and the segment of enamel rods is smaller and often misshapen, making a particular dark line seen by the use of a microscope (we referred this line as pathological stress line, PSL in short). Retzius reported that this line is incremental reflecting the layered apposition of enamel during amelogenesis (Retzius, 1937), and after that this line is termed as the Retzius line.[38,39] The line is conceptually akin to tree rings that are markers of environmental adversity in the tree's life.

Skinner and Anderson[40] reported that the Retzius lines corresponded well with the timing of the specific episode of stress, described in anecdotal reports and medical record. According to a chronology of the permanent dentition (van Bee, 1983),[41] the age band (epoch) reflected the timing of exposure to stress. PSL of the first, second, and third molar reflects the timing of exposure to stress at 3/4 months to 6 years of age, birth or slightly before the age of 3 years, and ages 7 to 14 years, respectively.[41] The factors affecting a disruption of ameloblast are unknown etiological/susceptibility factors, including genetic vulnerability to stress or specially weaken ameloblast. Nutritional factors, such as depleted nutrient stores and inefficient utilization and the availability of key nutrients, are hypothesized to contribute an additional 30% to ameloblast disruption.[42]

PSL in Schizophrenia: Findings of Yui

Subjects

We examined PSL in the third molar in 35 chronic paranoid schizophrenics: 25 males and 10 females, 41.9 ± 13.5 yeas old and 32 normal controls (5 males and 27 females, 28.3 ± 9.1 years old). The changes of density in potassium (P), calcium (Ca), and magnesium (Mg) in PSL were examined by a scanning microscope and an electron probe microanalyzer (EPMA). Since the rate of enamel is well known, PSL was assessed by its length and definition in each enamel at half years from 9 to 13 years old: 0, none; 1, slight; 2, mild; 3, moderate; 4, severe. Mineral changes were examined in PSL portions rated as 2–4 due to the extent of changes of mineral density in one area (1.5 × 1.5 mm^2): 0, none; 1, 1/4; 2, 1/3; 3, 1/2, 4, 2/3.

Results

PSL scores in the 35 schizophrenics were significantly higher than the 32 normal controls (4.8 ± 5.1 vs. 2.0 ± 2.3, $P < 0.01$). The 32 schizophrenics exhibited PSL indicative of stressors experienced at the ages of 10.5 to 11.5 years. Scores in potassium and calcium were significantly decreased and those of magnesium were significantly increased in PSL portions.

Discussion

These findings suggest that stress sensitization may be induced from 10.5 to 11.5 years, which is comparable to previous studies indicating that stress exposed at 10–13 years may be related to the development of adulthood schizophrenia. A decreased density of P and Ca and an increased density of Mg have been reported to be caused by severe emotional stress. The findings on EPMA suggest that severe emotional stress during childhood may induce stress sensitization.

STRUCTURAL BRAIN CHANGES UNDERLYING VULNERABILITY TO SCHIZOPHRENIA

Brain structural changes in schizophrenia are thought to represent a complex and dynamic process in which multiple brain regions are differentially involved. A strategy elaborating on the morphologic characteristics of the schizophrenia spectrum could shed light on the pathological process underlying schizophrenia.[43] Common neurobiological deviations among the schizophrenia spectrum may be essential for the pathogenesis of schizophrenia, but some additional pathological changes may also be required for the development of full-blown schizophrenia. Clarifying the neurobiological similarities and differences between established schizophrenia and schizotypal (personality) disorder, a schizophrenia spectrum disorder without manifestation of overt and sustained psychosis, would potentially discriminate the pathophysiological mechanisms underlying the vulnerability to schizophrenia from those associated with overt psychosis. We have made extensive comparisons of brain morphology by using magnetic resonance imaging (MRI)-based volumetry between patients with schizophrenia and those with schizotypal disorder.

METHODS

Subjects

Thirty-five patients (23 males, 12 females) with schizotypal disorder, 62 patients with schizophrenia (32 males, 30 females), and 63 control subjects (35 males, 28 females) were included in this study. The patients with schizotypal disorder all met the criteria for schizotypal disorder in ICD-10 as well as the criteria for schizotypal personality disorder in DSM-IV. At the time of MRI scanning, six patients were neuroleptic-naïve and 29 patients were being treated with low doses of antipsychotics. All subjects have received consistent clinical follow-up and none of them has developed overt schizophrenia to date. The patients with schizophrenia fulfilled both ICD-10 and DSM-IV criteria for schizophrenia. All schizophrenia patients apart from one female patient were receiving neuroleptic medication. The control subjects consisted of healthy volunteers who were excluded if they had a history of psychiatric, neurological, serious medical or surgical illness, or substance abuse disorder. They were also screened for history of psychiatric disorders in their first-degree relatives. All control subjects were given the Minnesota Multiphasic Personality Inventory to exclude subjects with abnormal profiles. The three groups were matched for age, sex, handedness, height, and parental education. After complete description of the study to the subjects, written informed consent was obtained. This study was approved by the local committee on medical ethics.

MRI Acquisition and Processing

MRI scans were acquired with a 1.5 T scanner (Vision, Siemens Medical System, Inc., Erlangen, Germany). A three-dimensional T1-weighted gradient-echo sequence FLASH (fast low-angle shots) with $1 \times 1 \times 1$ mm voxels was used. Imaging parameters were: TE = 5 msec; TR = 24 msec; flip angle = $40°$; field of view = 256 mm; matrix size = 256×256. The image data were processed on a Unix workstation with the software package Dr. View 5.0.

Volumetric Analysis of Regions of Interest (ROIs)

Volumetric measurements were performed in 23 regions of interest covering the entire cerebral cortex except the occipital lobe. The ROIs included: prefrontal cortices (superior, middle, and inferior frontal gyri, orbitofrontal gyrus, straight gyrus, and ventral medial prefrontal cortex), cingulate gyrus, insula (short and long insula), precentral gyrus, postcentral gyrus, parietal association cortices (superior parietal lobule, supramarginal gyrus, angular gyrus, and precuneus), temporal pole, superior, middle, and inferior temporal gyri, fusiform gyrus, parahippocampal gyrus, hippocampus, and amygdala. Parcellations of the cortex were made, in principle, according to the cerebral gyri/sulci. The detailed procedures for delineations of each ROI were described elsewhere.[43–48]

Statistical Analysis

Effect sizes (Cohen's d) of cortical gray matter changes in each ROI were calculated from relative volumes (100 × absolute ROI volume/ICV) in the schizotypal and schizophrenia patients compared with the controls.

RESULTS

FIGURE 1 shows effect sizes of the gray matter volume changes in 35 schizotypal patients and 62 schizophrenia patients compared with 63 healthy control subjects. In schizophrenia, large volume reductions (Cohen's d $\geq$ 0.8) were observed in the superior temporal gyrus, amygdala, and straight gyrus. Medium volume reductions (0.5 $\leq$ Cohen's d < 0.8) were present in the superior and inferior frontal gyri, cingulate gyrus, insula, fusiform gyrus, postcentral gyrus, inferior parietal lobule, and precuneus. Small volume reductions (0.2 $\leq$ Cohen's d < 0.5) were demonstrated in the hippocampus, middle frontal gyrus, precentral gyrus, and superior parietal lobule. Schizotypal patients showed volume reductions of similar magnitude to those in schizophrenia in the temporal lobe structures, while the prefrontal and posterior parietal cortices were relatively preserved. In addition, modest volume increase in the middle frontal gyrus was observed in schizotypal disorder.

DISCUSSION

Volume reductions common to schizophrenia and schizotypal disorder were revealed in the amygdala, hippocampus, superior temporal gyrus, and anterior parietal cortex. These changes may represent the vulnerability to schizophrenia. Volume decreases in the amygdala/hippocampus frequently reported in

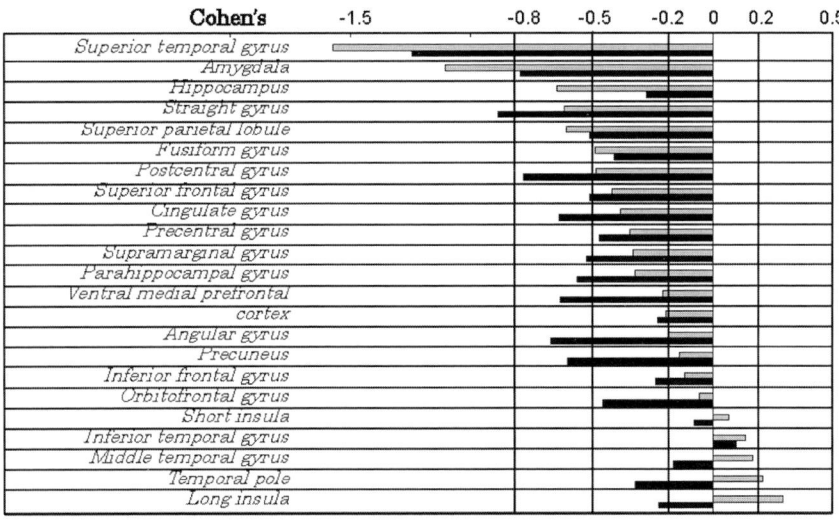

FIGURE 1. Bar graphs representing effect sizes of the gray matter volume changes in 35 schizotypal patients (gray) and 62 schizophrenia patients (black) compared with 63 healthy control subjects.

subjects at genetic risk for schizophrenia[49] support these notions. They are also consistent with the finding of volume reduction of the superior temporal gyrus in community-based sample of male schizotypal subjects.[50] In schizophrenia, further volume reductions were observed in the prefrontal cortex, posterior parietal cortex, cingulate, insula, and fusiform cortex. Functional deficits in these areas, especially the prefrontal cortex, may result in the loss of inhibitory control in other brain regions leading to manifestation of the subclinical temporal lobe dysfunction as positive psychotic symptoms in schizophrenia[43,51] On the other hand, preserved volume of these areas might be related to that schizotypal subjects are protected from overt psychosis. As to the stress vulnerability in schizophrenia, only a few studies have examined the relation of brain morphology with the reactivity to stress[52,53] Future studies are needed to clarify how structural alterations in the brain affect the reactivity to stress as well as how brain morphology is modulated by stress in schizophrenia.

REFERENCES

1. ZUBIN, J. & B. SPRING. 1977. Vulnerability-a new view of schizophrenia. J. Abnorm. Psychol. **86:** 103–126.
2. PERRY, B. 2002. Childhood experience and expression of genetic potential: what childhood neglect tells us about nature and nurse. Brain Mind **3:** 79–100.

3. MYIN-GDERMEYS, I., L. KRABBENDAM, P.A.E.G. DELESPAUL & J. VAN OS. 2003. Do life events have their effect on psychosis by influencing the emotional reactivity to daily life stress. Psychol. Med. **33**: 327–333.
4. READ, J., J. VAN OS, A.P. MORRISON & C.A. ROSS. 2005. Childhood trauma, psychosis and schizophrenia: a literature review with theoretical and clinical implications. Act. Psychiatr. Scand. **112**: 330–350.
5. GLASER, J-P., J. VAN OS, P.T.M. PORTEGIJIS & I. MYIN-GERMEYS. 2006. Childhood trauma and emotional reactivity to daily life stress in adult frequent attenders of general practitioners. J. Psychosom. Res. **61**: 229–236.
6. HABEL, U., I. KRASENBRIK, U. BOWI, et al. 2006. A special role of negative emotion in children and adolescents with schizophrenia and other psychoses. Psychiatry Res. **145**: 9–19.
7. HABEL, U., M. KLEIN, T. KELLERMANN, et al. 2005. Same or difference? Neural correlates of happy and sad mood in healthy males. Neuroimage **26**: 206–214.
8. MYIN-GERMEYS, I., L. KRABBENDAM, J. JOLLES, et al. 2002. Are cognitive impairments associated with sensitivity to stress in schizophrenia? An experience sampling study. Am. J Psychiatry **159**: 443–449.
9. MONROE, S. 1983. Major and minor life events as predictors of psychological distress: further issues and findings. J. Behav. Med. **6**: 189–205.
10. MYIN-GERMEYS, I., J. VAN OS. 2007. Stress-reactivity in psychosis: evidence for an affective pathway to psychosis. Clin. Psychol. Rev. **27**: 60–66.
11. SPAUWEN, J., L. KRABBENDAM, R. LIEB, et al. 2006. Impact of psychological trauma on the development of psychotic symptoms: relationship with proneness. Br. J. Psychiatry. **188**: 527–5330
12. POULTON, R., A. CAPSI, T.F. MOFFITT, et al. 2000. Children's self-reported psychiatric. symptoms and adult schizophreniform disorder: a 15-year longitudinal study. Arch. Gen. Psychiatry **75**: 1053–1058.
13. MARCELIS, M., F. NAVARRO-MAREU, R. MURRAY, et al. 1998. Urbanization and psychosis: a study of 1942–1978 birth cohorts in the Netherlands. Psycho. Med. **28**: 871–879.
14. VAN OS, J., C. PEDERSEN, P. MORTENSEN. 2004. Confirmation of synergy between urbanicity and familial liability in the causation of psychosis. Am. J. Psychiatry **161**: 2312–2314.
15. HOWES, O.D., C. MCDONALD, M. CANNON, et al. 2004. Pathways to schizophrenia: the impact of environmental factors. Int. J. Neuropsychopharm. **7**(Suppl 1): S7-S13.
16. CANNON, M., A. CASPI, T.E. MOFFITT, et al. 2002. Evidence for early-childhood, pan-developmental impairment specific to schizophreniform disorder: results from a longitudinal birth cohort. Arch. Gen. Psychiatry **59**: 449–456.
17. CANNON, T.D., I.M. ROSSO, M. HOLISTER, et al. 2000. A prospective cohort study of genetic and peripheral influences in the etiology of schizophrenia. Schizophr. Bull. **26**: 351–366.
18. CHOI, D.W. & S.M. ROTHMSAN. 1990. The role of glutamate neurotoxicity in hypoxic-ishemic neuronal death. Am. Rev. Neurosci. **13**: 171–182.
19. VYTHILINGA, M., C. HEIM, J. NEWPORT, et al. 2002. Childhood trauma associated with smaller hippocampal volume in women with major depression. Am. J. Psychiatry **159**: 2072–2080.
20. SCHULTS, K., C. MCDONASLD, S. FRANGOU, et al. 2003. Hippocampal volume in familial and nonfamilial schizophrenic probands and their unaffected relatives. Biol. Psychiatry **53**: 562–579.

21. HEIM, C. & C.B. NEMEROFF. 2001. The role of childhood trauma in the neurobiology if mood and anxiety disorders: preclinical and clinical studies. Biol. Psychiatry **49:** 1023–1039.
22. MOORE, H., A.R. WEST, A.A. GRACE. 1999. The regulation of forebrain dopamine transmission: relevance to the pathophysiology and psycopathology of schizophrenia. Biol. Psychiatry **46:** 40–55.
23. KAPUR, S. 2003. Psychosis as a state of aberrant salience: a framework linking biology, phenomenology, and pharmacology in schizophrenia. Am. J. Psychiatry **160:** 13–23.
24. ZELENA, D., A. FOLDES, Z. MERGL, et al. 2004. Effects of repeated restrain stress on hypothalamo-pituitary-adrenocortical function in vasopressin deficient Brattleboro rats. Brain Res. Bull. **63:** 521–530.
25. ZELENA, D., Z. MERGL & G.B. MAKARA. 2005. Glutamate agonists activate the hypothalamo-pituitary-adrenocortical axis through paraventricular nucleus but not through vasopressinergic neurons. Brain Res. 1031**:** 185–193.
26. MARCELIS, M., E. CAVALIER, J. GIELEN, et al. 2004. Abnormal response to metabolic stress in schizophrenia: marker of vulnerability or acquired sensitization? Psychol. Med. **34:** 1103–1111.
27. SHEITMAN, B.B. & J.A. LIEBERMAN. 1998. The natural history and pathophysiology of treatment resistant schizophrenia. J. Psychiat. Res. **32:** 143–150.
28. LIEBERMAN, J.A., D. PERKINS, A. BELGER, et al. 2001. The early stage of schizophrenia: speculations on pathogenesis, pathophysiology, and therapeutic approaches. Biol. Psychiatry **50:** 884–897.
29. YUI, K., S. GOTO, T. IKEMOTO, et al. 1999. Neurobiological basis of relapse prediction in stimulant-induced psychosis and schizophrenia: the role of sensitization. Mol. Psychiatry **4:** 512–523.
30. GLENTHØJ, B.Y. & R. HEMMONHSEN. 1997. Dopaminergic sensitization: implications for the pathogenesis of schizophrenia. Prog. Neuro-Psychopharm. Biol. Psychiat. **21:** 23–46.
31. BUKA, S.L., M.T. TSUANG, E.F. TOREY, et al. 2001. Maternal cytokine during pregnancy and adult psychosis. Brain. Behav. Immun. **15:** 411–420.
32. OLAF, D. & L. Alsan. 2007. Maternal intrauterine infection, cytokine, and brain damage in the preterm newborn. Pediat. Res. **42:** 1–8.
33. VAN, O.S., L. KRABBENDAM, I. MYRIN-GERMEY & P. DELESPAUL. 2005. The schizophrenia envirome. Curr. Opin. Psychiatry **18:** 141–145.
34. BROWN, A.S., J. HOOTON, C.A. SCHAEFER, et al. 2004. Elavated maternal interleukin-8 levels and risk of schizophrenia in adult offspring. Am. J. Psychiatry **161:** 889–895.
35. NANKOVA, B.B., A.W. TANK & E.L. SABBAN. 1999. Transient or sustained transcriptional activation of the genes encoding rat adrenomedullally catecholamine biosynthetic enzymes by different durations of immobilization stress. Neuroscience **94:** 803–808.
36. SABBAN, E.L., B. HIREMAGALUR, R. NANKOVA, et al. 1995. Molecular biology of stress-elicited induction of catecholamine biosynthetic enzymes. Ann. N.Y. Acad. Sci. **771:** 327–338.
37. LIU, Z., R. KVETNANSKY, L. SEROVA, et al. 2005. Increased susceptibility to transcriptional changes with novel stressors in adrenal medulla of rats exposed to prolonged cold stress. Mol. Brain Res. **141:** 19–29.
38. RPKYTOVAS, K. 1965. The effect of Retzius' striae on the progression of the carious process in the enamel. Cesk. Stomatol. **65:** 425–428.

39. RINES, S. 1998. Growth tracks in dental enamel. J. Hum. Evol. **35:** 331–350.
40. SKINNER, M. & G.S. ANDRRTSON. 1991. Individual and enamel histology: a case report in forensic anthropology. J. Forensi. Sci. **36:** 939–948.
41. VAN BEEK, G.C. 1983. Dental Morphology: an illustrated guide. II edition, Wright Publishing Oxford.
42. GOODMAM, A.H. & J.C. ROSE. 1990. Assessment of systematic physiological perturbations from dental enamel hypoplasia and associated histological structures. Yearbook Phys. Anthropol. **33:** 59–110.
43. SUZUKI, M., S.-Y. ZHOU, T. TAKAHASHI, et al. 2005. Differential contributions of prefrontal and temporolimbic pathology to mechanisms of psychosis. Brain **128:** 2109–2122.
44. ZHOU, S.-Y., M. SUZUKI, H. HAGINO, et al. 2005. Volumetric analysis of sulci/gyri-defined *in vivo* frontal lobe regions in schizophrenia: precentral gyrus, cingulate gyrus and prefrontal region. Psychiatry Res. Neuroimaging **139:** 127–139.
45. TAKAHASHI, T., M. SUZUKI, S.-Y. ZHOU, et al. 2005. Volumetric MRI study of the short and long insular cortices in schizophrenia spectrum disorders. Psychiatry Res. Neuroimaging **138:** 209–220.
46. TAKAHASHI, T., M. SUZUKI, S.-Y. ZHOU, et al. 2006. Morphologic alterations of the parcellated superior temporal gyrus in schizophrenia spectrum. Schizophr. Res. **83:** 131–143.
47. TAKAHASHI, T., M. SUZUKI, S.-Y. ZHOU, et al. 2006. Temporal lobe gray matter in schizophrenia spectrum: a volumetric MRI study of the fusiform gyrus, parahippocampal gyrus, and middle and inferior temporal gyri. Schizophr. Res. **87:** 116–126.
48. ZHOU, S.-Y., M. SUZUKI, T. TAKAHASHI, et al. 2007. Differential volume deficits in the parietal regions in schizophrenia spectrum disorders. Schizophr. Res. **89:** 35–48.
49. LAWRIE, S.M., H. WHALLEY, J.N. KESTELMAN, et al. 1999. Magnetic resonance imaging of brain in people a high risk of developing schizophrenia. Lancet **353:** 30–33.
50. DICKEY, C.C., R.W. MCCARLEY, M.M. VOGLMAIER, et al. 1999. Schizotypal personality disorder and MRI abnormalities of temporal lobe gray matter. Biol. Psychiatry **45:** 1393–1402.
51. KURACHI, M. 2003. Pathogenesis of schizophrenia: Part II. Temporo-frontal two-step hypothesis. Psychiatry Clin. Neurosci. **57:** 9–15.
52. BREIER, A., O.R. DAVIS, R.W. BUCHANAN, et al. 1993. Effects of metabolic perturbation on plasma homovanillic acid in schizophrenia: relationship to prefrontal cortex volume. Arch. Gen. Psychiatry **50:** 541–550.
53. MARCELIS, M., J. SUCKLING, P. HOFMAN, et al. 2006. Evidence that brain tissue volumes are associated with HVA reactivity to metabolic stress in schizophrenia. Schizophr. Res. **86:** 45–53.

Glucocorticoid Hyper- and Hypofunction

Stress Effects on Cognition and Aggression

JEANSOK J. KIM[a] AND JÓZSEF HALLER[b]

[a]*Department of Psychology and Program in Neurobiology and Behavior, University of Washington, Seattle, Washington, USA*

[b]*Institute of Experimental Medicine, Hungarian Academy of Science, Budapest, Hungary*

ABSTRACT: It is now well documented that both increased and decreased stress responses can profoundly affect cognition and behavior. This mini review presents possible neural mechanisms subserving stress effects on memory and aggression, particularly focusing on glucocorticoid (GC) hyper- and hypofunction. First, uncontrollable stress impedes hippocampal memory and long-term potentiation (LTP). Because the hippocampus is important for the stability of long-term memory and because LTP has qualities desirable of an information storage mechanism, it has been hypothesized that stress-induced alterations in LTP contribute to memory impairments. Recent evidence suggests a neural–endocrine network comprising amygdala, prefrontal cortex (PFC), and glucocorticoids may be involved in regulating stress effects on hippocampal mnemonic functioning. Second, antisocial aggressiveness correlates with chronically decreased glucocorticoid production, and this condition leads in rats to behavioral–autonomic deficits reminiscent of the human disorder. Glucocorticoid deficiency-induced antisocial aggressiveness results from functional changes in the PFC, medial and central amygdala, and altered serotonin and substance P neurotransmissions. Accordingly, a neurobiological understanding of how stress and glucocorticoid deficiency alter brain, cognition, and behavior is an important challenge facing modern neuroscience with broad implications for individual and social well-being.

KEYWORDS: learning; memory; aggression; corticosterone; hippocampus; amygdala; prefrontal cortex; synaptic plasticity

INTRODUCTION

All organisms experience "demands" in their environments, such as predation (threats, danger), dominance (status, relationship), and competition for

Address for correspondence: Jeansok J. Kim, Ph.D., Department of Psychology and Program in Neurobiology & Behavior, University of Washington, Seattle, WA 98020, USA. Voice: +1-206-616-2685; fax: +1-206-685-3157.

jeansokk@u.washington.edu

resources (food, shelter, mate). Through the process of natural selection, evolution has endowed organisms with stress response, which is a crucial component of the animal's defense mechanism.

Stress, as a modern scientific concept profoundly influenced by nearly half century of work by Hans Selye,[1] describes any significant distressing situations (real or perceived) that require necessary psychophysiological readjustment or adaptation for well-being of the individual.[2–5] In recent decades, there has been an enormous interest in the manifold effects of stress on brain and behavior, as uncontrollable (chronic and intense) stress has been implicated in myriad psychopathologies, including anxiety, depression, posttraumatic stress disorders (PTSDs), schizophrenia, and drug use relapse.

The pioneering work of Selye directed attention to the effects of glucocorticoid (GC) hyperfunction as a pathogenic factor. Recent evidence suggests, however, that the consequences of chronic glucocorticoid hypofunction are not less devastating, and this endocrine condition also leads to a series of psychopathologies including antisocial aggressiveness.

A common thread binding stress-associated disorders appears to be the glucocorticoid-associated alterations in gene expression. Because glucocorticoid effects on cognition and behavior are reliably observed in a diverse array of animals (ranging from fish to humans), animal models (such as rodents) provide valuable means to study common neurobiological substrates underlying behavioral changes linked to alterations in glucocorticoid production.[2–5] This admittedly selective mini-review serves to highlight particular viewpoints and findings (from the authors) concerning stress-induced alterations in the hippocampal memory system and glucocorticoid deficiency-related changes in aggressive behavior. The highlighted findings may provide insights into understanding various stress-related disorders that severely limit the quality of human life in today's increasingly hectic and long-living society.

Stress Effects on Memory and Plasticity

Stress and Hippocampal Memory

The hippocampus is crucial for the stable formation of declarative/explicit memory in humans and spatial/relational memory in rodents.[6–8] It is also involved in inhibiting stress-related hypothalamus-pituitary-adrenal (HPA) axis activity.[2–5] As a high concentration of receptors for corticosteroids—the principal glucocorticoid synthesized by the adrenal cortex and secreted in abundance in response to stress (*cortisol* in human; *corticosterone* in rodent; CORT)—is localized in the hippocampus, the structure is quite susceptible to stress. This susceptibility has implications for its non-stress-related mnemonic functioning. Supporting this view are findings that stress generally impairs hippocampal memory tasks. In humans, impairments in verbal recall tasks have been observed in (i) PTSD patients[9,10]; (ii) people with hypercortisolemia conditions, such as Cushing's disease and chronic depression[11]; and (iii) subjects administered high doses of CORT or stress.[12] In rodents, stress generally induces deficits in spatial memory tasks.[4,13]

Stress Effects on Hippocampal Plasticity

A number of stress-associated changes that can potentially influence mnemonic functioning have been identified in the hippocampus, such as (transient) alterations in the motivation-arousal-emotion systems, (relatively long-lasting) modifications in long-term potentiation (LTP), morphological changes, neuronal endangerment, and suppression of adult neurogenesis.[3] Among these, the stress-induced impairment in LTP has garnered a particular interest as LTP is the leading candidate cellular model of information storage in the mammalian brain.[14] Paralleling the behavioral data, *in vitro* and *in vivo* studies indicate that stress impairs LTP.[15–20] Originally, hippocampal slices from rats that experienced 30-min restraint + 30 intermittent tailshocks were shown to exhibit LTP deficits in the Schaffer collateral/commissural-CA1 pathway.[15] Subsequent studies established that the LTP impairment is mainly due to psychological, and not physical (e.g., pain), aspects associated with stress.[16,19] The LTP deficit is also observed in the dentate gyrus, and persists up to 48 h in rats (following an acute stress). There also seems to be a critical stress dosage requirement as 10 shocks (which robustly produce fear conditioning and elevate CORT) do not impair LTP.[16] Recent studies further indicate that stress produces a time-dependent biphasic effect on LTP (an immediate enhancing effect followed by a longer-lasting inhibitory effect on LTP),[21] and the same stress that impairs LTP enhances long-term depression (LTD; an additional synaptic model of memory) in the hippocampus.[17]

The discovery that stress impairs LTP is significant in two ways.[22] First, it offers a testable substrate to investigate the phenomenon of stress-induced memory deficits; that is, if the notion that changes in synaptic efficacy is essential for memory is correct [Hebb's postulate[23]], then LTP impairments associated with stress might explain stress-induced memory deficits (FIG. 1A). Second, the LTP impairment can serve as a standard neural marker to compare behavioral effects resulting from the use of diverse "putative" stressors across laboratories. Not all stress paradigms would be expected to alter LTP and behavior in similar manners. Thus, the problem of qualifying, quantifying, or scaling different stressors and their behavioral effects can be normalized by examining LTP.[22]

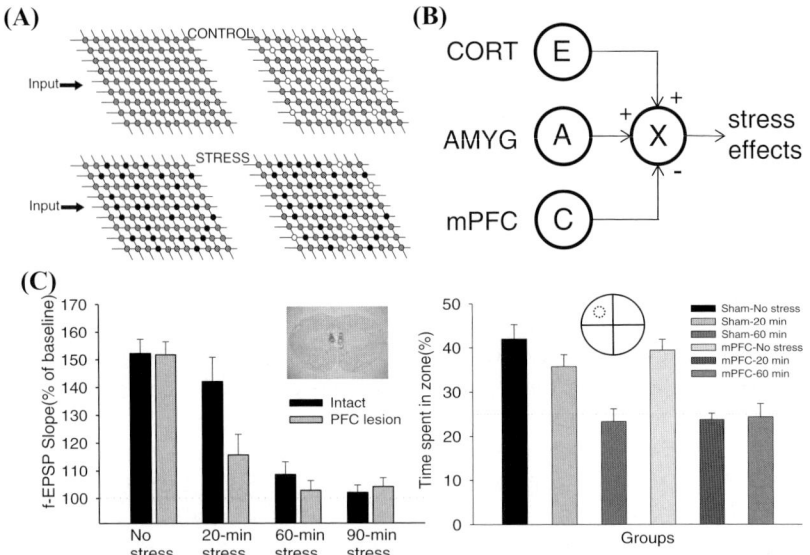

FIGURE 1. (**A**) An illustration of how stress-induced impairments in LTP can impede subsequent memory in the hippocampus. The gray circles on the matrices represent synapses with normal capacity to undergo plasticity (e.g., LTP), thereby supporting normal memory configuration (white circles). The black circles represent synapses with altered properties of plasticity (e.g., impaired LTP), which impair subsequent memory processing. (**B**) A simple connectionist model showing CORT (excitability, E), AMYG (aversiveness, A), and mPFC (controllability, C) interaction to produce stress effects. The model posits that CORT and AMYG exert excitatory (+) and mPFC exerts inhibitory (–) stress influences. X denotes a target structure, such as the hippocampus. (**C**) Electrolytic mPFC lesions and stress effects on LTP and spatial memory. In no stress condition, LTP (measured 40–60-min posttetanus) was robustly observed in hippocampal slices prepared from intact and mPFC-lesioned rats. Both 60-min and 90-min stress reliably impaired LTP in intact and mPFC-lesioned rats. While 20-min stress was ineffective in intact rats, it effectively reduced LTP in mPFC-lesioned animals. The inset shows an example of mPFC-lesioned brain section. Percentage of time spent swimming in the platform (represented by a dotted circle) quadrant zone during the 60-sec probe test. Sham-No stress, Sham-20-min (ineffective) stress, and mPFC-No stress groups exhibited reliable spatial memory, whereas Sham-60-min (effective) stress, mPFC-20-min stress, and mPFC-60-min stress groups did not.

Glucocorticoid Effects on Hippocampus

In most stress-hippocampus research, the usual strategy has been to relate the glucocorticoid level directly to stress effects, an approach that unveiled enormously useful information.[2–5] Seminal rodent work has shown that the hippocampus is enriched with both mineralocorticoid receptors (MRs) and glucocorticoid receptors (GRs),[24] and that CORT actions through these receptors mediate several stress effects on the hippocampus. Importantly, a dual relationship between the level of CORT and the magnitude of LTP has been described, where both low (via adrenalectomy, ADX) and high (via administration) levels of CORT are associated with impaired LTP.[25] Other studies showed that selective activation of MRs increases LTP while added activation of GRs attenuates LTP.[26] This suggests that basal (low) levels of CORT enhance potentiation through preferential stimulation of the high-affinity MRs and, during stress GR stimulation becomes important because levels of CORT become high enough to saturate low-affinity GRs.[2] Behavioral studies found similar results—spatial memory is impaired with GR, but not MR, activation.[27]

If CORT is the main contributing factor from behavioral stress, then removing it (during stress) and directly applying it (sans stress) should preclude and produce stress effects, respectively. However, there are data inconsistent with this simple linear CORT–stress effect notion. For instance, LTP is further reduced in ADX rats following stress and is not restored by CORT replacement.[28] In intact animals administered dexamethasone (a synthetic GC that binds to GRs and by mimicking feedback suppresses the pituitary-adrenal response to stress), stress-induced LTP impairments occurred nonetheless, suggesting that the elevated GC level is not the whole mechanism whereby stress suppresses LTP.[29] Notably, rats that experienced but were able to terminate shocks (i.e., control over shock cessation) did not show LTP impairments unlike "yoked" animals receiving identical shocks without control, although CORT levels were elevated equally in both groups.[16] Recently, male rats with stress equivalent levels of CORT—via administration of CORT or exposure to a sexually receptive female—were found not impaired in spatial memory.[30] Thus, CORT is unlikely an invariant physiological measure of stress since other sources (such as sex, exercise) can also significantly elevate it. Collectively, these data indicate that the elevated level of CORT is not a sufficient condition to reproduce stress effects on the hippocampus. Instead, it is likely that other factors (besides GC) constitute the central stress network that exerts cognitive influences.[3]

Amygdala and Prefrontal Cortex (PFC) Involvement in Stress

Evidence from anatomical, physiological, and behavioral studies indicates that the amygdala sends projections to the hippocampus, is involved in various stress-related behaviors (such as gastric erosion, analgesia, anxiety), alters the magnitude of perforant path-dentate gyrus LTP, and modulates the strength of hippocampal memory formation.[31,32] Consistent with these results, recent studies found that amygdalar lesions and inactivation effectively block stress-induced impairments in hippocampal LTP and spatial memory.[18]

It is also plausible that the massive reciprocal connection between frontal cortical regions to "core" limbic structures regulates stress–hippocampal interactions. Of particular interest are the projections from the medial PFC (prelimbic and dorsal anterior cingulate cortices) to the amygdala and hippocampus. Specifically, the mPFC sends projections to the amygdala.[33,34] The mPFC also sends projections to the hippocampal formation, via the entorhinal area and thalamus (nucleus reunions).[35,36] The functional implication of these connections is that during stress "higher level" processing occurring in the mPFC can influence the hippocampus through multiple routes. Consistent with this view are findings that mPFC activity closely correlates with controllability, that is, inhibition (or extinction) of aversively motivated behavior.[37,38] In humans, damage to this structure (or hyporesponsive mPFC) results in impulsiveness (loss of controllability) and inappropriate emotional behavior in the absence of intellectual deficits.[39]

We recently tested the notion that mPFC exerts inhibitory influences on stress. Rats with mPFC lesions received either 20-min restraint + 20 shocks (ineffective stress), 60-min restraint + 60 shocks (effective stress), or 90-min restraint + 90 shocks (>effective stress) (FIG. 1C). In intact animals, CA1 LTP was significantly observed in slices from control (No stress) and ineffective 20-min stress groups but not from effective 60-min and 90-min stress groups. In PFC-lesioned animals, the ineffective 20-min stress now reliably impaired LTP. Behaviorally, in rats with mPFC lesions, the same ineffective stress procedure also impaired long-term spatial memory. These findings suggest that mPFC normally exerts inhibitory stress influence (consistent with the view that a key function of mPFC is "controllability"), and that damage to this structure exacerbates stress effects on hippocampal LTP.

Need for a Systems Level Analysis of Stress

To fully understand the detrimental neurocognitive consequences of stress, it is vital to identify the basic elements (or controlling antecedent conditions) of stress and their neurobiological substrates.[22] Recently, Kim and Diamond[4] proposed that stress must satisfy three criteria: (i) it should generate heightened excitability/arousal (a variable denoted *"E"*); (ii) it should induce perceived aversiveness (a variable *"A"*); and (iii) it should produce uncontrollability (a variable *"U,"* an inverse of controllability). At the simplest conceptual level stress (S) can be represented: $S = E \times A \times U$. The magnitude of neurocognitive stress effects is then determined by the dynamic (and not necessarily orthogonal) interactions between E, A, and U, which can be adjusted by varying the level of each variable. FIGURE 2B illustrates a connectionist version with *CORT*, *AMYG* (amygdala), and *mPFC* representing neurobiological counterparts of *excitability (E)*, *aversiveness (A)*, and *controllability (C)*, respectively. As mentioned earlier, these biological designations are supported by the evidence that: (i) CORT levels correlate with excitability/arousal; (ii) amygdalar lesion/inactivation reduces aversive behavior while stimulation evokes

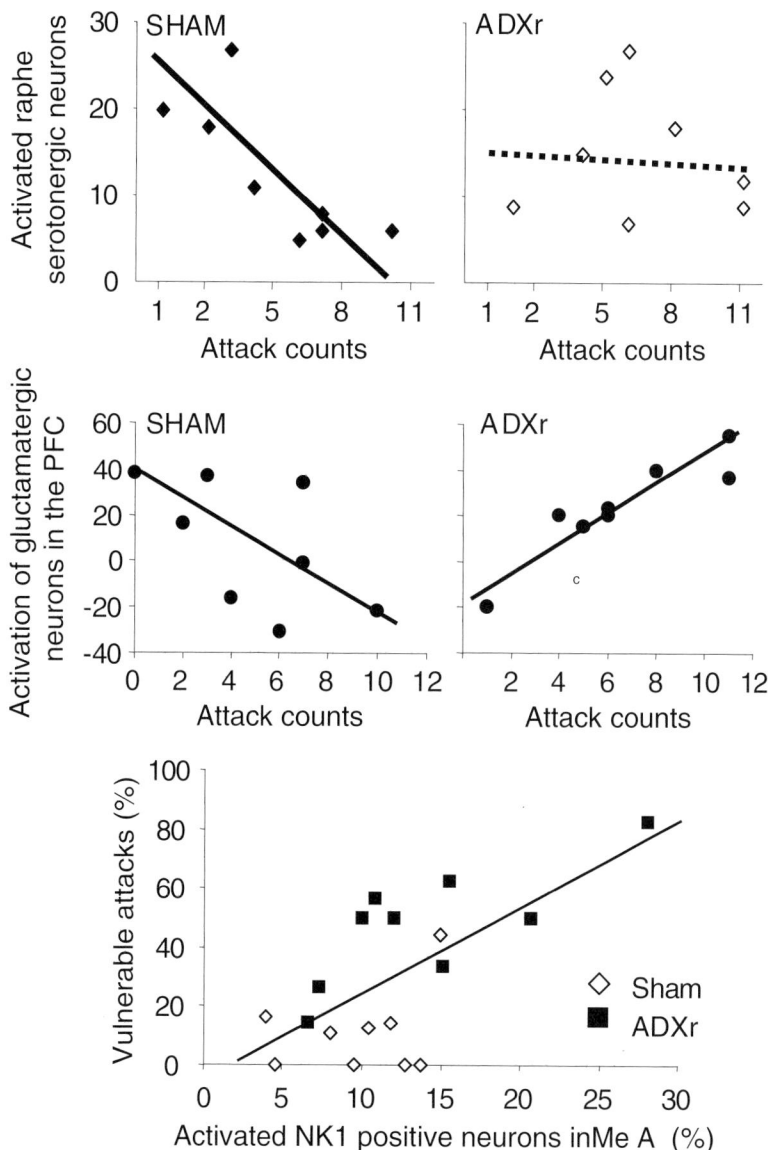

FIGURE 2. Correlations between neuronal activation and the execution of attacks. In controls (Sham), serotonergic- and marginally, prefrontal glutamatergic activation correlated negatively with the execution of attacks, whereas no significant correlation was seen between vulnerable attacks and NK1 receptor-expressing neurons of the medial amygdala (MeA). This picture changed dramatically in glucocorticoid-deficient ("violent", ADXr) rats: serotonergic neurons appeared to lose their role in controlling attacks, whereas PFC glutamatergic and MeA NK1 cell activation correlated positively with the execution of attacks and vulnerable attacks, respectively.

aversive behavior; (iii) the mPFC activity correlates with controllability while damaged (or hyporesponsive) mPFC results in impulsiveness or loss of controllability; and (iv) CORT, AMYG, and mPFC are all responsive to stress.

In summary, identifying and rigorously testing the rules and design features of the systems level model of stress could provide an important foundation for developing a more comprehensive (qualitative and quantitative) understanding of the way in which stress influences neurocognition and contributes to a variety of stress-related disorders.

Glucocorticoid Deficiency and Aggression

Glucocorticoid Status and Antisocial Aggressiveness

It is now widely accepted that not only increased but also decreased GC signaling leads to pathological conditions.[40] To date, reduced plasma GCs were associated with antisocial personality disorder and its childhood antecedent conduct disorder, PTSD, fibromyalgia, chronic fatigue, and burnout.[41–44] In contrast to PTSD, where the effects of GC deficiency appear to be mediated by a secondary upregulation of noradrenergic and CRH signaling,[45] antisocial behavior develops in conjunction with low autonomic responsiveness.[46] Thus, decreased plasma GCs may underlay a series of psychopathologies, and the nature of the disorder may also depend on associated neural changes (e.g., increased or decreased noradrenergic neurotransmission).

In an attempt to develop a model of antisocial aggressiveness, we studied adrenalectomized rats, in which low and stable levels of GRs were ensured by subcutaneous corticosterone pellets (ADXr). Mimicking in rats the endocrine condition associated with antisocial aggressiveness in humans (i.e., low GC levels) resulted in the development of three important symptoms of the disorder: (i) antisocial type of aggressiveness (attacks aimed at vulnerable body parts of opponents), (ii) low autonomic arousal during fights, and (iii) social deficits.[47,48]

Laboratory models are usually evaluated by investigating their face, construct, and predictive validity.[49] Considering the phenomenological similarities between human antisocial aggressiveness and the behavior of our rats, one can assume that the face validity of the model is acceptable. Construct validity is suggested by the shared GC background. Predictive validity is more problematic, as no reliably efficient treatments are known for antisocial personality disorder. Nevertheless, earlier findings suggested that serotonergic agents act in a rather similar fashion in antisocial personality-disordered people and rats submitted to our model.[50–53] This similarity indicates that the model has a certain degree of predictive validity.

The Neural Background of Antisocial Aggressiveness in Rats

By comparing the neural background of glucocorticoid deficiency-induced violent and normal (territory-related) aggression, we identified four areas that may be responsible for the development of abnormal attacks: the PFC, medial amygdala (MeA), central amygdala (CeA), and the raphe.

In the first series of experiments, we investigated aggression-induced neuronal activation by labeling the c-Fos protein. Aggression *per se* activated all the brain areas that were earlier believed to control aggression. As compared to controls, glucocorticoid deficiency-induced violent aggression induced a dramatic increase in the activation of the CeA that is tightly bound to the control of fear.[54] To our surprise, no significant differences were seen in other aggression-related brain areas. We assumed, however, that the changes seen in c-Fos studies were accompanied by more subtle alterations that cannot be revealed by a simple c-Fos staining. Therefore, the next series of experiments aimed at the neurochemical identification of neurons that were activated by the execution of attacks. This approach identified three other brain areas that may be responsible for the execution of violent attacks: the raphe, PFC, and MeA.

In rats submitted to resident/intruder conflicts (normal–territorial–aggression), the activation of serotonergic neurons showed a significant negative correlation with attack counts. In line with earlier findings, this suggested that the activation of serotonergic neurons downregulates attacks. This correlation, however, was lost in rats submitted to the violent aggression model, suggesting that the serotonergic control of aggression is lost in glucocorticoid-deficient rats[52] (FIG. 2, upper panel). This finding was confirmed later by pharmacological experiments, where the effect of the serotonergic anxiolytic buspirone was dramatically altered in GC-deficient, as compared to sham-operated, rats.[53]

In the PFC, we double labeled neurons for activation (e.g., c-Fos) and neurochemical markers.[55] Glucocorticoid deficiency markedly and specifically decreased the activation of CCK-containing GABAergic interneurons that control the inputs of glutamatergic principal cells, and ensure the fine-tuning of their function.[56] A multiple regression analysis showed that normal and violent attacks were associated with different patterns of principal cell activation when all the relevant PFC subareas were analyzed together (FIG. 2, middle panel). Moreover, the specific pattern seen in GC-deficient rats predicted the expression of vulnerable attack by a precision that exceeded 90%.

Finally, we investigated the activation of NK1 receptor-expressing neurons in brain areas relevant to aggression control. The reason for performing these studies was that substance P and its NK1 receptor was implicated in the defensive rage of cats.[57] The activation of NK1 receptor-expressing neurons was significantly stronger in GC-deficient rats and correlated significantly with the execution of violent attacks[58] (Fig. 2, lower panel). Moreover, NK1 receptor blockade specifically decreased violent aggression, without affecting more normal forms of aggressiveness.[58]

CONCLUSIONS

Psychopathologies may not only be associated with GC excess, but also with GC deficiency. In antisocial aggressiveness, GC hypofunction is associated with reduced autonomic activation and social deficits. Our laboratory findings suggest that GC deficiency and human antisocial aggressiveness are causally linked, as mimicking the former in rats resulted in behavioral symptoms strongly reminiscent of the latter. It occurs that chronic GC deficiency leads to neural changes that are tightly bound to the execution of antisocial attacks in rats. Such changes were identified in restricted brain regions, namely the raphe, PFC, as well as the central and medial amygdala. Noteworthy, these brain areas are strongly interrelated: serotonergic neurons have a strong effect on prefrontal functions, which tightly control the amygdala. We suggest that studying the impact of GC deficiency on behavior and brain function reveals important aspects of the mechanisms that underlay GC deficiency-associated behavioral disorders; moreover, such studies may identify novel treatment strategies.

REFERENCES

1. SELYE, H. 1936. A syndrome produced by diverse nocuous agents. Nature **138**: 32.
2. MCEWEN, B.S. & R.M. SAPOLSKY. 1995. Stress and cognitive function. Curr. Opin. Neurobiol. **5**: 205–216.
3. KIM, J.J. & K.S. YOON. 1998. Stress: metaplastic effects in the hippocampus. Trends Neurosci. **21**: 505–509.
4. KIM, J.J. & D.M. DIAMOND. 2002. The stressed hippocampus, synaptic plasticity and lost memories. Nat. Rev. Neurosci. **3**: 453–462.
5. DIAMOND, D.M., A.M. CAMBELL, C.R. PARK, et al. 2007. The temporal dynamics model of emotional memory processing: a synthesis on the neurobiological basis of stress-induced amnesia, flashbulb and traumatic memories, and the Yerkes-Dodson law. Neural Plasticity **2007**: 1–33.
6. SCOVILLE, W.B. & B. MILNER. 1957. Loss of recent memory after bilateral hippocampal lesions. J. Neurol. Neurosurg. Psychiatry **20**: 11–21.
7. O'KEEFE, J. & L. NADEL. 1978. The Hippocampus as a Cognitive Map. Oxford University Press, New York NY.
8. EICHENBAUM, H. 2000. A cortical-hippocampal system for declarative memory. Nat. Rev. **1**: 41–50.
9. BREMNER, J.D., T.M. SCOTT, R.C. DELANEY, et al. 1993. Deficits in short-term memory in posttraumatic-stress-disorder. Am. J. Psychiatry **150**: 1015–1019.

10. UTTO, M., J.J. VASTERLING, K. BRAILEY & P.B. SUTKER. 1993. Memory and attention in combat-related posttraumatic-stress-disorder (PTSD). J. Psychopathol. Behav. Assess. **15:** 43–52.
11. STARKMAN, M.N., S.S. GEBARSKI, S. BERENT & D.E. SCHTEINGART. 1992. Hippocampal formation volume, memory dysfunction, and cortisol levels in patients with Cushing's syndrome. Biol. Psychiatry **32:** 756–765.
12. NEWCOMER, J.W., S. CRAFT, T. HERSHEY, *et al.* 1994. Glucocorticoid-induced impairment in declarative memory performance in adult humans. J. Neurosci. **14:** 2047–2053.
13. DIAMOND, D.M., C.R. PARK, K.L. HEMAN & G.M. ROSE. 1999. Exposing rats to a predator impairs spatial working memory in the radial arm water maze. Hippocampus **9:** 542–52.
14. BLISS, T.V.P. & G.L. COLLINGRIDGE. 1993. A synaptic model of memory: long-term potentiation in the hippocampus. Nature **361:** 31–39.
15. FOY, M.R., M.E. STANTON, S. LEVINE & R.F. THOMPSON. 1987. Behavioral stress impairs long-term potentiation in rodent hippocampus. Behav. Neural Biol. **48:** 138–149.
16. SHORS, T.J., T.B. SEIB, S. LEVINE & R.F. THOMPSON. 1989. Inescapable versus escapable shock modulates long-term potentiation in the rat hippocampus. Science **244:** 224–226.
17. KIM, J.J., M.R. FOY & R.F. THOMPSON. 1996. Behavioral stress modifies hippocampal plasticity through N-methyl-D-aspartate receptor activation. Proc. Natl. Acad. Sci. **93:** 4750–4753.
18. KIM, J.J., J.W. KOO, H.J. LEE & J.-S. HAN. 2005. Amygdalar inactivation blocks stress-induced impairments in hippocampal long-term potentiation and spatial memory. J. Neurosci. **25:** 1532–1539.
19. DIAMOND, D.M., M. FLESHNER & G.M. ROSE. 1994. Psychological stress repeatedly blocks hippocampal primed burst potentiation in behaving rats. Behav. Br. Res. **62:** 1–9.
20. DIAMOND, D.M. & C.R. PARK. 2000. Predator exposure produces retrograde amnesia and blocks synaptic plasticity. Progress toward understanding how the hippocampus is affected by stress. Ann. N. Y. Acad. Sci. **911:** 453–455.
21. AKIRAV, R.-L.G. 1999. Biphasic modulation of hippocampal plasticity by behavioral stress and basolateral amygdalar stimulation in the rat. J. Neurosci. **19:** 10530–10535.
22. KIM, J.J., E.Y. SONG & T.A. KOSTEN. 2006. Stress effects in the hippocampus: synaptic plasticity and memory. Stress: Internatl. J. Biol. Stress **9:** 1–11.
23. HEBB, D.O. 1949. The Organization of Behavior: A Neuropsychological Theory. Wiley. New York, NY.
24. REUL, J.M H.M. & E.R. DE KLOET. 1985. Two receptor systems for corticosterone in rat brain: microdistribution and differential occupation. Endocrinology **117:** 2505–2511.
25. DIAMOND, D.M., M.C. BENNETT, M. FLESHNER & G.M. ROSE. 1992. Inverted-U relationship between the level of peripheral corticosterone and the magnitude of hippocampal primed burst potentiation. Hippocampus **2:** 421–430.
26. PAVLIDES, C., Y. WATANABE, A.M. MARGARINOS & B.S. MCEWEN. 1995. Opposing roles of type I and type II adrenal steroid receptors in hippocampal long-term potentiation. Neuroscience **68:** 387–394.

27. CONRAD, C.D., S.J. LUPIEN & B.S. MCEWEN. 1999. Support for a bimodal role for type II adrenal steroid receptors in spatial memory. Neurobiol. Learn. Mem. **72:** 39–46.
28. SHORS, T.J., S. LEVINE & R.F. THOMPSON. 1990. Effect of adrenalectomy and demedullation on the stress-induced impairment of long-term potentiation. Neuroendocrinology **51:** 70–75.
29. FOY, M.R., S. LEVINE & R.F. THOMPSON. 1990. Manipulation of pituitary-adrenal activity affects neural plasticity in rodent hippocampus. Psychol. Sci. **3:** 201–204.
30. WOODSON, J.C., D. MACINTOSH, M. FLESHNER & D.M. DIAMOND. 2003. Emotion-induced amnesia in rats: working memory-specific impairment, corticosterone-memory correlation, and fear versus arousal effects on memory. Learn. Mem. **10:** 326–336.
31. LEDOUX, J.E. 1996. The Emotion Brain. Simon & Schuster. New York, NY.
32. MCGAUGH, J.L. 2000. Memory–a century of consolidation. Science **287:** 248–251.
33. HEIDBREDER, C.A. & H.J. GROENEWEGEN. 2003. The medial prefrontal cortex in the rat: evidence for a dorsal-ventral distinction based upon functional and anatomical characteristics. Neurosci. Biobehav. Rev. **27:** 555–579.
34. MCDONALD, A.J. 1998. Cortical pathways to the mammalian amygdala. Prog. Neurobiol. **55:** 257–332.
35. HURLEY, K., H. HERBERT, M.M. MOGA & C.B. SAPER. 1991. Efferent projections of the infralimbic cortex of the rat. J. Comp. Neurol. **308:** 249–276.
36. SESACK, S.R., A.Y. DEUTCH, R.H. ROTH & B.S. BUNNEY. 1989. Topographical organization of the efferent projections of the medial prefrontal cortex in the rat: an anterograde tract-tracing study using *Phaseolus vulgaris* Leucoagglutinin. J. Comp. Neurol. **290:** 213–242.
37. WELLMAN, C.L., A. IZQUIERDO, J.E. GARRETT, *et al*. 2006. Impaired stress-coping and fear extinction and abnormal corticolimbic morphology in serotonin transporter knock-out mice. J. Neurosci. **27:** 684–691.
38. MAIER, S.F., J. AMAT, M.V. MARATTA, *et al*. 2006. Behavioral control, the medial prefrontal cortex, and resilience. Dialog. Clin. Neurosci. **8:** 397–406.
39. MCNALLY, R.J. 2006. Cognitive abnormalities in post-traumatic stress disorder. Trends Cog. Sci. **10:** 271–277.
40. RAISON, C.L. & A.H. MILLER. 2003. When not enough is too much: the role of insufficient glucocorticoid signaling in the pathophysiology of stress-related disorders. Am. J. Psychiatry **160:** 1554–1565.
41. CROFFORD, L.J., S.R. PILLEMER, K.T. KALOGERAS, *et al*. 1994. Hypothalamic-pituitary-adrenal axis perturbations in patients with fibromyalgia. Arthritis Rheum. **37:** 1583–1592.
42. DEMITRACK, M.A., J.K. DALE, S.E. STRAUS, *et al*. 1991. Evidence for impaired activation of the hypothalamic-pituitary-adrenal axis in patients with chronic fatigue syndrome. J. Clin. Endocrinol. Metab. **73:** 1224–1234
43. VIRKKUNEN, M. 1985. Urinary free cortisol secretion in habitually violent offenders. Acta Psychiatr. Scand. **72:** 40–44.
44. YEHUDA, R. 2001. Biology of posttraumatic stress disorder. J. Clin. Psychiatry. **62**(Suppl 17):41–46.
45. BAKER, D.G., S.A. WEST, W.E. NICHOLSON, *et al*. 1999. Serial CSF corticotropin-releasing hormone levels and adrenocortical activity in combat veterans with posttraumatic stress disorder. Am. J. Psychiatry **156:** 585–588.

46. RAINE, A. 1996. Autonomic nervous system factors underlying disinhibited, antisocial, and violent behavior. Biosocial perspectives and treatment implications. Ann. N. Y. Acad. Sci. **794:** 46–59.
47. HALLER, J., J. HALASZ, E. MIKICS & M.R. KRUK. 2005. Chronic glucocorticoid deficiency-induced abnormal aggression, autonomic hypoarousal, and social deficit in rats. J. Neuroendocrinol. **16:** 550–557.
48. HALLER, J. & M.R. KRUK. 2006. Normal and abnormal aggression: human disorders and novel laboratory models. Neurosci. Biobehav. Rev. **30:** 292–303.
49. WILLNER, P. 1986. Validation criteria for animal models of human mental disorders: learned helplessness as a paradigm case. Prog. Neuropsychopharmacol. Biol. Psychiatry **10:** 677–690.
50. KUTCHER, S.P., P. MARTON & M. KORENBLUM 1989. Relationship between psychiatric illness and conduct disorder in adolescents. Can. J. Psychiatry **34:** 526–529.
51. WHITEHEAD, P.D. 2003. Causality and collateral estoppel: process and content of recent SSRI litigation. J. Am. Acad. Psychiatry Law **31:** 377–382.
52. HALLER, J., M. TOTH & J. HALASZ. 2005. The activation of raphe serotonergic neurons in normal and hypoarousal-driven aggression: a double labeling study in rats. Behav. Brain Res. **161:** 88–94.
53. HALLER, J., Z. HORVATH & N. BAKOS. 2007. The effect of buspirone on normal and hypoarousal-driven abnormal aggression in rats. Prog. Neuropsychopharmacol. Biol. Psychiatry **31:** 27–31.
54. HALASZ, J., ZS. LIPOSITS, M.R. KRUK & J. HALLER. 2002. Neural background of glucocorticoid dysfunction-induced abnormal aggression in rats: involvement of fear- and stress-related structures. Eur. J. Neurosci. **15:** 561–569.
55. HALASZ, J., M. TOTH, I. KALLO, *et al*. 2006. The activation of prefrontal cortical neurons in aggression—A double labeling study. Behav. Brain Res. **175:** 166–175.
56. FREUND, T.F. 2003. Interneuron diversity series: Rhythm and mood in perisomatic inhibition. Trends Neurosci. **26:** 489–495.
57. SHAIKH, M.B., A. STEINBERG & A. SIEGEL. 1993. Evidence that substance P is utilized in medial amygdaloid facilitation of defensive rage behavior in the cat. Brain Res. **625:** 283–294.
58. HALASZ, J., M. TOTH, E. MIKICS, *et al*. 2007. The effect of NK1 receptor blockade on territorial aggression and in a model of violent aggression. Biol. Psych. In press.

Cognitive Activation Theory of Stress, Sensitization, and Common Health Complaints

HOLGER URSIN AND HEGE ERIKSEN

Unifob Health and Department of Education and Health Promotion, University of Bergen, Bergen, Norway

ABSTRACT: According to the cognitive activation theory of stress (CATS), a formal system of systematic definitions, the term "stress" is used for stress stimuli, the stress experience, the nonspecific, general stress response, and the experience of the stress response. The stress response is normal, healthy, and necessary alarm. If sustained there may be a risk of illness and disease. The level and duration of the alarm depend on the expectancy of the outcome of stimuli and the specific responses available for coping. The most common health complaints are subjective health complaints like muscle pain, tiredness and mood changes. These are normal aches of short duration and low intensity for most people. For some the pains and complaints are substantial and longlasting with serious implications for functioning. There are no sharp or obvious limits in the distribution of health complaints, separating "normal" and endurable pain and complaints, and intolerable complaints that need professional help. These conditions are most often unspecific, and are the most common reason for encounters with health professionals, and the most frequent reason for sick leave and disability. There is a striking comorbidity for all these conditions. This may be explained by psychobiological sensitization within neural loops, maintained by sustained activation, which has been suggested as a mechanism for these conditions.

KEYWORDS: cognitive activation theory of stress; subjective health complaints; sensitization

INTRODUCTION

The cognitive activation theory of stress (CATS)[1] offers a formal system of systematic definitions, aiming at reducing the reliance on words with imprecise meanings and usage. This is particularly important in a field where there are so many conflicting terms and attributions, among experts, and among the general public. This is not only a matter of concern for experts and theoretical

Address for correspondence: Holger Ursin, University of Bergen, Unifob Health, Christiesgt. 13, Bergen, Norway 5015. Voice: +47-5558-6227; fax: +47-5558-9872.
Holger.Ursin@psych.uib.no

discussions. Erroneous attributions may be very costly, for the individual, and for society. When expensive and ambitious preventive and curative initiatives are built on superstition the net results may be negative, or even disastrous.

The term "stress" is used for stress stimuli, for the stress experience, for the nonspecific, general stress response, and for the experience of the stress response.[2] For CATS the most basic assumption is that the stress response is a normal, healthy, and necessary alarm response. If this was not true, the stress response would not be present in all species, in all individuals, in all cultures. The stress response is a part of our biological inheritance, and it is by no means an outdated response. In principle, the response is simply an increase in arousal. However, if sustained it may contribute to illness and disease.

It seems to be a consensus that physical demands and psychological characteristics that produce the stress response have nothing in common. All stimuli are appraised[3] or filtered [2] before they gain access to the response system. The main "filters" are related to response outcome expectancy and stimulus expectancy.[4] It is the individual's experience of the demands and the expectancies of the outcome, which determine whether the demands will cause stress responses, which—if sustained—may cause illness and disease in man and animals.[2] Within this cognitive tradition CATS define coping as positive response outcome expectancies.[1] This means that the individual expects that he or she will be able to handle the situation with a positive result. In these situations, there is a low level of subjective health complaints and low levels of psychophysiological, psychoendocrine, and psychoimmune arousal.[1]

When it is impossible for the individual to establish coping, other expectancies may develop. When the individual learns that there are no relationships between anything the individual can do and the outcome, we refer to this as helplessness in CATS. Two of R.L. Solomon's students, Overmier and Seligman,[5] found that dogs with previous experience with inescapable shocks did not learn avoidance tasks. They found that this state of "helplessness" generalized to situations where control is possible. Translated to CATS, helplessness occurs when the perceived probability of avoiding the aversive stimulus with a response is the same as for no response. In other words, the response is without any perceived consequence for the occurrence of the aversive event. The organism has no control. This expectancy has been accepted as a model for anxiety and depression. Since the CATS formulations are valid for animals as well, this is a useful theoretical basis for animal models for depression and anxiety.[6]

Hopelessness is even worse. In CATS, this term is used for an acquired expectancy that most or all responses lead to negative results. Hopelessness is more directly opposite of coping than helplessness, since it is a negative response outcome expectancy. There is control, responses have effects, but they are all negative. The negative outcome is his or her fault since the individual has control. This introduces the element of guilt, which may make hopelessness a better model for depression than helplessness.[7]

CATS is a general and comprehensive stress theory, compatible with other theoretical positions. In working life in humans, the most influential model is the demand–control model.[8] It is the combination of psychological demands, task control, and skill use at work, which predicts stress-related ill health and behavioral correlates of work. Jobs with high demands, low control, and low social support carry the highest risk of illness and disease. Low psychological demands and high levels of control carry the lowest risk. Jobs with high psychological demands and high control, and low psychological demands and low control, carry an average risk. The model predicts disease, especially related to cardiovascular disease, and it is particularly the control dimension that is the most robust predictor.[9] Newer tradition emphasizes individual stress management, coping abilities, and subjective feelings of being in control or being able to cope.[9] In CATS, it is the expectancy of being able to cope that is the essence,[10] not the objective possibility of having control. Also, CATS is applicable directly to animal experimentation, and has a developed pathophysiological model built into the theory.

The effort–reward imbalance model[11] has its focus on reward and contractual fairness in employment, and is a strict model for human work relations, without a pathophysiological basis.

COMMON HEALTH COMPLAINTS

The most common health complaints are subjective health complaints like muscle pain, tiredness, and mood changes. These are normal aches and complaints of short duration and low intensity for most people. For some the pains and complaints are substantial and longlasting with serious implications for functioning. There are no sharp or obvious cut points between "normal" and "pathological" levels in subjective health complaints, separating "normal" and endurable pain and complaints, and intolerable complaints that need professional help. These conditions are most often unspecific, and are the most common reasons for encounters with health professionals, and the most frequent reason for sick leave and disability.

A variety of subjective illnesses with few or no objective findings have appeared at regular intervals as epidemics in our society under different labels. Examples are chronic fatigue syndrome, food intolerance, myalgic encephalitis, "yuppie flu," whiplash, fibromyalgia, postviral syndrome, and the Gulf War syndrome. Patients diagnosed with these illnesses complain of muscle pain, tiredness, depression, fatigue, headaches, sleep disturbances, concentration problems, memory lapses, flu-like symptoms, and "allergies." There are, however, few or no objective findings that might explain the "disease," or the complaints go beyond what is regarded as "reasonable" by the physician.

The prevalence of subjective health complaints is very high in national surveys. In a recent study conducted in Norway 96% reported that they had experienced at least one type of complaint during the preceding 30 days.[12] However, the prevalence of substantial complaints was moderate: Only 13% reported substantial musculoskeletal complaints, 5% "pseudoneurological" complaints (tiredness, mood changes), 4% gastrointestinal complaints, 2% allergy, and 18% flu-like complaints. The high prevalence makes the finding of such complaints "normal"; most people have them. It does not hinder them or make them seek medical advice or help from society; neither does it signal that they are in any inherent danger of developing dangerous and debilitating conditions. However, they may develop into conditions where complaints are so longlasting and intense that they require medical and social interventions, including sickness compensation. The transition from the "normal" complaint to the serious condition seems to be a continuous process, with no clear or objective thresholds to indicate a distinction. We suggest that this transition is due to a psychobiological sensitization.

It appears reasonable to assume that humans in every culture and environment experience health complaints like pains, fatigue, itching, dizziness etc., ranging from minor and transient to disabling and permanent. The interpretation and the meaning of complaints and sensations may be determined by the culture in which we live and our idiosyncratic attitudes. A headache can be interpreted as a sign from the body that it needs rest, but can also be a cause of worry: is this the first sign of a brain tumor, harmful radiation from mobile phone, or evil spirits released to harm me?

In media, as well as by professionals, it is often assumed that these complaints, as well as serious somatic disease, are a result of a misfit between our physiological constitution and the modern, civilized life.[13] Historical analyses argue against the idea that the subjective health complaints, or psychosomatic conditions, fatigue, or hysteria really are new phenomena, typical for our time.[14]

Similarly, it does not appear as if these complaints are specific for industrialized societies. In a comparative study of 120 aborigine Mangyans living in the jungle of Mindoro Island in the Philippines, the frequency of subjective health complaints was found to be more frequent than a representative sample from the Norwegian population, indicating that these complaints are not specific for industrialized societies.[15] A similar investigation of subjective health complaints in 320 Maasai people, living on the savannah in Eastern Africa, showed the same finding. These seminomadic people, living along the Great Rift Valley in southern Kenya and northern Tanzania, had significantly higher level of subjective health complaints than the standard Norwegian population. A life style very different from the "stressed" Nordic European population, with a diet of milk, meat, and blood from the cattle, and living in a primitive, but highly organized society, does not imply a low level of subjective health complaints, or "psychosomatic" complaints.[16]

SENSITIZATION: THE PATHOPHYSIOLOGY OF SUSTAINED ACTIVATION ("STRESS")

There is a striking comorbidity for subjective health complaints, or the unspecific conditions that constitute the main reasons for sickness absence and common health complaints. This may be explained by psychobiological sensitization within neural loops, which has been suggested as a mechanism for these conditions.[17]

Sensitization is an increased efficiency in a neural circuit, due to a change in the synapses from repeated use. This feed-forward mechanism increases the response to a stimulus. Sensitization is a typical feature of pain pathways, pain produces pain. Patients referred to a back pain clinic for low back pain do not have back pain only, they also complain about general pain, headaches, tiredness, anxiety, and depressed thoughts.[18] Patients hospitalized for irritable bowel disease have similar comorbidity.[19] The level of comorbidity is also a significant prognostic factor for spinal pain.[20]

This basic neurobiological process may be assumed to have a cognitive analogue. Brosschot et al.[21] have suggested that this cognitive correlate is an attentional bias, giving priority to thoughts and information related to fears and somatic complaints. They find that patients with subjective health complaints (unexplained medical complaints) show sensitization and extensive activation of cognitive networks related to illness and pain. Brosschot refers to this as the "night and day watch" of the sensitized organism.

We believe this is an acceptable theoretical basis for the design, and indeed for the effects, of cognitive behavioral treatment programs for low back pain,[22] as well as for other illnesses like fatigue.[23] Within CATS, this night and day watch is related to sustained activation, which is the motor sustaining the activation of specific pain and illness-related cognitive networks. Employees in work situations with high risk of developing sickness leave and muscle pain complain about "stress." The prevalence of subjective health complaints is high in populations that experience low job satisfaction[24] and low levels of coping.[17] These are all cases where CATS predicts high and sustained activation.

CONCLUSION

Stress complaints are very prevalent, so are subjective health complaints. It appears to be important to realize that many of our common aches and bodily sensations are normal phenomena, there are movements in the guts that may be felt, muscles and joints do hurt occasionally. Actually, almost all of us have had such experiences the last 30 days. In spite of this, the plurality of us is happy, in good health, and satisfied with our working conditions.

Only when these sensations become very strong do we need attention and care. Sickness absence is a major problem for those that are involved, and it is also a major economic and social problem in the modern welfare society.

When we need care, it is reasonable to demand that the interventions are based on rational thinking and evidence-based methods. There is far too much emphasis on interventions that do not have any proven effect. Very large interventions are being performed on a bogus theoretical basis, without the necessary research control to identify whether the intervention was effective, what part of the interventions that was effective, and for whom the effect was beneficial.

REFERENCES

1. URSIN, H. & H.R. ERIKSEN. 2004. The cognitive activation theory of stress. Psychoneuroendocrinology **29:** 567–592.
2. LEVINE, S. & H. URSIN. 1991. What is stress? *In* Stress. Neurobiology and Neuroendocrinology. M.R. Brown, C. Rivier & G. Koob, Eds.: 3–21. Marcel Decker. New York. NY.
3. FOLKMAN, S. & R.S. LAZARUS. 1990. Coping and emotion. *In* Psychological and Biological Approaches to Emotion. N.L. Stein, B. Leventhal & T. Trabasso, Eds.: 313–332. Lawrence Erlbaum, Hillsdale, NJ.
4. BOLLES, R.C. 1972. Reinforcement, expectancy and learning. Psychol. Rev. **79:** 394–409.
5. OVERMIER, J.B. & M.E.P. SELIGMAN. 1967. Effects of inescapable shock upon subsequent escape and avoidance responding. J. Comp. Physiol. Pychol. **63:** 28–33.
6. ERIKSEN, H.R., R. MURISON, A.M. PENSGAARD & H. URSIN. 2005. Cognitive activation theory of stress (CATS): from fish brains to the Olympics. Psychoneuroendocrinology **30:** 933–938.
7. PROCIUK, T.J., L.J. BREEN & R.J. LUSSIER. 1976. Hopelessness, internal-external locus of control, and depression. J. Clin. Psychol. **32:** 299–300.
8. KARASEK, R.A. & T. THEORELL. 1990. Healthy Work, Stress, Productivity, and the Reconstruction of Working Life. Basic Books. New York. NY.
9. THEORELL, T. & R. A. KARASEK. 1996. Current issues relating to psychosocial job strain and cardiovascular disease research. J. Occup. Health Psychol. **1:** 9–26.
10. ERIKSEN, H.R. & H. URSIN. 1999. Subjective health complaints: is coping more important than control? Work Stress **13:** 238–252.
11. SIEGRIST, J. & A. RODEL. 2006. Work stress and health risk behavior. Scand. J. Work Environ. Health **32:** 473–481.
12. IHLEBÆK, C., H.R. ERIKSEN & H. URSIN. 2002. Prevalence of subjective health complaints (SHC) in Norway. Scand. J. Pub. Health **30:** 20–29.
13. FOLKOW, B. 2000. Man's two environments and disorders of civilization: aspects on prevention. Blood Pressure **9:** 182–191.
14. SHORTER, E. 1992. From Paralysis to Fatigue. A History of Psychosomatic Illness in the Modern Era. The Free Press. New York. NY.
15. ERIKSEN, H.R., B. HELLESNES, P. STAFF & H. URSIN. 2004. Are subjective health complaints a result of modern civilisation? Int. J. Behav. Med. **11:** 122–125.
16. WILHELMSEN, I., S. MULINDI, D. SANKOK, *et al*. 2007. Subjective health complaints are more prevalent in Maasais than in Norwegians. Nordic J. Psychiatry In press.
17. ERIKSEN, H.R. & H. URSIN. 2002. Sensitization and subjective health complaints. Scand. J. Psychol. **43:** 189–196.

18. HAGEN, E.M., E. SVENSEN, H.R. ERIKSEN, *et al.* 2006. Comorbid subjective health complaints in low back pain. Spine **31:** 1491–1495.
19. VANDVIK, P.O., I. WILHELMSEN, C. IHLEBæK & P.G. FARUP. 2004. Comorbidity of irritable bowel syndrome in general practice: a striking feature with clinical implications. Aliment. Pharmacol. Therap. **20:** 1195–1203.
20. VON KORFF, M., P. CRANE, M. LANE, *et al.* 2005. Chronic spinal pain and physical-mental comorbidity in the United States: results from the national comorbidity survey replication. Pain **113:** 331–339.
21. BROSSCHOT, J.F., W. GERIN & J.F. THAYER. 2006. The perseverative cognition hypothesis: a review of worry, prolonged stress-related physiological activation, and health. J. Psychosom. Res. **60:** 113–124.
22. AIRAKSINEN, O., J-I. BROX, C. CEDRASCHI, *et al.* 2006. European guidelines for the management of chronic nonspecific low back pain. Eur. Spine J. **15:** S192–S300.
23. DEALE, A, K. HUSAIN, T. CHALDER & S. WESSELY. 2001. Long-term outcome of cognitive behavior therapy versus relaxation therapy for chronic fatigue syndrome: A 5-year follow up study. Am. J. Psychiatry **158:** 2038–2042.
24. SVENSEN, E., B.B. ARNETZ, H. URSIN & H.R. ERIKSEN. 2007. Health complaints and satisfied with the job? A cross-sectional study on work environment, job satisfaction and subjective health complaints. J. Occup. Environ. Med. **49:** 568–573.

The Catecholamine–Cytokine Balance

Interaction between the Brain and the Immune System

J. SZELÉNYI AND E.S. VIZI

Institute of Experimental Medicine, Laboratory of Neuroimmunology, Budapest, Hungary

ABSTRACT: Cytokines are involved both in various immune reactions and in controlling certain events in the central nervous system (CNS). In our earlier studies, it was shown that monoamine neurotransmitters, released in stress situations, represent a tonic sympathetic control on cytokine production and on the balance of proinflammatory/anti-inflammatory cytokines. Basic and clinical studies have provided evidence that the biophase level of monoamines, determined by the balance of their release and uptake, is involved in the pathophysiology and treatment of depression, while inflammatory mediators might also have a role in its etiology. In this work, we studied the role of changes in norepinephrine (NE) level on the lipopolysaccharide (LPS) evoked tumor necrosis factor (TNF)-α and interleukin (IL)-10 response both in the plasma and in the hippocampus of mice. We demonstrated that the LPS induced TNF-α response is in direct correlation with the biophase level of NE, as it is significantly higher when the release of NE of vesicular origin was completely inhibited in an animal model of depression (reserpine treatment) and it is significantly lower in the case of increasing biophase levels of NE by genetic (NET–KO) or chemical (desipramine) disruption of NE reuptake. IL-10 was changed inversely to TNF-α levels only in the desipramine-treated animals. Our results showed that depression is related both to changes in peripheral and in hippocampal inflammatory cytokine production and to monoamine neurotransmitter levels. Since several anti-inflammatory drugs also have antidepressant effects, we hypothesized that antidepressants are also able to modulate the LPS-induced inflammatory response, which might contribute to their antidepressant effect.

KEYWORDS: cytokine; catecholamine; neuroimmunomodulation; inflammation; depression

INTRODUCTION

Cytokines are low molecular weight proteins or glycoproteins that were initially thought to be restricted to the immune system as mediators of communication between immune cells. Recently, however, increasing attention has been paid to the fact that many cytokines also play a key role in the central nervous and endocrine systems. It was reported that cytokines were also produced by brain cells, and that they interact closely with neurohormones and neurotransmitters.[1,2] The regulatory pathways that control the immune system include mediators, most notably catecholamines (CA), from the brain, and various hormones produced by the endocrine system, for example, corticosteroids (FIG. 1).

A large number of external and/or internal stimuli, including bacterial endotoxin (LPS), viral infections, differentiating factors, prostaglandins, cytokines, etc., are able to activate the immune system via macrophages, resulting in increased production of proinflammatory cytokines, such as interleukin (IL)-1, IL-12, and tumor necrosis factor (TNF)-α.[3,4] As all normal immune responses are temporary, both antigen-specific and nonspecific immune responses are well controlled.[5] Sympathetic neurotransmitters have an important role in the fine-tuning of immune and inflammatory responses. Recent studies have also supplied ample evidence of the relationship between stress or other situations that influence catecholamine levels and cytokine production, and also between the production of inflammatory mediators and the development of some diseases of the central nervous system (CNS), for example, depression, Alzheimer disease, etc.[6,7]

OCCURRENCE OF CYTOKINES IN THE BRAIN

Cytokines are involved both in the immune response and in controlling various events in the CNS; that is, they are equally immunoregulators and modulators of neural functions and neuronal survival. There is ample evidence to indicate the constitutive expression of numerous Th1- and Th2-type cytokines and also their functionally active receptors in "normal" adult brain. There is also a diurnal rhythm in the expression of at least IL-1β and TNF-α that is under complex neuroendocrine control, probably mainly by physiological variations of the corticotropin-releasing factor (CRF) levels.[8]

Cytokines might interact with the CNS in numerous ways, in which it is not negligible whether these mediators originate from peripheral immunocompetent cells or are produced within the CNS. The peripheral and central cytokine compartments appear to be integrated, and their effect might either inhibit or synergize each other; however, it should always be taken into account that they are differentially regulated. On the other hand, cytokine production is under the tonic control of the peripheral nervous system and the CNS

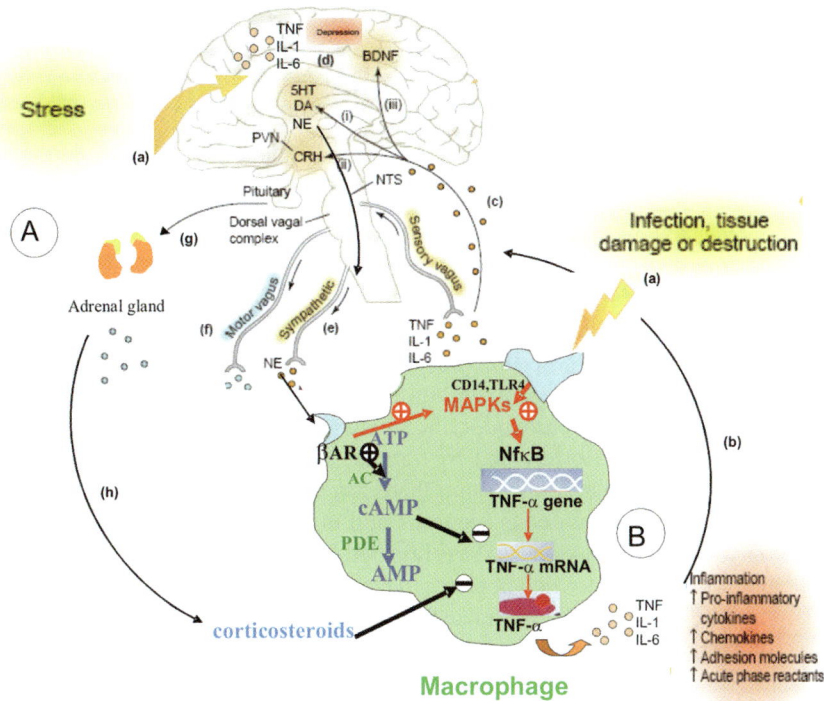

FIGURE 1. Bidirectional interactions between the central nervous system (CNS) and cytokines. Varieties of psychological or physical stimuli alter the biophase level of catecholamines and activate the production of cytokines, both in the immune- and the CNS. *Cognitive stimuli* like stress (a) induce both catecholamine (i) and cytokine (d) production in the brain (**A**). Catecholamines are able to modulate the immune response via receptors expressed on the immune cells (e). This regulation might be either attenuative or intensifying, depending on the costimulus. *Physical stimuli* (a) (e.g., infections, tissue damages, etc.) induce cytokine production, usually in the periphery (**B**), however, their effect might also influence the events in the CNS (b) (c). These cytokines also promote production of each other, change the level of stress hormones, and are under the regulation of the HPA axis via corticosteroids (g) (h). These interactions might also be either synergistic or attenuative. Many of these mediators (e.g., IL-1, NGF, TNF-α, NO, etc.) are upregulated by neuronal activity and promote behavioral changes and CNS diseases. The activity of the immune system and the production of catecholamines are also under diurnal control, therefore, the type and pattern of their biophase concentration provide feedback control on their own activities by tuning the levels of cytokines, hormones, and other mediators. The motor vagus might also influence these interactions (f). Collectively, these actions form a biochemical cascade involved in physiological and pathological psychological-, and immune regulation, as well as changes in their responses to external or internal challenges. CNS = central nervous system; HPA axis = hypothalamic-pituitary-adrenal axis; IL-1 = interleukin-1; IL-6 = interleukin-6; TNF-α = tumor necrosis factor-alpha; NE = norepinephrine; DA = dopamine; 5HT, 5-hydroxy-tryptamine; CRH = corticotropin-releasing hormone; βAR = beta-adrenergic receptor; MAPKs = mitogen-activated protein kinases.

and the cytokine balance can be modulated by the action of neurotransmitters released from nonsynaptic varicosities.[9]

The various cytokines affecting directly the CNS might originate from:
— Peripheral immune cells that are the main source of cytokines The cytokines produced by the immune cells in the periphery are able to cross the blood–brain barrier,[10] even in healthy, basal conditions. Active, saturable, and specific transport of certain cytokines across the blood–brain barrier was also demonstrated earlier.[11]
—Neuronal cells within the CNS are also able to produce cytokines even in the healthy state. Most of the cytokines and their receptors have been demonstrated and/or postulated in various cell types of the CNS, both under healthy and diseased circumstances. Cytokines produced by a cascade of neurons and glial cells within the brain may participate in the complex autonomic, neuroendocrine, metabolic, and behavioral responses to brain injuries originated by trauma, ischemia, infection, or inflammation.[12–14] Once in the brain, there is a CNS–cytokine network that is made up of its own cells (neurons and glial elements). These cells not only produce cytokines and express cytokine receptors, but also amplify cytokine signals, which in turn can have profound effects on neurotransmitter and corticotropin-releasing hormone (CRH) function, as well as on behavior.[12–16]

CONTRIBUTION OF CATECHOLAMINE AND CYTOKINE RECEPTORS IN THE INFLAMMATORY RESPONSE

The sympathetic nervous system (SNS) innervates immune organs and, when activated, releases its signaling molecules in the vicinity of immune cells. Accordingly, cytokine balance can be modulated by sympathetic neurotransmitters, both in the CNS and in the periphery. Immune cells express various neurotransmitter receptors that are sensitive to monoamines, and the production of cytokines (and other immune/inflammatory mediators (chemokines and free radicals) is modulated by activation of these receptors.[1] Once the neurotransmitters have reached the target cells, they occupy their appropriate receptors and the initiated signal transduction modulates the cytokine production of the cell.

The catecholamines norepinephrine (NE) and epinephrine (EPI) exert their effects by binding to 7 transmembrane spanning G-protein-coupled cell-surface receptors termed adrenoceptors. Adrenoceptors can be classified into three major groups: $\alpha 1-$, $\alpha 2-$, and β-adrenoceptor types.[17] Each of these three major types can be subdivided further into at least three subtypes: $\alpha 1A$, $\alpha 1B$, $\alpha 1C$; $\alpha 2A$, $\alpha 2B$, $\alpha 2C$; and $\beta 1$, $\beta 2$, and $\beta 3$ and at least the presynaptic $\alpha 2-$ and the peripheral $\beta 2$-adrenoceptors play major roles in the regulation of cytokine balance.

There is ample evidence that in the immunomodulatory effect of catecholamines, cAMP plays one of the key roles.[7,18–20] Thus, occupation of neurotransmitter receptors that stimulate or inhibit adenylate–cyclase also influence the cytokine profile of the system. NE and the adrenergic drugs may influence the immune response directly, through adrenergic receptors expressed on macrophages and also on other immunologically competent cells, as well as indirectly via alteration of endogenous NE levels by influencing the activity of release-regulating presynaptic α2-adrenoceptors (α2-AR) located on sympathetic nerve terminals.[9,21–24] Activation of the presynaptic α2–adrenoceptors results in a negative feedback effect on NE release, leading to decreased extracellular NE concentration.[25,26] The majority of the direct effect of NE prevails via the β2-adrenoceptors expressed by various immune cells.

CYTOKINES AND NEURONAL SURVIVAL

Cytokines can be expressed under resting physiological conditions in resident CNS cells, but are also induced during injury and development. In addition, under pathological conditions, cytokines can be expressed in infiltrating macrophages in the brain. Inflammatory processes have been implicated in both acute and chronic neurodegenerative conditions. The inflammatory response of the CNS is weaker and delayed over other tissues,[27] however, many inflammatory responses (e.g., activation of microglia and release of inflammatory mediators) can be induced rapidly. Physiologically, the cellular expression of cytokines in the CNS is strictly controlled; however, under certain pathological conditions, the expression of various cytokine genes may become spatially and temporally modified. Literary data indicate, for example, that the expression of tumor growth factor (TGF)-β1 by hippocampal CA1 neurons is upregulated during the first hours after ischemia.[28] Proinflammatory cytokine production and signaling results in important changes in the neurons long before their ultimate cell death.

Neurotoxic and neuroprotective mechanisms are both closely related to the balance between the proinflammatory and anti-inflammatory cytokines. The process of neurodegeneration is closely related to the shift of cytokine production toward the side of proinflammatory cytokines like IL-1 or TNF−α, regardless of the fact that they are produced within the CNS or have systemic origins. On the other hand, anti-inflammatory cytokines in the CNS maintain homeostasis and protect cell viability by inhibiting inflammatory responses.[29–31] Since there is an inhibitory cross-regulation between the two groups of cytokines that indirectly suppress the synthesis of each other, this provides a further fine-tuning of the balance between neurodegenerative and neuroprotective effects.

The role of proinflammatory cytokines like IL-1 or TNF-α in neurotoxicity is considerable but controversial. *In vivo* animal experiments have revealed that interferon (IFN)-γ/LPS administration into the rat hippocampus induces delayed neuronal apoptosis.[32] Concerning IL-1, it was demonstrated that this cytokine itself was not toxic to healthy brain tissue or to normal neurons, but even in very low concentrations, it augmented traumatic, ischemic, or inflammatory brain injuries.[31,33] This dual effect of cytokines expressed within the CNS is similar to the role played by cytokines in the periphery, that of helping to select populations of mature cell types by enhancing survival of some and eliminating others through apoptosis. In immune cells, and similarly, in neurons, the transcription factor nuclear factor (NF)-κB may play a pivotal role in these context-dependent effects on cell survival or death, acting much like a switch that can block death signals when induced and can allow death signal activation of the apoptotic pathway when suppressed or blocked. *In vitro* studies in mixed neuronal/glial cultures show that several cytokines play dual context-dependent maturation roles in either promoting or preventing apoptotic neuronal cell death.

Considering that proinflammatory cytokines are not neurotoxic per se, it is possible that they alter neuronal survival not only by intracellular receptor–receptor interactions but also by cell–cell interactions.[34] The primary source of IL-1 and TNF-α after brain injury is microglia, but astrocytes, oligodendrocytes, and other cells in the CNS may also produce cytokines, which can interact with each other. This is supported by another concept of neurodegeneration that implies the existence of a different level of regulation between pro- and anti-inflammatory cytokines in the brain. This concept is based on the intracellular cross-talk between heterologous receptors on a single cell that can either promote or inhibit receptor activity. If a receptor representing a death signal (e.g., TNF-α receptor) interacts with a receptor for survival signal (e.g., IGF1 receptor) it might inhibit IGF1-mediated neuroprotection; that is, the receptor–receptor interaction is a "silencing of the survival signal."[35]

Thus, activation of neurotransmitter receptors that stimulate adenylate–cyclase accompanied by LPS stimulus of the immune system leads to a shift toward T helper 2 (Th2)-type responses, which are both neuroprotective and anti-inflammatory, whereas downregulation of intracellular cAMP stimulates a T helper 1 (Th1)-type response, resulting in cell destructive effects and inflammation.

BIDIRECTIONAL COMMUNICATION BETWEEN CYTOKINES AND NEUROTRANSMITTERS

The modulation of cytokine balance by catecholamines is governed by the biophase concentration of these monamines. The amount of catecholamines in the extracellular space is a function of the balance between their vesicular release and their reuptake by the monoamine transporter system.

Stimulation of β_2-AR is the classical example of activation of adenylyl cyclase via stimulatory G proteins (Gs), resulting in the subsequent increase in intracellular cAMP. Although both α2- and β2-adrenoceptors are expressed on the surface of various immune cells,[23,36,37] macrophagic α2-adrenoceptors have only a minor role. Occupation of α2-adrenoceptors on macrophages results in the suppression of the intracellular cAMP level, because these receptors are associated with a Gi-type protein.[17] Since cAMP has generally been proven to suppress the inflammatory immune response, sympathetic control of the innate immune response is believed to be necessarily immunosuppressive,[1,38,39] while the exact mechanism is not yet fully understood.

As the immunoregulatory effect of catecholamines becomes effective practically as a costimulus, the most common way of studying it is to apply an agonist together with a known immune activator. Cytokine production by macrophages or macrophage-like cells (e.g., microglia) is most commonly induced by bacterial endotoxin (LPS). Evidence is available that the monocyte/macrophage system orchestrates the innate immune response to LPS by producing cytokines and other mediators, like TNF-α, IL-1β, IL-8, or nitric oxide (NO). However, no data are available on the effect of sympathetic signals on the cytokine production evoked by other known stimuli, such as cytokines, tumor promoters, or viral components.

In earlier studies, we investigated whether the immunomodulatory effect of the SNS would also be effective, together with stimuli other than LPS.[40] LPS stimulation via the CD14 receptor complex activates several intracellular pathways that include the IκB kinase NF-κB pathway and the MAPK pathways: ERK1 and 2, and p38.[41] We presented evidence that the β-adrenergic agonist isoproterenol had opposite immunomodulatory effects on TNF-α, IL-12, and NO production in macrophages stimulated by phorbol esters (PMA; a PKC activator) versus LPS (FIG. 1B). We also demonstrated that these opposite effects of β_2-AR stimulation on LPS- versus PMA-induced mediator production correlated with their effects on MAPK activation. These results show that the ERK and p38 MAPKs may act as molecular switches and that, depending on the applied stimulus, they can regulate sympathetic immunomodulation induced by β-adrenergic agonists/antagonists.[40,42]

Effects of cytokines on the noradrenergic system have also been described (FIG. 1A). In addition to the effects on neurotransmitter metabolism, it was shown that inflammatory cytokines exerted profound stimulatory effects on the hypothalamic-pituitary-adrenal (HPA) axis hormones as well as on CRH (mRNA and protein), both in the hypothalamus and amygdala, brain regions that have an important role in fear and anxiety.[43,44] Changes in the catecholamine metabolism in brain regions being essential to the regulation of emotion, including the limbic system (amygdala, hippocampus, and nucleus accumbens), might influence sickness behavior.[45,46] These effects are, in large part, mediated by the cross-talk of cytokines and their receptors within the HPA axis tissues that facilitate the integration of cytokine signals.[47] As an example,

IL-1 was shown to stimulate hypothalamic and preoptic noradrenergic neurotransmission,[48] similar to the effects observed after administration of various forms of IL-1.[45] There is inconsistent data for other proinflammatory cytokines and their influence on noradrenergic neurotransmission. IL-2 showed similar effects as IL-1, while other studies report that TNF-α inhibited NE release from the median eminence.[49] It is now well established that immune activation triggers the SNS to release the neurotransmitters noradrenaline (NA), adrenaline, and dopamine.[50–53]

THE CATECHOLAMINE–CYTOKINE INTERACTION IN STRESS AND DEPRESSION

Growing evidence suggests that overactivation of innate immune responses following stress and during depression might come at the expense of decreased cellular and humoral acquired immune responses.[15,54] Activation of the stress system might promote cytokine production through several mechanisms. Despite suppressing certain immune processes, activation of the SNS is linked in several studies to proinflammatory activation in the periphery, which might, in turn, influence inflammatory processes in the CNS. It was demonstrated that stress-induced activation of NF-κB in peripheral blood mononuclear cells appeared to be dependent on noradrenaline.[55]

A dysregulation of the cytokine balance could induce depressive symptoms, due to lower levels of anti-inflammatory cytokines and higher levels of proinflammatory cytokines. Anti-inflammatory cytokines are known to evoke an anti-inflammatory state, both on their own (IL-10, TGF-β) receptors and also by the blockade of the binding of proinflammatory stimuli to their cell-surface receptors (IL-1RA, soluble TNF receptors II)

In our earlier studies, it was shown that cytokine production was under tonic, sympathetic control.[1,7] We could demonstrate that the amount of catecholamines in the extracellular space was a function of the balance between their vesicular release and their reuptake by the monoamine transporter system. The release of NE, for example, is controlled by the negative feedback mechanism of presynaptic α2-ARs,[25,26] as discussed above. For the rapid removal of noradrenaline released from sympathetic neurons, its reuptake via noradrenaline-transporter (NET) is responsible.

Low levels of extracellular NE can be achieved experimentally by reserpine treatment. This results in an animal model of depression, where we could demonstrate that the TNF-α response was significantly higher than that in the healthy mice.[21] On the other hand, the extracellular NE level could be increased either by genetic removal or by a chemical blockade of the NET. Recently, it was demonstrated that genetically modified mice, which became NET-deficient (NET–KO)[56] resulted in a significant decrease in the LPS-evoked TNF-α response, confirming the assumption that the balance between

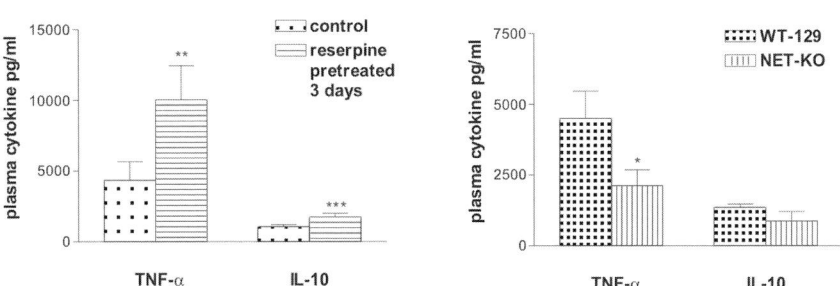

FIGURE 2. Extracellular catecholamine level balance as one of the key modulators of inflammatory mediator production. Cytokines were measured by enzyme-linked immunosorbent assay (ELISA) technique (R&D kits) after LPS induction (10 mg/kg ip. 90 min) in the plasma of wild-type (WT) and that of noradrenaline transporter deficient (NET–KO) mice.

NE release and uptake is one of the key modulators of inflammatory mediator production (FIG. 2).[6,57]

According to the mostly accepted theory of depression, the extent and/or the duration of monoamine neurotransmitter action are key points in the development and therapy of depression. A number of effective antidepressants have been reported to inhibit the activity of monoamine transporters that resulted in an increased biophase level of monoamines. A similar effect could be achieved by genetic removal of monoamine transporter genes, as it was realized in NET–KO,[56] DAT–KO,[58] and 5HTT–KO mice.[59] These animals behave like chronically antidepressant treated ones, exhibiting high extracellular monoamine levels. Long-term inhibition of the NET by antidepressants has been reported to change the density and function of pre- and postsynaptic α2-ARs, which may contribute to the antidepressant effects of NET inhibitors, such as desipramine. In the NET–KO animals, it was demonstrated that density of α2-AR was upregulated in the brainstem, hippocampus, and striatum.[60] In these mice, the α2-AR autoreceptors are not desensitized and the inhibitory tone on NE release is stronger as a consequence of elevated extracellular NE concentration.[61]

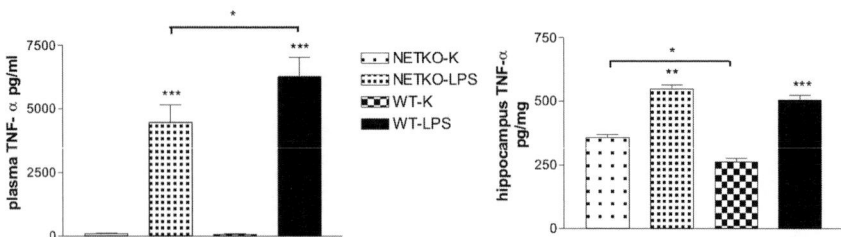

FIGURE 3. Comparison of the TNF-α production in the plasma and in the hippocampus of WT and that of NET–KO mice after LPS induction (10 mg/kg ip. 90 min).

Since the extracellular level of monoamines is highly dependent on the activity of their transporters, it was assumed that, in acute treatment of mice, not only desipramine but also other monoamine transporter inhibitors, exhibiting antidepressant characteristics (dopamine transporter and serotonin transporter inhibitors), had modulatory effects on the inflammatory immune response. Whereas an approximately 3-week long chronic treatment is necessary to start the therapeutic antidepressant effect of these drugs, it was studied whether the immunomodulatory effect was also present after chronic treatments.[62,63]

Concerning the crucial role that cAMP plays in cytokine production, it may be assumed that tricyclic antidepressants (TCAs) produce their immunomodulatory effects through this signaling mechanism. The antidepressant effects of phosphodiesterase inhibitors (like rolipram) [64] and the overexpression of the cAMP response element-binding protein (CREB) in the hippocampus after chronic antidepressant administration,[65] might support this assumption. However, an increasing amount of evidence is available that chronic treatment with TCAs has pleiotropic effects in addition to blocking the monoamine uptake systems, which contribute to exert their antidepressive action in the CNS.

In our recent experiments, we demonstrated significant differences between peripheral and hippocampal cytokine production. As is shown in FIGURE 3, there are differences in both basal and induced TNF-α production. An induction of TNF-α production with LPS resulted in a significantly higher TNF-α production in the plasma of wild-type (WT) than that in the NET–KO animals, while there was no difference in their basal levels. In the hippocampus of NET–KO animals, the basal TNF-α content was significantly higher than that of the WT mice, while upon induction, both strains responded similarly.

In conclusion, recognition of the interactions between the catecholamine- and inflammatory systems might be important, since certain anti-inflammatory drugs also have antidepressant effects, and most of the antidepressant drugs can modulate the immune response. Considering the essential role of sympathetic neurotransmitters and the fact that β-adrenergic agonists/antagonists are widely used in the therapy of many diseases, the recognition that inhibitors of the monoamine uptake system might have multiple targets and are also able to modulate the inflammatory response is highly important, both in future therapy and in the development of new drugs.

ACKNOWLEDGMENT

This work was supported by OTKA Grant No. T046896 and by ETT Grant No. 298/2006.

REFERENCES

1. ELENKOV, I.J., R.L. WILDER, G.P. CHROUSOS & E S. VIZI. 2000. The sympathetic nerve–an integrative interface between two supersystems: the brain and the immune system. Pharmacol. Rev. **52:** 595.
2. SZELENYI, J. 2001. Cytokines and the central nervous system. Brain Res. Bull. **54:** 329.
3. BEUTLER, B. 1995. TNF, immunity and inflammatory disease: lessons of the past decade. J. Investig. Med. **43:** 227.
4. BENDTZEN, K. 1994. Cytokines and natural regulators of cytokines. Immunol. Lett. **43:** 111.
5. DINARELLO, C.A. 1992. Anti-cytokine strategies. Eur. Cytokine Netw. **3:** 7.
6. SZELENYI, J. & Z. SELMECZY. 2002. Immunomodulatory effect of antidepressants. Curr. Opin. Pharmacol. **2:** 428.
7. VIZI, E.S. 1998. Receptor-mediated local fine-tuning by noradrenergic innervation of neuroendocrine and immune systems. Ann. N. Y. Acad. Sci. **851:** 388.
8. BREDOW, S., N. GUHA-THAKURTA, P. TAISHI, et al. 1997. Diurnal variations of tumor necrosis factor alpha mRNA and alpha-tubulin mRNA in rat brain. Neuroimmunomodulation **4:** 84.
9. VIZI, E.S. & J.P. KISS. 1998. Neurochemistry and pharmacology of the major hippocampal transmitter systems: synaptic and nonsynaptic interactions. Hippocampus **8:** 566.
10. WATKINS, L.R., S.F. MAIER & L.E. GOEHLER. 1995. Cytokine-to-brain communication: a review & analysis of alternative mechanisms. Life Sci. **57:** 1011.
11. BANKS, W.A. & A.J. KASTIN. 1997. Relative contributions of peripheral and central sources to levels of IL-1 alpha in the cerebral cortex of mice: assessment with species-specific enzyme immunoassays. J. Neuroimmunol. **79:** 22.
12. BREDER, C.D., C. HAZUKA, T. GHAYUR, et al. 1994. Regional induction of tumor necrosis factor alpha expression in the mouse brain after systemic lipopolysaccharide administration. Proc. Natl. Acad. Sci. USA **91:** 11393.
13. STERNBERG, E.M. 1997. Neural-immune interactions in health and disease. J. Clin. Invest. **100:** 2641.
14. WOICIECHOWSKY, C., K. ASADULLAH, D. NESTLER, et al. 1998. Sympathetic activation triggers systemic interleukin-10 release in immunodepression induced by brain injury. Nat. Med. **4:** 808.

15. RAISON, C.L. & A.H. MILLER. 2003. When not enough is too much: the role of insufficient glucocorticoid signaling in the pathophysiology of stress-related disorders. Am. J. Psychiatry **160**: 1554.
16. DANTZER, R. 2004. Cytokine-induced sickness behaviour: a neuroimmune response to activation of innate immunity. Eur. J. Pharmacol. **500**: 399.
17. BYLUND, D.B., D.C. EIKENBERG, J.P. HIEBLE, et al. 1994. International Union of Pharmacology nomenclature of adrenoceptors. Pharmacol. Rev. **46**: 121.
18. BOURNE, H.R., L.M. LICHTENSTEIN, K.L. MELMON, et al. 1974. Modulation of inflammation and immunity by cyclic AMP. Science **184**: 19.
19. KAMBAYASHI, T., C.O. JACOB, D. ZHOU, et al. 1995. Cyclic nucleotide phosphodiesterase type IV participates in the regulation of IL-10 and in the subsequent inhibition of TNF-alpha and IL-6 release by endotoxin-stimulated macrophages. J. Immunol. **155**: 4909.
20. KAMINSKA, B. 2005. MAPK signalling pathways as molecular targets for anti-inflammatory therapy–from molecular mechanisms to therapeutic benefits. Biochim. Biophys. Acta **1754**: 253.
21. SZELENYI, J., J.P. KISS & E.S. VIZI. 2000. Differential involvement of sympathetic nervous system and immune system in the modulation of TNF-alpha production by alpha2- and beta-adrenoceptors in mice. J. Neuroimmunol. **103**: 34.
22. SZELENYI, J., J.P. KISS, E. PUSKAS, et al. 2000. Contribution of differently localized alpha 2- and beta-adrenoceptors in the modulation of TNF-alpha and IL-10 production in endotoxemic mice. Ann. N. Y. Acad. Sci. **917**: 145.
23. SZELENYI, J., J.P. KISS, E. PUSKAS, et al. 2000. Opposite role of alpha2- and beta-adrenoceptors in the modulation of interleukin-10 production in endotoxaemic mice. Neuroreport **11**: 3565.
24. ELENKOV, I.J. & E.S. VIZI. 1991. Presynaptic modulation of release of noradrenaline from the sympathetic nerve terminals in the rat spleen. Neuropharmacology **30**: 1319.
25. VIZI, E.S. 1979. Presynaptic modulation of neurochemical transmission. Prog. Neurobiol. **12**: 181.
26. KISS, J.P., G. ZSILLA, A. MIKE, et al. 1995. Subtype-specificity of the presynaptic alpha 2-adrenoceptors modulating hippocampal norepinephrine release in rat. Brain Res. **674**: 238.
27. PERRY, V.H., M.D. BELL, H.C. BROWN & M.K. MATYSZAK. 1995. Inflammation in the nervous system. Curr. Opin. Neurobiol. **5**: 636.
28. ZHU, Y., S. ROTH-EICHHORN, N. BRAUN, et al. 2000. The expression of transforming growth factor-beta1 (TGF-beta1) in hippocampal neurons: a temporary upregulated protein level after transient forebrain ischemia in the rat. Brain Res. **866**: 286.
29. ROTHWELL, N.J. 1999. Annual review prize lecture cytokines killers in the brain? J. Physiol. **514**(Pt 1): 3.
30. FEUERSTEIN, G.Z., X. WANG & F.C. BARONE. 1998. The role of cytokines in the neuropathology of stroke and neurotrauma. Neuroimmunomodulation **5**: 143.
31. ALLAN, S.M. & N.J. ROTHWELL. 2001. Cytokines and acute neurodegeneration. Nat. Rev. Neurosci. **2**: 734.
32. MATSUOKA, Y., Y. KITAMURA, H. TAKAHASHI, et al. 1999. Interferon-gamma plus lipopolysaccharide induction of delayed neuronal apoptosis in rat hippocampus. Neurochem. Int. **34**: 91.

33. ROTHWELL, N., S. ALLAN & S. TOULMOND. 1997. The role of interleukin 1 in acute neurodegeneration and stroke: pathophysiological and therapeutic implications. J. Clin. Invest. **100:** 2648.
34. JEOHN, G.H., L.Y. KONG, B. WILSON, *et al.* 1998. Synergistic neurotoxic effects of combined treatments with cytokines in murine primary mixed neuron/glia cultures. J. Neuroimmunol. **85:** 1.
35. VENTERS, H.D., R. DANTZER & K.W. KELLEY. 2000. A new concept in neurodegeneration: TNFalpha is a silencer of survival signals. Trends Neurosci. **23:** 175.
36. SPENGLER, R.N., R.M. ALLEN, D.G. REMICK, *et al.* 1990. Stimulation of alpha-adrenergic receptor augments the production of macrophage-derived tumor necrosis factor. J. Immunol. **145:** 1430.
37. ABRASS, C.K., S.W. O'CONNOR, P.J. SCARPACE & I.B. ABRASS. 1985. Characterization of the beta-adrenergic receptor of the rat peritoneal macrophage. J. Immunol. **135:** 1338.
38. CHONG, Y.H., Y.J. SHIN & Y.H. SUH. 2003. Cyclic AMP inhibition of tumor necrosis factor alpha production induced by amyloidogenic C-terminal peptide of Alzheimer's amyloid precursor protein in macrophages: involvement of multiple intracellular pathways and cyclic AMP response element binding protein. Mol. Pharmacol. **63:** 690.
39. VAN DER POLL, T., J. JANSEN, E. ENDERT, *et al.* 1994. Noradrenaline inhibits lipopolysaccharide-induced tumor necrosis factor and interleukin 6 production in human whole blood. Infect. Immun. **62:** 2046.
40. SZELENYI, J., Z. SELMECZY, A. BROZIK, *et al.* 2006. Dual beta-adrenergic modulation in the immune system: stimulus-dependent effect of isoproterenol on MAPK activation and inflammatory mediator production in macrophages. Neurochem. Int. **49:** 94.
41. GUHA, M. & N. MACKMAN. 2001. LPS induction of gene expression in human monocytes. Cell Signal. **13:** 85.
42. MAGOCSI, M., E.S. VIZI, Z.S. SELMECZY & A.J.S. BROZIK. 2007. Multiple G protein-coupling specificity of beta-adrenoceptor in macrophages. Immunology. In press.
43. CAPURON, L. & A.H. MILLER. 2004. Cytokines and psychopathology: lessons from interferon-alpha. Biol. Psychiatry **56:** 819.
44. BESEDOVSKY, H.O. & A. DEL REY. 1996. Immune-neuro-endocrine interactions: facts and hypotheses. Endocr. Rev. **17:** 64.
45. DUNN, A.J., J. WANG & T. ANDO. 1999. Effects of cytokines on cerebral neurotransmission. Comparison with the effects of stress. Adv. Exp. Med. Biol. **461:** 117.
46. GAO, H.M., J. JIANG, B. WILSON, *et al.* 2002. Microglial activation-mediated delayed and progressive degeneration of rat nigral dopaminergic neurons: relevance to Parkinson's disease. J. Neurochem. **81:** 1285.
47. SILVERMAN, M.N., B.D. PEARCE, C.A. BIRON & A.H. MILLER. 2005. Immune modulation of the hypothalamic-pituitary-adrenal (HPA) axis during viral infection. Viral Immunol. **18:** 41.
48. DUNN, A.J. 1988. Systemic interleukin-1 administration stimulates hypothalamic norepinephrine metabolism parallelling the increased plasma corticosterone. Life Sci. **43:** 429.
49. ELENKOV, I.J., K. KOVACS, E. DUDA, *et al.* 1992. Presynaptic inhibitory effect of TNF-alpha on the release of noradrenaline in isolated median eminence. J. Neuroimmunol. **41:** 117.

50. AKIYOSHI, M., Y. SHIMIZU & M. SAITO. 1990. Interleukin-1 increases norepinephrine turnover in the spleen and lung in rats. Biochem. Biophys. Res. Commun. **173:** 1266.
51. BESEDOVSKY, H., E. SORKIN, M. KELLER & J. MULLER. 1975. Changes in blood hormone levels during the immune response. Proc. Soc. Exp. Biol. Med. **150:** 466.
52. DUNN, A.J. 1992. Endotoxin-induced activation of cerebral catecholamine and serotonin metabolism: comparison with interleukin-1. J. Pharmacol. Exp. Ther. **261:** 964.
53. SHIMIZU, N., T. HORI & H. NAKANE. 1994. An interleukin-1 beta-induced noradrenaline release in the spleen is mediated by brain corticotropin-releasing factor: an *in vivo* microdialysis study in conscious rats. Brain Behav. Immun. **8:** 14.
54. RAISON, C.L., L. CAPURON & A.H. MILLER. 2006. Cytokines sing the blues: inflammation and the pathogenesis of depression. Trends Immunol. **27:** 24.
55. BIERHAUS, A., J. WOLF, M. ANDRASSY, *et al*. 2003. A mechanism converting psychosocial stress into mononuclear cell activation. Proc. Natl. Acad. Sci. USA **100:** 1920.
56. XU, F., R.R. GAINETDINOV, W.C. WETSEL, *et al*. 2000. Mice lacking the norepinephrine transporter are supersensitive to psychostimulants. Nat. Neurosci. **3:** 465.
57. SELMECZY, Z., J. SZELENYI & E.S. VIZI. 2003. Intact noradrenaline transporter is needed for the sympathetic fine-tuning of cytokine balance. Eur. J. Pharmacol. **469:** 175.
58. MORON, J.A., A. BROCKINGTON, R.A. WISE, *et al*. 2002. Dopamine uptake through the norepinephrine transporter in brain regions with low levels of the dopamine transporter: evidence from knock-out mouse lines. J. Neurosci. **22:** 389.
59. ALEXANDRE, C., D. POPA, V. FABRE, *et al*. 2006. Early life blockade of 5-hydroxytryptamine 1A receptors normalizes sleep and depression-like behavior in adult knock-out mice lacking the serotonin transporter. J. Neurosci. **26:** 5554.
60. GILSBACH, R., A. FARON-GORECKA, Z. ROGOZ, *et al*. 2006. Norepinephrine transporter knockout-induced up-regulation of brain alpha2A/C-adrenergic receptors. J. Neurochem. **96:** 1111.
61. VIZI, E.S., G. ZSILLA, M.G. CARON & J.P. KISS. 2004. Uptake and release of norepinephrine by serotonergic terminals in norepinephrine transporter knock-out mice: implications for the action of selective serotonin reuptake inhibitors. J. Neurosci. **24:** 7888.
62. SZELENYI, J., Z. SELMECZY & E.S. VIZI. 2004. Effect of monoamine transporter inhibitors on the LPS-induced cytokine production. **18:** 95.
63. SZELENYI, J. & E.S. V. 2007. Neuroimmune correlates of sleep in depression: role of cytokines. *In* Neuroimmunology of Sleep, S.R.C. Pandi-Perumal & P. Daniel; Ed.: Dr Chrousos, Georgios), 295. Springer. New York, NY.
64. ZHU, J., E. MIX & B. WINBLAD. 2001. The antidepressant and antiinflammatory effects of rolipram in the central nervous system. CNS Drug Rev. **7:** 387.
65. CHEN, A.C., Y. SHIRAYAMA, K.H. SHIN, *et al*. 2001. Expression of the cAMP response element binding protein (CREB) in hippocampus produces an antidepressant effect. Biol. Psychiatry **49:** 753.

Chronic Stress and Social Changes

Socioeconomic Determination of Chronic Stress

MÁRIA S. KOPP,[a] ÁRPÁD SKRABSKI,[a] ANDRÁS SZÉKELY,[a] ADRIENNE STAUDER,[a] AND REDFORD WILLIAMS[b]

[a]*Institute of Behavioral Sciences, Semmelweis University, Budapest, Hungary*

[b]*Behavioral Medicine Research Center, Duke University, Durham, North Carolina, USA*

> ABSTRACT: In the last decades in the transforming societies of Central and Eastern Europe, premature mortality increased dramatically, especially among men. Increasing disparities in socioeconomic conditions have been accompanied by a widening socioeconomic gradient in mortality among men. Social cohesion and meaning in life may help to counterbalance the widening gap in material circumstances. Not the difficult social situation in itself, but the subjective experience of relative disadvantage, the prolonged negative emotional state, that is, chronic stress seems to be the most important risk factor. The health consequences of a low socioeconomic situation among men might be mostly explained by chronic stress caused by work and close-partner–related factors, and the toxic components of this interaction are depression and hopelessness. In the case of women, the broader personal and family relations are the most important health-related factors. Weekend workload, low social support at work and low control at work accounted for a large part of variation in male premature cardiovascular mortality rates, whereas job insecurity, high weekend workload, and low control at work contribute most markedly to variations in premature cardiovascular mortality rates among women. There are two general approaches that scientists and practitioners might take: train individuals and groups to use skills that will enable them to cope better with the stressful conditions that are damaging their health; and lobby governments to adopt policies that will result in decreased chronic stress on the societal level.
>
> KEYWORDS: chronic stress; transforming societies; premature mortality; depression; gender differences; socioeconomic differences, coping, stress management

Address for correspondence: Maria S. Kopp, M.D., Ph.D., Institute of Behavioral Sciences, Semmelweis University, Budapest, H-1089, Nagyvarad ter 4, Hungary. Voice: 36-1-210-2953; fax: 36-1-210-2955.

kopmar@net.sote.hu

INTRODUCTION

In the last decades in the transforming societies of Central and Eastern Europe (CEE), premature mortality increased dramatically, especially among men. For example in Hungary between 1960 and 2005 in the 40- to 69-year-old male population the mortality rate increased from 12 to 16 per thousand by 33%, while in the same age group among women it decreased by 4%. Thus in 2005, an additional 11.395 men deceased from this age group than in 1960 (20.736 men in 1960, 32.131 men in 2005).[1] Up to the end of the 1960s the middle-aged (40- to 69-year-old) male mortality rates in Hungary had been similar to the Western European levels. Subsequently, mortality rates continued to decline in Western Europe, whereas in Hungary and in other CEE countries this tendency reversed, especially among middle-aged men.[2-12] In the late 1980s, the mortality rates among 40–69-year-old men in Hungary rose to higher levels than they were in the 1930s, while the mortality rates in the older age groups were comparable to the worst in Western Europe. Among Hungarian men the risk of early cardiovascular death (before age 65) was 3.13 times higher than the European average in 1993, while it was only 1.27 times higher in 1970.[13] Between 1965 and 1992, the life expectancy declined considerably among men in Hungary and in the former Soviet Union, while it increased in Western Europe, in the United States, and in Canada.[14]

Since the breakup of the Soviet Union in the late 1980s to early 1990s, a marked increase in annual mortality and shortening of life expectancy has been observed in Russia, Ukraine, and in the three Baltic countries. The annual national mortality in Russia had increased by 50% since 1989, from about 6 deaths per 1,000 to about 9 deaths per 1,000 in 2002. Most of this increased mortality was due to cardiovascular disease and occurred disproportionately in persons of lower socioeconomic status.[15,16] Similar to the situation in Hungary, this dramatic increase in annual mortality could not be accounted for by increases in standard cardiovascular disease risk factors such as lipids, smoking, excessive alcohol use, or obesity. Nor was reduced access to medical care a likely cause, because the increase in mortality in Moscow, where the medical system continued to be sound, was higher than in outlying areas in eastern Russia, where medical care delivery had never been at the Moscow level. Chronic stress and increased levels of psychosocial risk factors are likely contributors to these increased mortality rates. Surveys revealed increases in psychosocial tension, stress, vital exhaustion, and depression. Among patients visiting medical outpatient clinics a marked rise, to 45%, was observed in the proportion reporting signs of depression. Psychosocial factors arising out of the social, economic, and political upheavals since 1989–1990 are prime candidates to account for a significant part of the increased mortality.[15]

It is surprising that these deteriorations in the health of populations in CEE have not received more attention, given the magnitude of the problem. If the 33% increase in annual premature mortality of Hungarian men before 69 years of age or 50% increase of mortality in Russia, for example, were the result of some viral agent, such as "bird flu," there would be a worldwide mobilization of great proportions to deal with what would be widely recognized as a public health disaster. But that has not happened. The rising death rates and decreased life expectancy in CEE have received some attention, and several research groups[3,5,17-19] have mobilized efforts to understand the reasons for the deterioration in health in CEE, but there has not been the massive response one would think a public health problem of this magnitude would call forth, most of all for health politicians.

Why not? Could it be because it appears that psychosocial factors are playing a major role, leading to the perception that the problem is "only due to stress" and the attendant worsening of psychosocial factors like depression, anxiety, hostility, increased strain and decreased reward at work, declining social support, and the like--all factors that behavioral medicine research has clearly documented to have significant impacts on morbidity and mortality, but factors that the public health establishment finds harder to recognize as serious threats to health and lacks the means to confront?

CHRONIC STRESS AS A PUBLIC HEALTH ISSUE

What is the explanation for this public health disaster during a period of rapid economic change in Hungary? This deterioration cannot be ascribed to deficiencies in health care, because during these years there was a significant decrease in infant and old-age mortality and improvements in other dimensions of health care. Furthermore, between 1960 and 1989 there was a constant increase in the gross domestic product in Hungary. Thus the worsening health status of the Hungarian male population cannot be explained by a worsening material situation.

A growing polarization of the socioeconomic situation occurred in the CEE countries, especially in Hungary between 1960 and 1990 and in Russia and the Baltic countries in the last decades.[19] The vast majority of the population lived at a similarly low level economically in 1960, with practically no income inequality, and there were no mortality differences between socioeconomic strata. Since that time, increasing disparities in socioeconomic conditions have been accompanied by a widening socioeconomic gradient in mortality, especially among men.[2,20]

The theory of relative deprivation hypothesizes says that chronic stress can arise out of situations in which there is rapid improvement in living standards for some but not for others. Relative deprivation may be deleterious to both psychological and physical health, mediated by stress-related coping responses (e.g., more smoking, heavier drinking) as well as invidious social comparisons. Status syndrome is the name given to this phenomenon by Michael Marmot, which is especially important among the sudden hierarchy disruptions in changing societies.[21] Conversely, social cohesion and meaning in life may help to counterbalance the widening gap in material circumstances.[22–24] This might be the explanation for the observation in some Far-East countries such as Japan and Singapore that life expectancy increased considerably between 1965 and 1992 in spite of fundamental societal changes.

One of the most interesting features of the so-called Central Eastern European health paradox is the gender difference in worsening mortality, in spite of the fact that men and women share the same socioeconomic and political circumstances. In Hungary, the male–female difference in life expectancy in 2005 was 8.3 years, which is considerably higher than the average difference found in countries of Western Europe, for example 5.7 years in neighboring Austria and 4.6 years in Denmark and Great Britain.[1] The mortality ratio comparing the low to high educational stratum was 1.8 for Hungarian males while it was 1.2 for females.[19]

On the basis of the data of national representative surveys conducted in the Hungarian population (Hungarostudy 1983, 1988, 1995, 2002),[6–11,22–26] we found that a worse socioeconomic situation is linked to higher morbidity and mortality rates in Hungary as well. According to multivariate analyses, however, a relatively poor socioeconomic situation in itself does not cause higher morbidity and mortality rates, only through the mediation of depressive symptoms. Consequently, not only the difficult social situation in itself, but the subjective experience of relative disadvantage, the prolonged negative emotional state, that is, chronic stress, engenders to be the most important health risk factor.[19]

A self-destructive cycle develops from the enduring relatively disadvantageous socioeconomic situation and depressive symptoms. This cycle, which results in chronic stress, plays a significant role in the increase of morbidity and mortality rates in the lower socioeconomic groups of the population. Until the 1970s, with the uniformly low living standards, Hungarian health statistics showed more favorable trends than in several Western countries, such as in Great Britain or in Austria. During rapid socioeconomic changes, the disadvantaged continuously blame themselves or their environment, consider their future hopeless, experience permanent loss of control and helplessness, because they cannot afford a car and better living conditions, or have higher income, while others around them are able to achieve these. They constantly rate their own situation negatively, feel helpless, and experience a loss of control. This experience becomes widespread when society becomes rapidly polarized and social cohesion, trust, reciprocity, and social support decrease dramatically.[22,23]

Although the relationship is true in general, the significance of the different factors varies according to periods and to environmental processes. In relatively stable societies, existing without great social shocks, the social factors and the psychological coping with these factors have less significance. In a region like Hungary and the other CEE countries, dramatic changes have occurred in the last decades. During this time period, depression and premature cardiovascular and overall mortality increased in parallel, primarily among men.

Men were found to be more susceptible to the effects of relative income inequality and GDP deprivation, but the pathway of this relationship is yet to be explained. Animal experiments have shown males to be more sensitive than females to loss of dominance position, that is loss of position in hierarchy.[27] Most animal studies on social rank examine males, where social rank is the best predictor of quality of life and health. The relationship between social inequality and health applies to women as well as to men in several respects according to several studies, although the income and occupation of women are not as powerful predictors of mortality as they are for men.[5] There are significant gender differences in ways of coping during the sudden changes of the political–economic system, and male morbidity and mortality seem to be more affected by the socioeconomic changes.[9,10]

Middle-aged men are more vulnerable to the socioeconomic risks of their society, but this is closely connected with the different male–female roles in the society. Men are affected not only by their own social situation but by the subjective evaluation of social status of women as well.[25] The subjective social status and education of women were strongly and inversely correlated with male middle-aged mortality, which means that in subregions where women hold more negative appraisal of the social standing and have lower education, there is greater male health deterioration. In preventing chronic stress and the consequent high male premature mortality in CEE, women might play an important role.

The improvement of higher education of women seems to be beneficial both for male and female longevity. Educated women accept more the responsibility for the socioeconomic situation of their family. The feeling of relative socioeconomic deprivation among women in the relatively deprived regions, on the contrary, might result in a vicious cycle of relative deprivation among men.[25]

Social distrust and the rival attitude are important predictors of middle-aged mortality differences among men.[22,23] This indicates that in a suddenly changing socioeconomic situation, relative economic deprivation, rival attitude, and social distrust are all more important risk factors for men while the strong collective efficacy could be a protective factor, even in the case of men. Rival attitude was in highly significant negative association with participation in civic organizations; consequently, the protective effect of participation in civic associations might influence health through a lower rival, competitive attitude in members of civic networks among men.

The existing and broad socioeconomic differences are less important regarding the middle-aged female mortality differences. Higher neighborhood cohesion, religious involvement, and reciprocity of women were not so much influenced by sudden socioeconomic changes in the last decades; therefore the protective network of women remained relatively unchanged.[22,23]

WORK-RELATED CHRONIC STRESS AND HEALTH DETERIORATION

Variations in middle-aged mortality rates in a rapidly changing society in CEE are largely accounted for by distinct unfavorable working and other psychosocial stress conditions. The transformation in CEE countries created conditions of loss of control over life, economic deprivation, and social isolation that might undermine the health status of the population. This disproportionately affects the working-age population because their work and family support roles rendered them more vulnerable to socioeconomic disruption, which, in the CEE countries, were not buffered by a well-functioning civic society.[22,28] The factors that might undermine health differ according to gender and socioeconomic status.[25,28] Working conditions may play an important role in this regard, given the centrality of work in midlife.[18,26,29,30]

Hungary and the other CEE countries had witnessed a major change in the labor market in the last decades. Earlier, everybody had to have one working place and there was no possibility for more than one job. Afterwards, the first development of the "new economic mechanism" was the initiation of the so-called second or third economy, which meant the possibility of better economic situation by accepting or creating second or third jobs, but at the expense of considerable overwork. These developments were more frequent among groups in lower socioeconomic strata, as indicated by lower educational degree or lower income, because they often dispose of fewer choices or alternatives. It went along with a high degree of job instability and associated loss of control.[10,26]

According to the Hungarostudy 2002, high weekend workload, low social support at work, and low control at work accounted for a large part of variation in male premature cardiovascular (CV) mortality rates, whereas job insecurity, high weekend workload, and low control at work contribute most markedly to variations in premature CV mortality rates among women. Low social support from friends, depression, anomie, hostility, alcohol abuse, and cigarette smoking can also explain a considerable part of variations of premature CV mortality differences.[26]

FIRST RESULTS OF THE HUNGAROSTUDY EPIDEMIOLOGICAL PANEL FOLLOW-UP STUDY

In the Hungarostudy 2002 survey 12,600 persons above age 18 were interviewed in their homes. They represented the Hungarian population classified according to age, sex, and 150 subregions.[22,23] Socioeconomic, psychosocial, work- and family-related factors, and behavioral and self-reported health measures were recorded. From the latest Hungarostudy Epidemiological Panel (HEP) 2006 follow-up study, we analyzed the data of people who in 2002 were between the age of 40 and 69.[31]

A total of 99 men (8.8%) and 53 women (3.6%) died from the 40–69-year age groups till 2006. In this age group in both genders, oncological disorders were the most prevalent causes of death, 36.5% among men and 41.5% among women. The rate of cancer mortality is the highest is Hungary among the European countries today.[1] The rate of cardiovascular and cerebrovascular death was 35.1% among men and 29.3% among women; 12.1% of men and 19.5% of women died because of external causes; 4.1% of men because of hepatic cirrhosis.

Among the characteristics of the deceased persons, subjective health status was an important predictor of early death in both genders, more so among men than among women. The odds ratio of early death for self-rated disability was 5.84 (CI: 3.15–10.81), $P = 0.000$, for men and 2.02 (CI: 1.09–3.75), $P = 0.02$, for women. Self-rated health also predicted mortality, among men OR = 2.98 (CI: 1.94–4.56), $P = 0.000$, among women OR = 1.98 (CI: 1.11–3.52), $P = 0.02$. Of those women who died till 2006, 13.7% suffered from oncological disorders in 2002; among men unidentified cardiovascular disorders were significantly more prevalent in 2002. Although it is natural that people with poor health status are at a higher risk for early death, the question arises, which factors might explain this early health deterioration. TABLE 1 shows those health-related, socioeconomic, work-related, and other psychosocial and behavioral factors that significantly predicted early death among men and women.

Education (lower or higher than secondary) predicted only male premature mortality, the odds ratio was 1.84 for men. Among men, subjective social status and personal income were also significant predictors of mortality. Among women only the family-related socioeconomic measures were significant predictors of mortality, namely no car and no personal computer in the family. These measures were significant protective factors among men as well.

The question arises, which are the toxic components of the lower socioeconomic situation? Among men, depression, especially severe depression, anxiety, and work-related factors, first of all work insecurity, low control in work and anomie, that is demoralization significantly predicted the premature mortality. Interestingly, living with spouse (OR: 2.74, CI: 1.45–5.17, $P = 0.001$) was a highly significant protective factor only among men, but not among women. We controlled the effect of these measures according to age. After this control, severe depression, low self-efficacy, low well-being, low cheerfulness, not living with spouse, and no active sport significantly predicted mortality among men, independently of age (TABLE 1). These results mean that among men the most important chronic stress factors are work insecurity and living alone. Depression and anxiety are closely connected with these chronic stress factors, but because both had been true in 2002, we cannot determine which was the cause and which the consequence. It can be hypothesized that more depressed men regard their work situation as more insecure and that

TABLE 1. Significant predictors of death till 2006 among 40- to 69-year-old men and women from Hungarostudy 2002

	Male ORs, CIs, and significance	Male ORs adjusted for age, CIs, and significance	Female ORs, CIs, and significance	Female ORs adjusted for age, CIs, and significance
Self-reported health measures				
Self-reported disability	5.84 (3.15–10.81),000	4.27 (2.22–8.22),000	2.02 (1.09–3.75),02	NS
Low self-rated health	2.98 (1.94–4.56),000	3.88 (2.40–6.29),000	1.98 (1.11–3.52),02	2.68 (1.39–5.18),003
Oncological disorder in 2002	NS		3.19 (1.71–5.94),000	2.70 (1.22–5.95)
Mental health				
Severe depression (BDI score higher than 24)	5.12 (2.99–8.78),000	3.68 (1.98–6.85),000	NS	NS
Negative affect	2.12 (1.38–3.26),000	NS	3.58 (1.98–6.50),000	2.92 (1.50–5.69),002
Unhappiness	NS	NS	2.80 (1.58–4.96),000	NS
Hospital anxiety score (HAS)	2.80 (1.82–4.32),000	NS		NS
Low self-efficacy	2.30 (1.45–3.64),000	2.63 (1.57–4.40),000	2.42 (1.37–4.28),002	3.79 (1.75–8.20),000
Low WHO well-being	1.99 (1.32–3.03),001	2.57 (1.58–4.19),000	NS	NS
Low cheerfulness	1.73 (1.12–2.67),01	2.27 (1.41–3.67),001	1.97 (1.11–3.49),02	3.05 (1.49–6.25),002
Socioeconomic measures				
Education lower/higher than secondary	1.84 (1.14–2.96),011	2.39 (1.33–4.31),004	NS	NS
Low subjective social status	1.92 (1.13–3.24),014	NS	NS	NS
Subjective poverty	1.79 (1.02–3.13),04	NS	NS	NS

Continued

TABLE 1. Continued

	Male ORs, CIs, and significance	Male ORs adjusted for age, CIs, and significance	Female ORs, CIs, and significance	Female ORs adjusted for age, CIs, and significance
"Ontological insecurity," anomie				
No car in the family	2.95 (1.92–4.55),000	4.45 (2.59–7.64),000	2.30 (1.28–4.13),004	2.84 (1.44–5.64),003
No personal computer in the family	2.82 (1.55–5.13),000	3.25 (1.66–6.35),000	2.41 (1.02–5.70),04	2.59 (1.02–6.57),045
Anomie (no point in making plans for the future)	1.86 (1.11–3.14),02	NS	NS	NS
Hopelessness	2.28 (1.50–3.47),000	NS	2.84 (1.63–4.49),000	2.79 (1.29–6.06),009
Rival attitude	2.14 (1.14–4.03),016	NS	NS	NS
No meaning in life	1.92 (1.25–2.92),002	2.62 (1.61–4.25),000	NS	NS
Work-related stress measures				
Work insecurity	3.33 (1.95–5.69),000	3.60 (2.04–6.36),000	NS	NS
Low work control	2.06 (1.16–3.67),012	2.17 (1.19–3.95),011	NS	NS
Low personal income	1.65 (1.07–2.56),024	2.43 (1.41–4.19),001	NS	NS
Low family income	2.26 (1.30–3.95),003	2.47 (1.33–4.64),005	NS	NS
Low employment	1.86 (1.09–3.16),02	NS	NS	NS
Low social support from coworkers	NS	2.03 (1.19–3.48),000	2.04 (1.00–4.19),05	NS
Other social support measures				
Not living with spouse	2.20 (1.37–3.55),001	3.58 (2.07–6.16),000	NS	NS
Low social support from spouse	2.74 (1.45–5.17),001	4.03 (2.17–7.46),000	NS	NS
Not satisfied with personal relations	NS	NS	2.15 (1.21–3.83),008	3.90 (1.76–8.62),000
Family problems	NS	NS	2.40 (1.26–4.55),006	NS
Behavioral measures				
Smoking	1.88 (1.13–3.13),013	NS	NS	NS
Sport activity	2.99 (1.82–4.94),000	2.96 (1.72–5.09),000	NS	4.57 (2.37–8.79),000
Suicide attempt (lifetime)	NS	NS	3.30 (1.43–7.61),008	NS

OR = Odds ratios; CI = confidence intervals and significance; NS = non-significant.

depression influences personal partner connections, but the opposite may be true as well. Among the deceased men, the prevalence of severe depression was 24% in 2002, among the other men 5.8%. This result means that from the prevention point of view it would be fundamental to care for severely depressed middle-aged men.

Among 40–69-year-old women, the personal socioeconomic factors and the work-related measures were not significant predictors of early mortality. We controlled the risk factors according to age as well. After this control, among women dissatisfaction with personal relations, negative affect, low self-efficacy, low cheerfulness, and hopelessness remained significant predictors of early mortality. Interestingly, the lifetime occurrence of suicide attempts was also an independent predictor of early mortality, but only among women.[31]

These results mean that the health consequences of low education among men might be mostly explained by chronic stress caused by work- and close-partner-related factors, and the toxic components of this interaction are depression, hopelessness, and anxiety. In the case of women, the broader personal and family relations are the most important health-related factors. Interestingly among women, hopelessness and suicide attempts are the most important predictors of early death.

WHAT SORT OF STEPS COULD BE TAKEN TO DECREASE THE TOXIC EFFECTS OF CHRONIC STRESS?

There are two general approaches that scientists and practitioners might take: (a) Train individuals and groups in our populations to use coping skills that will enable them to cope better with the stressful conditions that are damaging their health. (b) Lobby our governments to adopt policies that will result in decreased chronic stress on society level. This is the case in the latest Swedish National Public Health programme. The key points of the public health plan are strengthening social power and participation, strengthening social and economic security, ensuring secure and good growing-up conditions, and ensuring healthy working environments.[32]

Coping skills training based on cognitive behavior therapy has been shown to reduce both psychosocial risk factors and the accompanying biological mechanisms that likely mediate their effects on health. One approach to such training, cognitive behavioral stress management (CBSM), has been shown in rigorously conducted randomized clinical trials to reduce anxiety, distress, and 24-h urine excretion levels of norepinephrine and cortisol in symptomatic HIV-infected gay men.[33,34] These are effects that would be expected to benefit people in CEE who are suffering from the psychological and biological consequences of the changes in social, economic, and political conditions in the region.

Another program, very similar to CBSM, the LifeSkills program,[35] provides training in ten coping skills and has now been evaluated in two randomized clinical trials with heart patients and shown to reduce both psychosocial risk factors and biomarkers of stress. One of the studies,[36] conducted in patients following a myocardial infarction, showed that patients randomized to coping skills training to reduce hostility showed significant decreases in both hostility and resting diastolic blood pressure in comparison to patients randomized to usual care. These benefits of the training were even greater at the 2-month follow-up. The other clinical trial[37] randomized patients who had undergone coronary bypass surgery to either LifeSkills training over six sessions or a 1-h lecture on stress effects on the heart and tips on how to cope with it. Compared to patients who received information only, those in the LifeSkills arm showed significant improvements in anger, anxiety, depression, social support, and blood pressure and heart rate, both at rest and in response to an anger recall task.

Results like these obtained with CBSM and LifeSkills coping skills training suggest that it may be possible, in a manner analogous to vaccination against disease-causing viruses, to "inoculate" people against the health-damaging effects of stressors like those being experienced by the populations of CEE.[38] While encouraging, we have to keep in mind, however, that these programs used a delivery system consisting of six to eight small-group sessions for 8–12 participants spread out over several weeks. This is not a delivery system that is likely to be able to reach the large numbers of people at risk—for example, the middle-aged men who are most affected in Hungary, or the lower socioeconomic groups in Russia—who could benefit from the training. Vaccinations and drug interventions can be delivered to the populace on a large scale, but what about behavioral interventions that provide training in coping skills?

One way to get around this problem would be to use electronic means, such as video or Internet-based training, to deliver coping skills training on a mass basis.[39,40]

ACKNOWLEDGMENT

This study was supported by the National Research Fund (OTKA) projects OTKA TS-40889 (2002) and TS-049785 (2004) Scientific School grants and NKFP 1/002/2001 and NKFP 1b/020/2004.

REFERENCES

1. Demographic Yearbook. 2005. Central Statistical Office. Budapest, Hungary.
2. BLACK, D., J.N. MORRIS, C. SMITH, et al. 1992. Inequalities in health: the Black report P. Townsend & N. Davidson, Eds.: 294–295. Health Divide. Penguin, London.
3. BOBAK, M. & M. MARMOT. 1996. East-West mortality divide: proposed research agenda. Br. Med. J. **312:** 421–425.

4. CORNIA, G.A. & R. PANICCIA. Eds. 2000. The Mortality Crisis in Transitional Economies. Oxford University Press. Oxford.
5. MARMOT, M. & R. WILKINSON. 1999. Social Determinants of Health. Oxford University Press. Oxford.
6. KOPP, M.S., Á. SKRABSKI & S. SZEDMÁK. 1995. Socioeconomic factors, severity of depressive symptomatology and sickness absence rate in the Hungarian population. J. Psychosom. Res. **39:** 1019–1029.
7. KOPP, M.S. & Á. SKRABSKI. 1996. Behavioural Sciences Applied to a Changing Society. Bibl Septem Artium Liberalium. Budapest.
8. KOPP, M.S., P. FALGER, A. APPELS & S. SZEDMÁK. 1998. Depressive symptomatology and vital exhaustion are differentially related to behavioural risk factors for coronary artery disease. Psychosom. Med. **60:** 752–758.
9. KOPP, M.S. 2000. Cultural transition. *In* Encyclopedia of stress. E.G. Fink, Ed.: 611–615. Volume 1. Academic Press. San Diego.
10. KOPP, M.S., Á. SKRABSKI & S. SZEDMÁK. 2000. Psychosocial risk factors, inequality and self-rated morbidity in a changing society. Soc. Sci. Med. **51:** 1350–1361.
11. KOPP, M.S., Á. SKRABSKI & A. SZÉKELY. 2002. Risk factors and inequality in relation to morbidity and mortality in a changing society. *In* Heart Disease: Environment, Stress and Gender. NATO Science Series, Life and Behavioural Sciences. G. WEIDNER, M.S. KOPP & M. KRISTENSON, Eds.: 101–113. Volume 327. IOS Press. Amsterdam.
12. KRISTENSON, M. & Z. KUCINSKIENE. 2002. Possible causes of the differences in coronary heart disease mortality between Lithuania and Sweden: the LiuViCordia study. *In* Heart Disease: Environment, Stress and Gender. NATO Science Series, Life and Behavioural Sciences. G. WEIDNER, M.S. KOPP, M. KRISTENSON, Eds.: 328–340. Volume 327. IOS Press. Amsterdam.
13. VARGÁNÉ HAJDÚ, P. & R. ÁDÁNY. 2000. Early death because of cardiovascular disorders in Hungary and in the European union, 1970–1997. Orvosi Hetilap **141:** 601–607.
14. CHAZOV, E. 2004. International Conference of Behavioral Medicine, Mainz (invited speaker).
15. MARMOT, M.G. 1996. The social pattern of health and disease. *In* Health and Social Organization. D. Blane, E. Brunner & R. Wilkinson, Eds.: 42–70. Routledge. London.
16. WEIDNER, G. 2002. The role of stress and gender related factors in the increase in heart disease in Eastern Europe: overview. *In* Heart Disease: Environment, Stress and Gender. NATO Science Series, Life and Behavioural Sciences. G. Weidner, M.S. Kopp & M. Kristenson, Eds.: 1–14. Volume 327. IOS Press. Amsterdam.
17. WILKINSON, R.G. 1996. Unhealthy societies: the afflictions of inequality. Routledge. London.
18. PIKHART, H., M. BOBAK, J. SIEGRIST, *et al.* 2001. Psychosocial work characteristics and self-rated health in four post-communist countries. J. Epidemiol. Commun. Health **55:** 624–630.
19. KOPP, M.S. & J. RÉTHELYI. 2004. Where psychology meets physiology: chronic stress and premature mortality—the Central-EE health paradox. Brain Res. Bull. **62:** 351–367.
20. MACKENBACH, J.P., A.E. KUNST, F. GROENHOF, *et al.* 1999. Socioeconomic inequalities in mortality among women and among men: an international study. Am. J. Public Health **89:** 1800–1806.

21. MARMOT, M. 2004. The Status Syndrome: how Social Standing Affects Our Health and Longevity. Times Books. New York.
22. SKRABSKI, Á, M.S. KOPP & I. KAWACHI. 2003. Social capital in a changing society: cross sectional associations with middle aged female and male mortality rates. J. Epidemiol. Commun. Health **57:** 114–119.
23. SKRABSKI, Á., M.S. KOPP & I. KAWACHI. 2004. Social capital and collective efficacy in Hungary: cross sectional associations with middle aged female and male mortality rates. J. Epidemiol. Commun. Health **58:** 340–345.
24. SKRABSKI, Á., M.S. KOPP, S. RÓZSA, et al. 2005. Life meaning: and important correlate of health in the Hungarian population. Int. J. Behav. Med. **12:** 78–85.
25. KOPP, M.S., Á. SKRABSKI, I. KAWACHI & N.E. ADLER. 2005. Low socioeconomic status of the opposite gender is a risk factor for middle aged mortality. J. Epidemiol. Commun. Health **59:** 675–678.
26. KOPP, M.S., Á. SKRABSKI, ZS SZÁNTÓ & J. SIEGRIST. 2006. Psychosocial determinants of premature cardiovascular mortality differences within Hungary. J Epidemiol. Commun. Health **60:** 782–788.
27. RÉTHELYI, J. & M.S. KOPP. 2004. Hierarchy disruption: women and men. Behav. Brain Sci. **27:** 17–18.
28. HERTZMAN, C., A. SIDDIQI & M. BOBAK. 2002. The population health context for gender, stress and cardiovascular disease in Central and Eastern Europe. *In* Heart Disease: Environment, Stress and Gender. G. Weidner, M.S. Kopp & M. Kristenson, Eds.: 15–25. IOS Press. Amsterdam.
29. SIEGRIST, J. 2002. Effort-reward imbalance at work and health. *In* Historical and Current Perspectives on Stress and Health. P. Perrewe & D. Gater, Eds.: 261–291. Elsevier. Oxford.
30. KARASEK, R.A. & T. THEORELL. 1990. Healthy work. Stress, Productivity, and the Reconstruction of Working Life. Basic Books. New York.
31. KOPP, M.S, A. SZÉKELY & Á. SKRABSKI. 2007. Public health burden of work stress in a transforming society. American Psychosomatic Society Conference, Budapest (invited speaker).
32. ÅGREN, GUNAR. 2000. National Public Health Programme. National Institute of Public Health, Sweden.
33. ANTONI, M.H., D.G. CRUESS, S. CRUESS, et al. 2000. Cognitive-behavioral stress management intervention effects on anxiety, 24-hr urinary norepinephrine output, and T-cytotoxic/suppressor cells over time among symptomatic HIV-infected gay men. J. Consult. Clin. Psychol. **68:** 31–45.
34. ANTONI, M.H., S. CRUESS, D.G. CRUESS, et al. 2000. Cognitive-behavioral stress management reduces distress and 24-hour urinary free cortisol output among symptomatic HIV-infected gay men. Ann. Behav. Med. **22:** 29–37.
35. WILLIAMS, R. & V. WILLIAMS. 1997. LifeSkills: 8 Simple Ways to Build Stronger Relationships, Communicate More Clearly, Improve Your Health, and Even the Health of Those Around You. Times Books/Random House. New York.
36. GIDRON, Y., K. DAVIDSON & I. BATA. 1999. The short-term effects of a hostility-reduction intervention on male coronary heart disease patients. Health Psychol. **18:** 416.
37. BISHOP, G.D., M. KAUR, V.L.M. TAN, et al. 2005. Effects of a psychosocial skills training workshop on psychophysiological and psychosocial risk in patients undergoing coronary artery bypass grafting. Am. Heart J. **150:** 602–609.

38. STAUDER, A., V.P. WILLIAMS, R.B. WILLIAMS & M.S. KOPP. 2006. Hungarian adaptation of the Williams LifeSkills program: preliminary results among psychosomatic patients (abstract). J. Psychosom. Res. **60:** 663.
39. BRENNER, S.L., S.B. HEAD, M.J. HELMS, *et al.* 2003. A videotape module to teach assertion skills. J. Appl. Soc. Psychol. **33:** 1140–1152.
40. KIRBY, E.D., V.P. WILLIAMS, M.C. HOCKING, *et al.* 2006. Psychosocial benefits of three formats of a standardized behavioral stress management program. Psychosom. Med. **68:** 816–823.

Attitude toward Death: Does It Influence Dental Fear?

GÁBOR FÁBIÁN,[a] ORSOLYA MÜLLER,[a] SZILVIA KOVÁCS,[b]
MINH TÚ NGUYEN,[c] TIBOR KÁROLY FÁBIÁN,[c] PÉTER CSERMELY,[d]
AND PÁL FEJÉRDY[c]

[a]*Clinic of Pediatric Dentistry and Orthodontics, Faculty of Dentistry, Semmelweis University, Szentkirályi utca 47, 1088, Budapest, Hungary*

[b]*Department of Neurosis, Budai Pediatric Hospital, Budapest, Hungary*

[c]*Clinic of Prosthetic Dentistry, Faculty of Dentistry, Semmelweis University, Budapest, Hungary*

[d]*Faculty of Medicine, Institute of Medical Chemistry Molecular Biology and Pathobiochemistry, Semmelweis University, Budapest, Hungary*

> ABSTRACT: The possible influence of fear of death and attitude toward death were studied related to dental anxiety in Hungarian elementary and secondary school subjects ($n = 277$; 114 males, 163 females; age between 8 and 18 years). Dental fear and anxiety scores were DAS: 10.8 ± 3.6; DFS: 40.6 ± 15.6; STAI-S: 38.0 ± 11.0; STAI-T: 40.3 ± 10.0. Lester's Attitude Toward Death Scale scores were 6.3 ± 1.3. Girls scored higher on DAS, STAI-S, and STAI-T scales ($P \leq 0.05$). Age influenced STAI-S, STAI-T, and Lester's Scale scores ($P \leq 0.05$). Lester's Scale scores influenced the expectations of the subjects about the dental fear of their surrounding people (parents, brother, sister, friends) ($P \leq 0.05$). A percentage of 7.22 of the subjects indicated a rather strong connection between dental fear and fear of death. These subjects had significantly higher dental fear and anxiety scores as compared to others ($P \leq 0.01$). Death-related content was found in 4.3% of drawings and in 10.5% of free associations (couplings) related to teeth (in 12.6% either in drawings or in couplings). The appearance of death-related content was higher with higher age, and higher expected dental fear of surrounding people ($P \leq 0.01$). Our data indicate a detectable influence of fear of death on dental fear, especially in subjects with higher dental fear scores.
>
> KEYWORDS: dental fear; attitude toward death; fear of death

Address for correspondence: Tibor Károly Fábián, D.M.D., Ph.D., Clinic of Prosthetic Dentistry, Faculty of Dentistry, Semmelweis University, Szentkirályi utca 47, 1088, Budapest, Hungary. Voice: +36-1-317-1622; fax: +36-1-317-1094.

fab@fok.usn.hu

INTRODUCTION

One of the supposedly basic conflict of existential psychotherapy is transitoriness[1] (mortal nature, death). The conflict of transitoriness seems to be highly important in dentistry as well. An important coupling of mouth and teeth to the conflict of transitoriness is rooted in the early phase of psychological development.[2,3] The early experience of the infant related to losing somebody (the baby must be taken off the breast following tooth eruption) and destroying something (intake of food with the use of the teeth) is strongly coupled to the oral region during the so-called oral-sadistic stage[4] of psychological development. Another important coupling is that teeth have a strong symbolic meaning related to strength and aggressiveness, whereas tooth loss, especially edentulousness, symbolizes the loss of vigor and vitality and evokes a symbolic meaning of growing old, evanescence, and death.[5,6] Consequently, the conflict of transitoriness may play an important role in most of the psychological events coupled to mouth and teeth, including fear of dental treatment.

Only two publications were found to investigate a representative number of subjects related to this topic.[7,8] One of them investigated the relationship between the different types of fear (Fear Survey Schedule II; FSS-II[9]) and dental fear in adult phobic patients.[7] Seeing a fight ($r = 0.33$; $P \leq 0.01$) or blood ($r = 0.32, P \leq 0.01$)—and in the case of men the cemeteries ($r = 0.5; P \leq 0.01$) and dead bodies ($r = 0.41; P \leq 0.01$)—were found among the individual fear items having the highest and most significant Pearson correlation with Dental Anxiety Scale (DAS) total scores.[7] In the other paper,[8] FSS-II and some additional items were used, and a factor model of fear was developed. The Pearson correlation of factor "illness and death" with DAS total scores was $r = 0.28$ ($P \leq 0.05$). The Pearson correlation of the same factor with Dental Fear Survey (DFS) total scores was $r = 0.25$ ($P \leq 0.05$).[8] The highest correlation was found in the case of factor "physical injuries" with both DAS ($r = 0.38, P \leq 0.05$) and DFS ($r = 0.46, P \leq 0.05$) total scores in this study.[8]

The above data indicate that the conflict of transitoriness may play an important role in the appearance of dental fear, although the relationship seems to be more complex rather than a direct one. In the present article, the authors would like to confirm this hypothesis.

METHODS

Hungarian primary and grammar school subjects from Budapest participated in this study ($n = 277$, 114 males, 163 females; age between 8 and 18 years, mean: 13.97 ± 2.77 years). The subjects participated voluntarily, after the appropriate information about the study had been given. Agreements of the students' parents were also obtained.[10]

To measure anxiety level the Hungarian version[11] of Spielberger's State and Trait Anxiety Inventory[12] (STAI-S, STAI-T) was used. Dental fear was measured by the Hungarian versions[13,14] of the Dental Anxiety Scale (DAS)[15] and Dental Fear Survey (DFS).[16,17] To measure the subjects' expectations in terms of dental fear of their surrounding people (parents, brother, sister, friends), the Expectation Scale[18] was used. This Scale[18] contains the last four items of the early version (first, nonreduced version) of the DFS scale.[16] To score this scale, the mean value is calculated from the items, which were answered[18] (i.e., subjects having no brothers or sisters do not answer that particular item). Chronbach alpha values of the scales in this study were STAI-S: 0.89; STAI-T: 0.87; DAS: 0.86; DFS: 0.92; Expectation Scale (in case all items were scored): 0.81.

Following administration of the scales, subjects were asked to make drawings[19] and write free associations (couplings)[20,21] as to the teeth. For drawing, a 15×15-cm square on a sheet of paper was used, and the participant was asked to draw a tooth (or teeth) in it.[19] For writing free associations (couplings), another sheet of paper was given and the participant was asked to write any thoughts that came into his or her mind about teeth.[20,21]

As the next step, the Hungarian version[22] of the Lester's Attitude Toward Death Scale[23] was administered. This scale contains 21 individual statements (items) about death, and the participant indicated which ones he or she believed to be true. The nominated items could be analyzed individually or the median value of the item scores can be used as a measure of fear of death as well.[23] Although there are some other, more complex scales[24–26] in the literature to measure fear of death, the items of the Lester's Scale are much less "drastic" and more convenient for children.

Finally participants were asked to indicate if they believed any of the following two statements to be true: (a) "Dental treatment may induce fear of death." (b) "Dental treatment never induces fear of death."

Following the collection of data, scales (or survey and inventories) were evaluated as usual, and the indications related to the above two statements were also registered. The drawings and written free associations were evaluated against the death-related contents. For this purpose, the content criteria of Gottschalk–Gleser[27] were used, but content related directly to a description of an actual dental treatment was not regarded as positive. For example, a description (or drawing) of a real drilling of a tooth was not taken into account, but a description of a dentist who is "eating his patients during the dental treatment" was taken as positive. The drawings and written free associations (couplings) were evaluated by three investigators, and only those were regarded as positive in which a minimum of one content (meeting the above criteria) was found by at least two of the investigators.

For statistical analysis, the SPSS/PC 10.0 software (SPSS, Inc., Chicago, IL) was used, the level of significance was $P \leq 0.05$.

TABLE 1. Mean score and standard deviation of the scales related to gender and age

	DAS	DFS	EXPECT	STAI-S	STAI-T	Lester
Gender						
m ($n = 114$)	10.2 ± 3.5	39.2 ± 14.3	2.3 ± 0.9	36.4 ± 10.4	38.6 ± 9.6	6.4 ± 1.3
fm ($n = 163$)	11.1 ± 3.7	41.6 ± 16.3	2.5 ± 0.8	39.2 ± 11.4	41.4 ± 10.2	6.2 ± 1.3
Age (year)						
8 ($n = 10$)	9.3 ± 6.0	40.8 ± 14.8	1.9 ± 0.9	38.5 ± 8.0	37.3 ± 8.7	6.8 ± 0.7
9 ($n = 18$)	10.7 ± 3.7	40.2 ± 17.3	2.4 ± 1.0	34.3 ± 7.7	39.7 ± 1.5	6.6 ± 0.1
10 ($n = 16$)	10.8 ± 2.5	39.4 ± 14.9	1.8 ± 0.5	34.4 ± 12.8	39.9 ± 9.9	7.2 ± 0.9
11 ($n = 16$)	11.1 ± 4.6	42.0 ± 15.3	2.1 ± 0.9	34.0 ± 12.2	37.9 ± 11.4	6.5 ± 1.2
12 ($n = 17$)	9.6 ± 3.9	36.2 ± 13.6	2.2 ± 0.8	30.6 ± 7.5	33.6 ± 7.6	6.7 ± 1.4
13 ($n = 19$)	10.9 ± 3.1	38.7 ± 13.4	2.3 ± 1.0	35.8 ± 10.9	36.5 ± 9.2	6.8 ± 1.5
14 ($n = 39$)	10.9 ± 3.7	40.9 ± 16.6	2.4 ± 0.8	40.3 ± 11.2	40.5 ± 11.3	6.1 ± 1.3
15 ($n = 47$)	10.8 ± 4.0	39.9 ± 14.6	2.5 ± 0.8	38.1 ± 10.7	41.1 ± 10.3	6.0 ± 1.0
16 ($n = 38$)	11.4 ± 3.1	42.3 ± 17.3	2.5 ± 0.8	38.1 ± 10.9	41.9 ± 10.6	6.2 ± 1.1
17 ($n = 42$)	10.8 ± 3.6	41.9 ± 17.4	2.7 ± 0.9	43.0 ± 11.4	41.0 ± 9.3	6.0 ± 1.7
18 ($n = 15$)	10.3 ± 2.4	41.3 ± 12.9	2.4 ± 0.8	41.4 ± 9.5	45.9 ± 9.5	6.5 ± 1.7
$\sum (n = 277)$	10.8 ± 3.6	40.6 ± 15.6	2.4 ± 0.9	38.0 ± 11.0	40.3 ± 10.0	6.3 ± 1.3

DAS = Dental Anxiety Scale; DFS = Dental Fear Survey; EXPECT = Expectation Scale; STAI-S and STAI-T = State and Trait Anxiety Inventory; Lester = Median Value of Lester's Attitude Toward Death Scale; m = male; fm = female; $\sum$ = total sample.

RESULTS

The gender and age of the subjects influenced the values obtained on most of the scales (TABLE 1). Female participants scored lower on Lester's Scale and higher on all other scales than male participants. The difference was significant in the case of DAS, Expectation Scale, STAI-S, and STAI-T scores (*t*-probe, $P \leq 0.05$). There was a general increasing tendency of most of the scales with age, with the exception of the Lester's scale, which decreased with age. The influence of age was significant in the case of STAI-S, STAI-T, and Lester's Scale (one-way analysis of variance [ANOVA], $P \leq 0.05$).

Higher Pearson correlation values (TABLE 2) were between dental fear scores (DAS and DFS) and between anxiety scores (STAI-S and STAI-T). Although Pearson correlation values were around zero between Lester's scale and all other scales (TABLE 2), one-way ANOVA indicated that the median value of the Lester's Scale score significantly influenced ($P \leq 0.05$) the Expectation Scale, and was influenced by the STAI-S and STAI-T scores (one-way ANOVA, $P \leq 0.05$). Age also influenced the median value of Lester's Scale score (one-way ANOVA; $P \leq 0.05$), whereas gender did not.

In many cases, value differences of several measured parameters were found between participants marking and participants not marking an individual item of Lester's Scale (TABLE 3). These differences may indicate relations between the individual items and the measured parameter. Such relations were found between gender and six individual items (nominator Ú nonnominator; Fisher's

TABLE 2. Pearson correlation of the scales

	DAS	DFS	EXPECT	STAI-S	STAI-T	Lester
DAS	1.00					
DFS	0.77**	1.00				
EXPECT	0.36**	0.37**	1.00			
STAI-S	0.34**	0.35**	0.26**	1.00		
STAI-T	0.36**	0.39**	0.20**	0.66**	1.00	
Lester	0.08^{ns}	0.07^{ns}	−0.01^{ns}	−0.07^{ns}	−0.03^{ns}	1.00

DAS = Dental Anxiety Scale; DFS = Dental Fear Survey; EXPECT = Expectation Scale; STAI-S and STAI-T = State and Trait Anxiety Inventory; Lester = Median Value of Lester's Attitude Toward Death Scale; * = $P \leq 0.01$; ** = $P \leq 0.01$; ns = not significant.

TABLE 3. Relations between several measured parameters and marking an item of Lester's Attitude Toward Death Scale

No. of item	Gender	Age	DAS	DFS	EXPECT	STAI-S	STAI-T
1 (110)	−	−	−	−	−	−	−
2 (64)	−	++	−	−	−	−	−
3 (116)	+	++	−	−	−	−	−
4 (38)	+	−	−	−	−	−	−
5 (164)	−	++	−	−	−	−	−
6 (47)	−	−	−	−	−	−	−
7 (104)	−	+	−	−	−	−	−
8 (129)	−	++	−	−	+	−	−
9 (152)	−	++	−	−	−	−	−
10 (190)	−	−	−	−	+	−	−
11 (126)	−	++	−	−	−	−	−
12 (145)	+	++	−	−	−	−	−
13 (129)	+	++	−	−	−	−	−
14 (124)	−	++	−	−	+	−	−
15 (131)	−	++	−	−	−	−	−
16 (162)	−	++	−	−	−	+	−
17 (36)	−	+	−	−	−	+	+
18 (161)	−	++	−	−	−	−	−
19 (80)	−	++	−	−	−	−	−
20 (100)	++	++	−	−	+	−	−
21 (99)	+	++	+	+	−	−	−

DAS = Dental Anxiety Scale; DFS = Dental Fear Survey; EXPECT = Expectation Scale; STAI-S and STAI-T = State and Trait Anxiety Inventory; () = number of subjects marked an item; + = $P \leq 0.05$; ++ = $P \leq 0.01$; − = not significant.

exact; $P \leq 0.05$); age and most of the items (nominator ⇔ nonnominator; t-probe; $P \leq 0.05$); dental fear scores (DAS and DFS), and one certain item with the most refusing attitude toward death (nominator ⇔ nonnominator; t-probe; $P \leq 0.05$); Expectation Scale and four individual items (nominator ⇔ nonnominator; t-probe; $P \leq 0.05$); anxiety (STAI-S and STAI-T), and a few individual items (nominator ⇔ nonnominator; t-probe; $P \leq 0.05$). It should also be mentioned that nominators of the only item being in relation with dental fear (with the most refusing attitude toward death) scored significantly higher (t-probe; $P \leq 0.05$) on dental fear scales (DAS and DFS) as compared to nonnominators.

Twenty participants (7.22%) asserted that "dental treatment may induce fear of death." (TABLE 4). Participants asserting this scored significantly higher (t-probe; $P \leq 0.01$) on dental fear scales (DAS and DFS) and anxiety inventories (STAI-S and STAI-T).

Similarly, 49 participants (17.69%) did not assert that "dental treatment never induces fear of death." (TABLE 5). Participants not asserting this statement scored significantly higher (t-probe; $P \leq 0.01$) on dental fear scales (DAS and DFS) and anxiety inventories (STAI-S and STAI-T).

There were 257 participants making drawings, of which 12 drawings (4.3% of *all* 277 participants) had death-related content (TABLE 6). Written free associations (couplings) were prepared in 181 cases, of which 29 couplings (10.5% of *all* 277 participants) contained death-related content (TABLE 6). In 35 cases (12.6% of *all* 277 participants), there was a death-related content in at least one of the works (either in the drawing or in the written free association, or in both) of participants. In the case of males, death-related contents were expressed more frequently in drawings (Fisher's exact; $P \leq 0.05$). Age, state-anxiety (STAI-S), and Expectation Scale sores were higher in the case of participants expressing death-related contents in written free associations (t-probe; $P \leq 0.01$, $P \leq 0.05$). Similarly, age and Expectation Scale scores were higher (t-probe; $P \leq 0.01$, $P \leq 0.05$) in the case of participants expressing death-related content either in drawings or in written associations, or in both.

DISCUSSION

In general, it is accepted that fear of death scores are somewhat higher in females than males,[28] although there are some studies that do not confirm this difference.[24,29,30] Age seems to have no significant influence on fear of death levels in adults and older-aged population.[28,31] However, younger-aged participants scored somewhat higher,[32] and there was a slightly rising tendency from 13- to 26-year-old adolescent and young adult populations.[29] Our data (TABLE 1) concerning the level of fear of death (median value of Lester's Scale) indicated no significant difference related to gender; however, they showed slightly downward tendency from 8 to 18 years.

Data in the literature indicate that there is a slightly positive Pearson correlation between anxiety,[28] dental fear,[7,8] and fear of death. Our results did not indicate such correlation (TABLE 2), although some nonlinear relations (ANOVA) were detected. Fear of death level was significantly influenced by anxiety scores. An interesting finding was that fear of death level significantly influenced the expectations of the participants about the dental fear of their

TABLE 4. Influence of gender, age, and several scale scores on the subjects' assertion of relationship between dental fear and fear of death

	Gender distrib	Mean age	DAS	DFS	EXP	STAI-S	STAI-T	Lest
Relationship asserted ($n = 20$)	m 12 fm 8	14.4 ± 3.1	13.2 ± 4.1	57.3 ± 22.0	2.7 ± 0.9	45.2 ± 11.5	47.3 ± 9.2	6.4 ± 1.0
Relationship not asserted ($n = 257$)	m 102 fm 155	14.0 ± 2.7	10.6 ± 3.5	39.3 ± 14.2	2.4 ± 0.9	37.5 ± 10.8	39.7 ± 9.9	6.3 ± 1.3
Sig. level	n.s.	n.s.	++	++	n.s.	++	++	n.s.

Gender distrib = gender distribution; DAS = Dental Anxiety Scale; DFS = Dental Fear Survey; EXP = Expectation Scale; STAI-S and STAI-T = State and Trait Anxiety Inventory; Lest = Median value of Lester's Attitude Toward Death Scale; () = number of assertions; Sig. level = level of significance; + = $P \leq 0.01$; ++ = $P \leq 0.01$; − = not significant.

TABLE 5. **Influence of gender, age, and several scale scores on the subjects' assertion of absence of relationship between dental fear and fear of death**

	Gender distrib	Mean age	DAS	DFS	EXP	STAI-S	STAI-T	Lest
Absence of relationship asserted ($n = 228$)	m 88 fm 140	18.9 ± 2.8	10.5 ± 3.5	38.9 ± 13.5	2.4 ± 0.9	37.0 ± 10.7	39.3 ± 9.7	6.3 ± 1.3
Absence of relationship not asserted ($n = 49$)	m 26 fm 23	14.6 ± 2.6	12.1 ± 4.1	48.5 ± 21.4	2.5 ± 0.9	42.6 ± 11.2	44.6 ± 10.7	6.5 ± 1.4
Sig. level	+	n.s.	++	++	n.s.	++	++	n.s.

Gender distrib = gender distribution; DAS = Dental Anxiety Scale; DFS = Dental Fear Survey; EXP = Expectation Scale; STAI-S and STAI-T = state and trait anxiety inventory; Lest = median value of Lester's Attitude Toward Death Scale; () = number of assertions; Sig. level = level of significance; + = $P \leq 0.01$; ++ = $P \leq 0.01$; − = not significant.

TABLE 6. Influence of gender, age, and several scale scores on the appearance of death-related contents in drawings and written free associations (couplings)

	Gen	Age	DAS	DFS	EXPE	STAI-S	STAI-T	Lest
Drawing ($n = 12$)	+	–	–	–	–	–	–	–
Coupling ($n = 29$)	–	++	–	–	+	++	–	–
Drawing and/or coupling ($n = 35$)	–	++	–	–	+	–	–	–

Gen = gender; DAS = Dental Anxiety Scale; DFS = Dental Fear Survey; EXPE = Expectation Scale; STAI-S and STAI-T = State and Trait Anxiety Inventory, Lest = Median value of Lester's Attitude Toward Death Scale; () = number of drawings and couplings containing death-related contents; + = $P \leq 0.01$; ++ = $P \leq 0.01$; – = not significant.

surrounding people (Expectation Scale), although dental fear level (DAS and DFS) was not influenced by fear of death scores.

It was also interesting to find that dental fear scores (DAS and DFS) had a relation to only one certain item of Lester's Scale, with the most refusing attitude toward death (TABLE 3), and participants with higher dental fear scores more frequently nominated this item as believed to be true. More pronounced relation was found between the individual items of Lester's Scale and the Expectation Scale (TABLE 3) supporting the above finding that expectations of the participants about the dental fear of their surrounding people is more closely related to the attitude toward death than dental fear itself.

A percentage of 7.22 of the participants asserted that "dental treatment may induce fear of death." (TABLE 4), and a percentage of 17.69 of the participants did not assert that "dental treatment never induces fear of death." (TABLE 5). On the basis of these data, we can conclude that a not quite negligible percentage of the participants experience some relation between fear of death and dental fear. The experience of such a relation seems to be stronger in participants with higher anxiety and dental fear scores (TABLES 4 and 5).

Death-related contents were found in 12.6% of the participants either in drawings or in written free associations or in both (TABLE 6). This finding renders a relation between tooth (teeth) and fear of death at the unconscious level of imaginations (notions) as well. Interestingly, no relation was found between the appearance of death-related contents and dental fear scores (DAS and DFS), whereas expectations of the participants (Expectation Scale) influenced the appearance of such contents significantly (TABLE 6).

Summarizing the data, the authors conclude that there is a significant relation between the conflict of transitoriness (mortal nature, death) and the tooth/teeth-related psychological events including dental fear. This relation is detectable at both conscious and unconscious level, especially in participants with higher dental fear scores.

REFERENCES

1. YALOM, D.I. 1980. Existential Psychotherapy. Basic Books. New York.
2. FÁBIÁN, T.K., G. FÁBIÁN & P. FEJÉRDY. 2007. Dental stress. *In* Encyclopedia of Stress. Second edition. G. Fink, Ed. in chief: 733–736. Academic Press. Oxford.
3. FÁBIÁN, T.K., SZ. KOVÁCS, O. MÜLLER, *et al.* 2006. Some aspects of existential psychotherapy in dentistry. Fogorv. Szle. **99:** 246.
4. ABRAHAM, K. 1924. Versuch einer Entwicklungsgeschichte der Libido auf Grund der Psychoanalyse seelischen Störungen 1–96. Internat. Psychoanalyt. Verlag. Leipzig. Germany.
5. FÁBIÁN, T.K. & G. FÁBIÁN. 1998. Stress of life, stress of death: anxiety in dentistry from the viewpoint of hypnotherapy. Ann. N.Y. Acad. Sci. **851:** 495–500.
6. FÁBIÁN, T.K. & G. FÁBIÁN. 2000. Dental stress. *In* Encyclopedia of Stress. G. Fink, Ed. in chief: 657–659. Academic Press. San Diego.
7. BERGGREN, U. 1992. General and specific fears in referred and self-referred adult patients with extreme dental anxiety. Behav. Res. Ther. **30:** 395–401.
8. BERGGREN, U., S.G. CARLSSON, J.E. GUSTAFSSON & M. HAKEBERG. 1995. Factor analysis and reduction of a Fear Survey Schedule among dental phobic patients. Eur. J. Oral Sci. **103:** 331–338.
9. GEER, J.H. 1965. The development of a scale to measure fear. Behav. Ther. Res. **3:** 45–53.
10. World Medical Association. 2001. World Medical Association Declaration of Helsinki. Ethical principles for medical research involving human subjects. Bull. World Health Organ. **79:** 373–374.
11. SIPOS, K. & M. SIPOS. 1978. The development and validation of the Hungarian form of the STAI. *In* Cross-Cultural Anxiety 2. C.D. Spielberger & R. DiazGuerro, Eds.: 51–61. Hemisphere Publishing Corporation. Washington and London.
12. SPIELBERGER, C.D., R.L. GORSUCH & R.E. LUSHENE. 1970. Manual for the Strait-Trait Anxiety Inventory. Consulting Psychologist Press. Palo Alto, CA.
13. FÁBIÁN, T.K., P. KELEMEN & G. FÁBIÁN. 1998. A Dental Anxiety Scale hazai bevezetése. Magyar populáción végzett fogászati szorongás-epidemiológiai vizsgálatok. Fogorv. Szle. **91:** 43–52.
14. FÁBIÁN, T.K., T. HANDA, M. SZABÓ, *et al.* 1999. A "Dental Fear Survey" (a "Fogászati vélemény kérdőív") magyar fordítása, hazai populáción végzett mérések eredményei. Fogorv. Szle. **92:** 307–315.
15. CORAH, N.L. 1969. Development of a Dental Anxiety Scale. J. Dent. Res. **48:** 596.
16. KLEINKNECHT R.A., R.K. KLEPACH & L.D. ALEXANDER. 1973. Origins and characteristics of fear of dentistry. J. Am. Dent. Assoc. **86:** 842–848.
17. KLEINKNECHT, R.A., R.M. THORNDIKE, F.D. MCGLYNN & J. HARKAVY. 1984. Factor analysis of the Dental Fear Survey with cross-validation. J. Am. Dent. Assoc. **108:** 59–61.
18. FÁBIÁN, G., L. FEJÉRDY, CS. FÁBIÁN, *et al.* 2003. Fogászati kezeléstõl való félelem epidemiológiai vizsgálata általános iskolás (8–15 éves) korcsoportban. Fogorv. Szle. **96:** 129–133.
19. TÓTH, ZS., L. FEJÉRDY, CS. FÁBIÁN, *et al.* 2006. Fogat ábrázoló rajzok alapparamétereinek vizsgálata normál populáción, 8–18 éves korcsoportban. Fogorv. Szle. **99:** 47–52.

20. FÁBIÁN, G, L. FEJÉRDY, B. KAÁN, et al. 2004. Adatok általános iskolás (8–15 éves) gyermekek fogászati kezeléssel kapcsolatos félelmeinek hátteréről. Fogorv. Szle. **97:** 128–132.
21. FEJÉRDY, L., B. KAÁN, G. FÁBIÁN, et al. 2005. Adatok budapesti középiskolások fogászati kezeléssel kapcsolatos félelmeinek hátteréről. Fogorv. Szle. **98:** 9–13.
22. KULCSÁR, ZS. 1998. Egészségpszichológia. 135–140. ELTE Eötvös Kiadó. Budapest, Hungary.
23. LESTER, D. 1991. The Lester attitude toward death scale. Omega **23:** 67–75.
24. LARRABEE, M.J. 1978. Measuring fear of death: a reliability study. J. Psychol. **100:** 33–37.
25. LOO, R. 1984. Personality correlates of the fear of death and dying scale. J. Clin. Psychol. **40:** 120–122.
26. TEMPLER, D.I. 1970. The construction and validation of a death anxiety scale. J. Gen. Psychol. **82:** 163–177.
27. GOTTSCHALK, L.A. & G.C. GLESER. 1969. The Measurement of Psychological States through the Content Analysis of Verbal Behavior. University of California Press. Berkeley, CA.
28. KASTENBAUM, R. 2000. Death anxiety. In Encyclopedia of Stress. G. Fink, Ed. in chief: 645–651. Academic Press. San Diego.
29. HARDT, D.V. 1975. Development of an investigatory instrument to measure attitudes toward death. J. Sch. Health **45:** 96–99.
30. LOO, R. & L. SHEA. 1996. Structure of the Collett-Lester fear of death and dying scale. Death Stud. **20:** 577–586.
31. CHRIST, A. 1961. Attitudes toward death among a group of acute geriatric psychiatric patients. J. Gerontol. **16:** 56–59.
32. KASTENBAUM, R. & R. AISENBERG. 1972. Psychology of Death, pp. 107. Springer. New York.

Stress, Immune Function, and Women's Reproduction

PABLO A. NEPOMNASCHY,[a] EYAL SHEINER,[b]
GEORGE MASTORAKOS,[c] AND PETRA C. ARCK[d]

[a]*Epidemiology Branch, NIEHS, Research Triangle Park, North Carolina, USA*

[b]*Soroka University Medical Center, Ben-Gurion University of the Negev, Beer-Sheva, Israel*

[c]*Second Department of Obstetrics Gynecology, Aretaieion Hospital, Athens University Medical School, Greece*

[d]*Division of PsychoNeuroImmunology Charité, University of Medicine, Berlin, Germany*

> ABSTRACT: Only 23% of women will begin a successful pregnancy during the first menstrual cycle in their attempt to conceive.[1] A large number of these failed reproductive attempts are attributed to a broad set of pathologies, but across studies an important proportion of unsuccessful cycles is consistently left unexplained. Stress has become a commonly cited factor when discussing unexplained reproductive failures. Early research on the effect of stress on reproduction was plagued with methodological problems and lacked a solid theoretical framework. However, recent experimental, clinical and population-based research provides new evidence and suggests novel biological mechanisms, which merit a fresh evaluation of the purported association. Here we briefly review the latest advancements in the study of the interplay between stress, the immune system and women's reproduction, discuss a proposed evolutionary origin for their relationship and examine the biological pathways that may mediate the connection between these three systems.
>
> KEYWORDS: stress; immune function; reproductive function; evolution

STRESS AS A MODULATOR OF REPRODUCTIVE FUNCTION

Reproduction is a particularly onerous endeavor for females of eutherian mammal species. Intrauterine gestation demands significant physiologic and immune changes, which result in increased health risks for the mother. Furthermore, in several species maternal investment can continue considerably after

Address for correspondence: Pablo A. Nepomnaschy, Ph.D., PO Box 12233, MD A3-05, Rm 309 111 TW Alexander Dr. RTP, NC 27709-2233, USA. Voice: 001-541-7812; fax: 001-919-541-2511.
nepomnaschyp@niehs.nih.gov

parturition. In the case of humans, maternal investment continues for almost the entire life span. Also important, a mother's ability to invest in reproduction also affects the survival and future reproduction of her offspring.

Given the extensive costs involved, the timing of each reproductive venture can critically affect females' lifetime reproductive success. Therefore, being able to time reproductive events would be a valuable adaptation. Such ability requires two mechanisms: one to ascertain the quality of the current reproductive environment relative to its potential quality in the near future; another to suppress reproductive function when convenient. The convenience of reproducing now versus later might be assessed through environmental cues. Increases in environmental unpredictability or deteriorations of the social or physical environment, such as the loss of a social network member, a period of negative energetic balance due to famine, a natural disaster, or the development of an infection, are all challenges that will increase the risks associated with pregnancy as well as threaten a female's ability to invest in reproduction postnatally. The organism perceives all of those challenges as stressors. Consequently, stress has been proposed to be one of the mechanisms used by organisms to assess the appropriateness of their current context for initiating a reproductive venture. Consistent with the proposed hypothesis, the activation of the stress axis has been found to trigger reproductive suppression mechanisms both in humans and nonhuman mammals.[2–9]

STRESS AND REPRODUCTION IN HUMANS

Current understanding of the relationship between stress and reproductive function in women is mainly based on fertility of patients. Various studies based on interview data have shown psychological stress to be associated with reduced fecundability (probability of conceiving in a given cycle), increased risk of early pregnancy loss, and even infertility. For example, some prospective studies focused on women undergoing fertility treatment suggest that those with perceived or actually higher workload are less likely to conceive, and among those who conceive the likelihood of successful pregnancy completion is reduced.[10] In contrast, a small prospective study of six cycles among Danish women failed to find a relation between work demands and control, and pregnancy outcome. However, when restricting the sample to those with idiopathic infertility, job strain did predict miscarriage.[11] Importantly, consistent with what has been observed among fertility patients, healthy, nulliparous women with low scores of psychosomatic symptoms, few negative life events, no fluctuations in body weight prior to pregnancy, and regular religious practice have been reported to prospectively predict higher than average fertility.[12]

Until recently hormonal evidence for the relationship between stress and reproductive suppression was mainly restricted to clinical studies of individual stressors or studies focused on subgroups of women affected by similar stressors, such as athletes[13] or individuals who shared a professional context.[14] In the last few years, however, Nepomnaschy *et al.* began reporting results from a long-term, prospective study including periodical physiologic data on both stress and reproductive biomarkers. They followed the daily lives of 61 participant Mayan women for 1 year. This society frequently endures psychological, immunologic, and energetic challenges, such as intervals of restricted food supply, infectious diseases, and social violence. Longitudinal analyses of their data uncovered several interesting relationships. First, participants' most critical concerns, identified through the analyses of open-ended interviews, were found to be accurate predictors of increases in each woman's cortisol levels.[15] As discussed below, cortisol is a key mediator in the body's response to stress and, consequently, is frequently used as a marker of stress.[6,16] Second, increases in cortisol levels were associated with significant changes in the profiles of the participants' reproductive hormones during their menstrual cycles.[8] Specifically, increased cortisol levels were tied to increases in gonadotrophin and progestin levels during the follicular phase. Increased cortisol levels were also associated with significantly lower progestin levels during the middle of the luteal phase. These untimely changes in gonadotrophins and gonadal steroids have been previously found to affect a female's chances to conceive.[17,18] Finally, in the case of conceptive cycles, increases in cortisol during the first 3 weeks of gestation were predictive of miscarriage.[3] Their results are consistent with the hypothesis that stress levels may be used by a woman's body to assess the quality of her environment and regulate her investment on reproduction accordingly.[3,8]

BIOLOGICAL PATHWAYS

Three "super-systems"—the endocrine, immune, and nervous systems—engage in multiple interactions during the body's response to acute and chronic stress. Each of these systems is also individually vulnerable and responds to stressors.[19,20] Communication between systems is possible because they use a common "chemical language," sharing respective ligands and their cognate receptors.[21,22] Next, we briefly discuss some of the salient aspects of what has been learned so far regarding each one of these three systems and their interplay.

Physiologically, the term stress describes the response of an organism to challenges to its dynamic equilibrium or homeostasis. Stress activates the hypothalamic-pituitary-adrenal axis (HPA), the adrenergic and the autonomic nervous systems. The principal central nervous system regulators of the HPA axis are corticotrophin-releasing hormone (CRH) and antidiuretic hormone (AVP). Further, peripheral mediators of the stress response, such as glucocorticoids (GCs), catecholamines, and neurotrophins are activated.[23,24] Apart from the central nervous system, CRH has been found in the adrenal

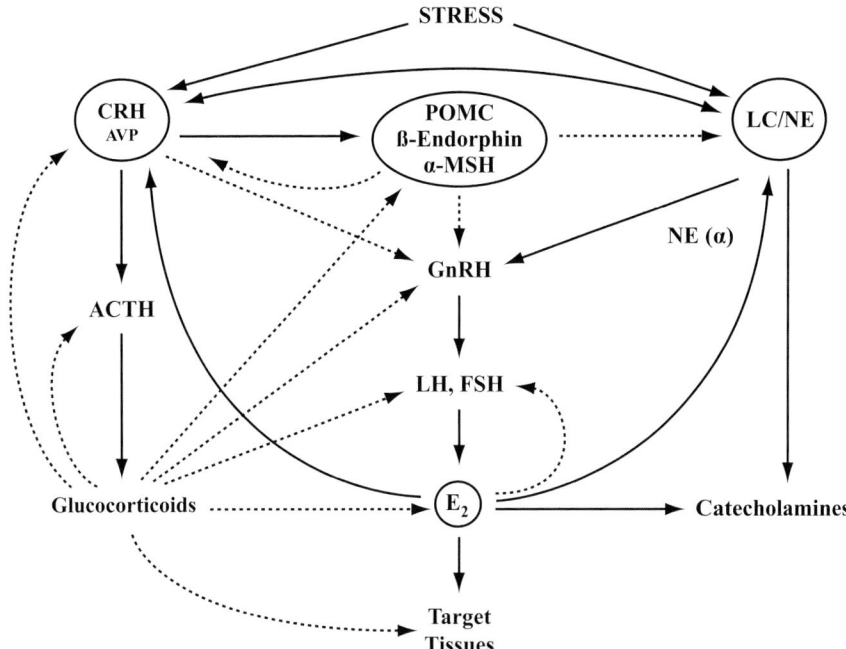

FIGURE 1. Heuristic representation of the interplay among the HPA axis, the LC/NE sympathetic system, and the HPG axis. POMC = proopiomelanocortin; α-MSH = α-melanocyte-stimulating hormone. The dotted lines represent inhibition while the solid lines represent stimulation.

medulla, ovaries, myometrium, endometrium, placenta, testis, and elsewhere. The "stress system" in the brain (the CRH neurons in the paraventricular nucleus of hypothalamus and other brain areas; locus ceruleus/norepinephrine system, central sympathetic system in the brainstem) collaborates with its peripheral components (HPA, and peripheral sympathetic nervous system).[25] CRH and its receptors are found in many extrahypothalamic sites in the brain. CRH secretion is complex and is based on reciprocal interactions among the various parts of the stress system. HPA axis activation inhibits the reproductive axis at all levels[26] either directly or through β-endorphin secreted from the arcuate proopiomelanocortin (POMC) neuron as well as through catecholamines. CRH suppresses the GnRH neurons of the arcuate and preoptic nuclei in the medial preoptic area (FIG. 1).[27,28] Glucocorticoids exert inhibitory effects on GnRH neurons, the pituitary gonadotrophs, influencing primarily the secretion of LH, and the gonads themselves while they render target tissues of sex steroids resistant to these hormones (FIG. 1).[29] The interaction between CRH and the reproductive axis is bidirectional, probably exerted via estrogen-responsive elements (ERE) in the promoter area of the CRH gene.[30] In the monkey, systemic administration of CRH rapidly decreases plasma LH levels.

The response of the hypothalamic-pituitary-gonadal (HPG) axis to stress is potentially biphasic. Humans and intact rodents exposed to acute stress respond with a small and often short-lived increase in plasma LH levels while prolonged stress inhibits LH release and blocks ovulation.

In premenstrual syndrome urinary-free cortisol excretion is normal in both phases of the cycle but adrenocorticotropic hormone (ACTH) responses to ovine (o)CRH stimulation are abnormal. Thus, in these women the HPA axis is perturbed but retains the ability to normalize its time-integrated function. Furthermore, suppression of gonadal function, caused by chronic HPA axis activation that is stimulated by the inflammatory cytokines (IL-1, TNF-α, IL-6), is observed in highly trained individuals or those sustaining anorexia nervosa or starvation.[25,31] These subjects have increased evening plasma cortisol and ACTH levels, increased urinary-free cortisol excretion, blunted ACTH responses to exogenous CRH, and present hypogonadotrophic hypogonadism and oligoamenorrhea.[32]

The actions of the stress system and the female reproductive system are bidirectionally interrelated.[26] Gonadal dysfunction and deregulation of the HPG axis are very common in Cushing syndrome and congenital adrenal hyperplasia.[33] In Cushing syndrome hypercortisolism-induced suppression of the HPG axis in women can lead to secondary amenorrhea. Furthermore, increased activity of the HPA axis, could probably be involved in the pathogenesis of the polycystic ovary syndrome (PCOS) characterized by chronic anovulation and hyperandrogenism (ovarian and adrenal). In PCOS increased activity of the enzyme (P450c17a) responsible for the conversion of progesterone to 17-hydroprogesterone and Δ4-androstedione within the adrenal cortex is associated with increased activity of the same enzyme within the ovaries. Increased adrenal production of 17-hydroxysteroids associated with the increased levels of the adrenarche indicator DHEA-S accounts only for 20% of the patients.[31]

Additionally, the presence of CRH has been demonstrated in the theca and stroma cells as well as in cells of the corpora lutea of rat and human ovaries.[34] CRH exerts an inhibitory effect on ovarian steroidogenesis, mediated through CRH-and interleukin-1-receptors, and may be linked to follicular atresia and luteolysis.[35] Also, the epithelial cells of human and rodent endometrium produce CRH throughout the menstrual cycle, while the stroma needs to undergo decasualization in order to produce CRH.[36]

Maternal HPA Axis during Pregnancy, Parturition, and Postpartum

Circulating CRH in plasma, produced by the placenta, decidua, and fetal membranes, increases exponentially in the last 2 months of gestation.[37] Amniotic fluid CRH-binding protein (CRHbp) levels fall approaching term.[38] The unbound CRH stimulates maternal ACTH secretion, which rises within normal limits.[39] The circadian rhythm of maternal CRH is maintained probably

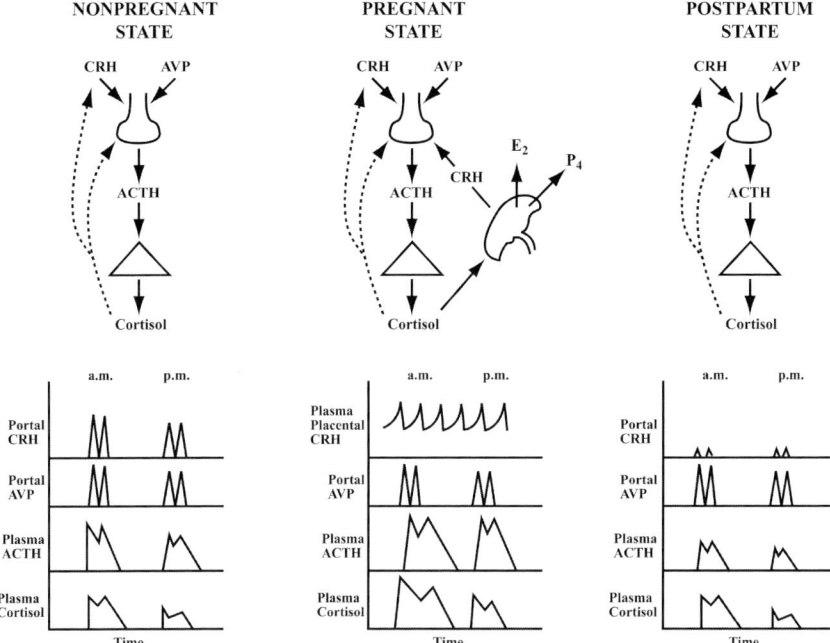

FIGURE 2. Heuristic model of the hypothalamic-pituitary-adrenal axis in the nonpregnant, pregnant, and postpartum state. The dotted lines represent inhibition while the solid lines represent stimulation.

due to the circadian AVP secretion by the parvicellular neuron of the PVN.[40] Maternal adrenal glands gradually become hypertrophic and free plasma cortisol rises during the third trimester reaching three times nonpregnant values (FIG. 2).

In sheep, placental CRH stimulates fetal ACTH production at term that in turn leads to a surge of fetal cortisol secretion precipitating parturition.[41] Activation of the HPA axis and increase of CRH during parturition have also been observed in primates. Cortisol competes with the action of progesterone in the regulation of placental CRH gene at the end of gestation. The fetal adrenal responds to the fetal pituitary ACTH and placental CRH with DHEAS production. The latter is aromatized in the placenta to estrogen promoting myometrial contractility. When labor does not progress satisfactorily in primiparous women delivering vaginally (spontaneously) ACTH levels increase. Women with preterm labor show significantly higher IL-1 levels than those of same gestational age with normal pregnancies. In women with preterm labor IL-1 and CRH levels are positively correlated suggesting that IL-1 acts directly and/or indirectly as a biological effector on placental CRH release. It seems, however, that critical levels of CRH should be achieved for the initiation of labor.[41] The HPA axis during pregnancy may function as a biological clock, "counting" from the early stages of gestation with placental CRH presumed to be the timing "starter," determining the course of pregnancy.[42]

During postpartum maternal plasma cortisol levels show a decline toward normal levels. Dynamic testing of the HPA axis shows transiently suppressed hypothalamic CRH secretion until 6 weeks postpartum. Although suppressed ACTH responses to oCRH stimulation are noted during postpartum, total plasma cortisol levels remain within normal range, probably due to the adrenal glands hypertrophy. The latter results from the maternal HPA axis hyperactivity during pregnancy.[32] Almost half of postpartum women develop a short-lived dysthymic disorder, called the "postpartum blues," while overt postpartum depression is common and occurs in up to 18% of newly delivered mothers.[43] Women who have postpartum blues or postpartum depression show more blunted ACTH responses to oCRH stimulation testing than euthymic women. Thus, the gradual recuperation of the HPA axis in the postpartum may be implicated in this period's mood disorders. Although the causal mechanism of this attenuated HPA axis response has yet to be elucidated, a role of estrogen on CRH expression can be speculated, exerted via EREs on the promoter area of the human CRH gene.[44] Estrogen given at high doses during the postpartum period is effective as an antidepressant possibly acting to reestablish normal CRH and norepinephrine responses to stressors.[32,45] In animal studies prolactin was shown to suppress HPA axis responses to stress. The HPA axis seems to influence the mother's psychological status and the mother–infant relationship/bonding.

Stress, Immune Function, and Pregnancy

Why women's immune systems generally do not reject the fetus in spite of spatial adjacencies of fetal "histoincompatible" tissue is still unclear. Such spatial adjacencies at the interface of fetal and maternal tissues—the so-called fetomaternal interface—guarantee nourishment of the fetus.[46] Hence, fetal tolerance is fundamental for successful pregnancy maintenance. Plural tolerance at the fetomaternal interface may be mediated via immune adaptation mechanisms evolving during early pregnancy. Such mechanisms include the predominance of anti-inflammatory, Th2 cytokines over proinflammatory Th1 cytokines in the decidua[47,48]; the expression of indoleamine 2, 3-dioxygenase (IDO), an enzyme that famishes immune rejection by depriving the T cells of tryptophan and by inhibiting lymphocyte proliferation[49,50]; and the presence of $CD4^+CD25^{bright}$ regulatory T (Treg) cells, which suppress an aggressive allogeneic response directed against the fetus in humans[51] and mice.[52] Further, an elaborate homeostasis between stimulatory and inhibitory signals promotes immune privilege at the fetomaternal interface.[53,54] Additionally, signal transducers and activators of transcription (STAT)3[55] and transforming growth factor (TGF)-β1[56] are involved in the regulation of fetal immune tolerance, likely by inhibiting the expression of proinflammatory mediators and promoting the synthesis of immunosuppressive factors. In addition, dendritic cells (DC) may be an essential cell subset for the regulation of the innate immune response mediating tolerance at the fetomaternal interface.[57] Furthermore, early in pregnancy, endometrial CRH at the implantation sites induces the expression of Fas ligand, which induces apoptosis, in the invading embryonic trophoblast and the maternal decidual cells on the fetal–maternal interface promoting thus, apoptosis of activated T lymphocytes participating in both the implantation and the tolerance process of pregnancy.[58] Pyrrolopyridine compounds have been developed as CRH receptor antagonists. Antalarmin (a pyrrolopyridine antagonist) prevented implantation in rats, by reducing the inflammatory-like reaction of the endometrium to the invading blastocyst. Consequently, antalarmin and its analogues might represent a new class of nonsteroidal inhibitors of early pregnancy.[58]

In view of the enormous complexity of the regulatory immune mechanisms involved in pregnancy maintenance, it is evident that pregnancy failure is most likely the result of complex deregulation. This deregulation can be initiated or aggravated by stress.[9] Rodent models have been particularly instructive to understand failing immune adaptation in response to stress during pregnancy.[58] Failure to sustain fetal immune tolerance during pregnancy in response to stress must be seen in the context of a neuroendocrine–immune disequilibrium. As previously outlined, stress activates the HPA axis. The upregulation of stress hormones, such as CRH and ACTH and peripheral mediators of stress response, such as GCs, catecholamines, and neurotrophins, may in turn strongly alter the immune response.[19,21,22] For example, GCs inhibit the production of proinflammatory cytokines, such as interleukin (IL)-12, interferon- (IFN-)γ, and tumor necrosis factor (TNF) and upregulate the production of IL-4, IL-10, and IL-13 by Th2 cells,[59] subsequently inducing a selective suppression of the Th1-mediated cellular immunity and a skew toward Th2-mediated humoral immunity. Interestingly, it has been postulated that this Th2 shift may actually protect the organism from systemic 'overshooting' with Th1/proinflammatory cytokines with tissue-damaging potential.[60]

The notion that stress represents a threat to pregnancy maintenance may appear to be in contradiction with the understanding that high levels of GC promote fetal tolerance by protecting pregnant women from overshooting abortogenic Th1/proinflammatory cytokines. However, this hypothesis can be rejected: besides the often-quoted immunosuppressive effects of GCs, relevant examples of proinflammatory actions of CRH —which triggers the release of GC–have been introduced.[61] In addition, as already described, neuroendocrine responses to stress also include activation of the sympathetic nervous system.[23] Lymphoid organs are prominently innervated by noradrenergic nerves fibers[62] and the immune system appears to be regulated via the sympathetic nervous system/catecholamines at regional, local, and systemic levels.[19–22,63] Lymphocytes express adrenergic receptors, and respond to stress-induced catecholamines with lymphocytosis, and distinct changes in lymphocyte trafficking, circulation, proliferation, and production of proinflammatory Th1-like, all of which can interfere with fetal tolerance.[64,65]

Besides the cardinal stress mediators, neurotrophin nerve growth factor (NGF) is progressively appreciated as a pivotal regulator involved in the stress response.[66] In addition to functioning as a trophic factor for peptidergic and sympathetic neurons and axon sprouting, NGF acts as a strong immunomodulator, endorsing interaction between neuronal, glia, and immune cells and facilitating cell migration through vascular endothelium.[67] Moreover, stress-triggered fetal rejection has been prevented by neutralizing NGF in mice.[68]

Apart from neurohormones and neurotrophins, progesterone mediates the onset, development, and maintenance of pregnancy.[69,70] Stress has been linked to decreased levels of progesterone in human[8] and nonhuman mammals.[70–73] Progesterone replacement abrogates effects of stress exposure by decreasing the levels of the abortogenic proinflammatory cytokines.[69] Such endocrine-immune cross-talk is exceedingly dependent on a specific $CD8^+$ T cell population, since depletion of CD8 leads to termination of the protective effect of progesterone on pregnancy.[69]

Uterine DC may serve as a switchboard between fetal rejection and tolerance.[57,74,75] DC are the most potent antigen-presenting cells (APCs) involved in the defense of the body and in the maintenance of the immune tolerance. The endogenous regulation of DC function is still poorly understood, yet their maturation, migration, and their expression of stimulatory and costimulatory molecules have major consequences on the immune response.[57] Vasoactive intestinal peptide (VIP), produced by decidual lymphocytes during the early postimplantation period[76] has been shown to decrease in mucosal tissue in response to stress in rats.[77] VIP presents potent anti-inflammatory actions and affects the early stages of DC differentiation and results in the generation of DC that cannot mature following inflammatory stimuli.[78] These DC exhibit a tolerogenic phenotype and are characterized by low expression of co-stimulatory molecules (CD40, CD80, and CD86), low production of pro-inflammatory, Th1-like cytokines, increased production of anti-inflammatory cytokines such as IL-10, and capacity to induce regulatory T cells with suppressive actions, all of which will promote fetal tolerance.[57] What remains to be elucidated is whether stress alters VIP expression at the fetomaternal interface.

Immature DC reside in the decidua during early pregnancy[57,74,75] and possibly serve as sentinel cells of the tissue environment for potential danger signals. In mice, the majority of decidual DC during early gestation is immature, and the highest percentage of immature DC occurs during blastocyst adhesion and early implantation.[79] However, in pregnancies with high abortion risk, for example, induced by exposure to experimental stressors, an increase of mature APC can be observed.[57] By blocking crucial ligands required on APC to induce T cell activation, mechanisms of fetal tolerance are restored in stress-challenged pregnancies.[54,57] Besides VIP, stress hormones and/or progesterone may initiate a considerable plasticity of DC phenotype and future research is likely to provide detailed insights on the impact of hormones on DC phenotype, for example, at the fetomaternal interface.

CONCLUDING REMARKS

Evidence continues to accumulate indicating that stress can lead to reproductive suppression. The relationship between stress, immune function, and reproduction may have arisen through natural selection due to its value in preventing or stopping pregnancy in dire circumstances. However, in modern industrialized environments stress-triggered mechanisms of reproductive suppression may have lost some of their original adaptive value and may even result in, or aggravate, reproductive pathologies.

Our understanding of the biologic mechanisms mediating the interplay between the stress, immune, and reproductive functions has advanced. There are, however, various aspects of their relationship that still require further research. Animal research is revealing some of the neuroendocrine–immune pathways linking reproductive suppression and stress. Translation of those results to human applications is, however, a complex process. The patient population in clinical studies is generally self-selected and stress "quantification" is difficult. Tissue collection can often only be performed retrospectively, leaving much room for discussions of cause versus effect. Clinicians and basic scientists should join efforts to elucidate hierarchical, temporal, and spatial interactions of key parameters during central and peripheral responses to stress, so that a list of candidate targets for clinically useful therapeutic interventions could be identified. Also, in order to understand the relationship between stress and reproduction in healthy women, more population-based prospective studies will be needed. We still have much to learn about basic important issues, such as how the effects of stress may change as gestation progresses or the role of the HPA axis during the transition between lactation amenorrhea and eumenorrhea. Only longitudinal studies will provide answers to those questions.

ACKNOWLEDGMENTS

We thank Jukic AM, Nguyen RH, and Berry NS for their helpful comments and suggestions. We thank the NIEHS (USA), the German Research Foundation, and the Charité for their support. P.C.A. is a partner of the EMBIC Network of Excellence, cofinanced by the European Commission throughout the FP6 framework program "Life Science, Genomics and Biotechnology for Health."

REFERENCES

1. BAIRD, D. & B. STRASSMANN. 2000. Women's fecundability and factors affecting it [invited review]. *In* Women and Health. MB Goldman, H.M., Ed.: 126–137 Academic Press. San Diego, CA.
2. ARENA, B. *et al*. 1995. Reproductive hormones and menstrual changes with exercise in female athletes. Sports Med. **19:** 278–287.
3. NEPOMNASCHY, P.A. *et al*. 2006. Cortisol levels and very early pregnancy loss in humans. Proc. Natl. Acad. Sci. USA **103:** 3938–3942.
4. LAATIKAINEN, T.J. 1991. Corticotropin-releasing hormone and opioid peptides in reproduction and stress. Ann. Med. **23:** 489–496.
5. WASSER, S.K. & D.P. BARASH. 1983. Reproductive suppression among female mammals: implications for biomedicine and sexual selection theory. Q. Rev. Biol. **58:** 513–538.
6. SAPOLSKY, R.M., L.M. ROMERO & A.U. MUNCK. 2000. How do glucocorticoids influence stress responses? Integrating permissive, suppressive, stimulatory, and preparative actions. Endocr. Rev. **21:** 55–89.
7. TILBROOK, A.J., A.I. TURNER & I.J. CLARKE. 2000. Effects of stress on reproduction in non-rodent mammals: the role of glucocorticoids and sex differences. Rev. Reprod. **5:** 105–113.
8. NEPOMNASCHY, P.A. *et al*. 2004. Stress and female reproductive function: a study of daily variations in cortisol, gonadotrophins, and gonadal steroids in a rural Mayan population. Am. J. Hum. Biol. **16:** 523–532.
9. ARCK, P.C. *et al*. 2001. Stress and immune mediators in miscarriage. Hum. Reprod. **16:** 1505–1511.
10. BARZILAI-PESACH, V. *et al*. 2006. The effect of women's occupational psychologic stress on outcome of fertility treatments. J. Occup. Environ. Med. **48:** 56–62.
11. HJOLLUND, N.H. *et al*. 1998. Job strain and time to pregnancy. Scand. J. Work Environ. Health **24:** 344–350.
12. VARTIAINEN, H. *et al*. 1994. Psychosocial factors, female fertility and pregnancy: a prospective study–Part I: Fertility. J. Psychosom. Obstet. Gynaecol. **15:** 67–75.
13. BONEN, A. 1994. Exercise-induced menstrual cycle changes. A functional, temporary adaptation to metabolic stress. Sports Med. **17:** 373–392.
14. SCHENKER, M.B. *et al*. 1997. Self-reported stress and reproductive health of female lawyers. J. Occup. Environ. Med. **39:** 556–568.
15. NEPOMNASCHY, P.A. 2005. Stress and Female Reproduction in a Rural Mayan Population, The University of Michigan. Ann Arbor, MI.
16. PIKE, I.L. & S.R. WILLIAMS. 2006. Incorporating psychosocial health into biocultural models: preliminary findings from Turkana women of Kenya. Am. J. Hum. Biol. **18:** 729–740.
17. BAIRD, D.D. *et al*. 1999. Preimplantation urinary hormone profiles and the probability of conception in healthy women. Fertil. Steril. **71:** 40–49.

18. FERIN, M. 1999. Clinical review 105: Stress and the reproductive cycle. J. Clin. Endocrinol. Metab. **84:** 1768–1774.
19. FLESHNER, M. & M.L. LAUDENSLAGER. 2004. Psychoneuroimmunology: then and now. Behav. Cogn. Neurosci. Rev. **3:** 114–130.
20. MCEWEN, B.S. 2004. Protection and damage from acute and chronic stress: allostasis and allostatic overload and relevance to the pathophysiology of psychiatric disorders. Ann. N.Y. Acad. Sci. **1032:** 1–7.
21. STEINMAN, L. 2004. Elaborate interactions between the immune and nervous systems. Nat. Immunol. **5:** 575–581.
22. WEIGENT, D.A. & J.E. BLALOCK. 1987. Interactions between the neuroendocrine and immune systems: common hormones and receptors. Immunol. Rev. **100:** 79–108.
23. BENSCHOP, R.J., M. RODRIGUEZ-FEUERHAHN & M. SCHEDLOWSKI. 1996. Catecholamine-induced leukocytosis: early observations, current research, and future directions. Brain Behav. Immun. **10:** 77–91.
24. WEBSTER, J.I., L. TONELLI & E.M. STERNBERG. 2002. Neuroendocrine regulation of immunity. Annu. Rev. Immunol. **20:** 125–163.
25. CHROUSOS, G.P. 1992. Regulation and dysregulation of the hypothalamic-pituitary-adrenal axis. The corticotropin-releasing hormone perspective. Endocrinol. Metab. Clin. North Am. **21:** 833–858.
26. MASTORAKOS, G., M.G. PAVLATOU & M. MIZAMTSIDI. 2006. The hypothalamic-pituitary-adrenal and the hypothalamic- pituitary-gonadal axes interplay. Pediatr. Endocrinol. Rev. 3(Suppl 1): 172–181.
27. HABIB, K.E., P.W. GOLD & G.P. CHROUSOS. 2001. Neuroendocrinology of stress. Endocrinol. Metab. Clin. North Am. **30:** 695–728; vii-viii.
28. RIVIER, C. & S. RIVEST. 1991. Effect of stress on the activity of the hypothalamic-pituitary-gonadal axis: peripheral and central mechanisms. Biol. Reprod. **45:** 523–532.
29. MAGIAKOU, M.A. *et al.* 1997. The hypothalamic-pituitary-adrenal axis and the female reproductive system. Ann. N.Y. Acad. Sci. **816:** 42–56.
30. TSIGOS, C. & G.P. CHROUSOS. 1994. Physiology of the hypothalamic-pituitary-adrenal axis in health and dysregulation in psychiatric and autoimmune disorders. Endocrinol. Metab. Clin. North Am. **23:** 451–466.
31. EVANS, J.J. 1999. Modulation of gonadotropin levels by peptides acting at the anterior pituitary gland. Endocr. Rev. **20:** 46–67.
32. CHROUSOS, G.P., D.J. TORPY & P.W. GOLD. 1998. Interactions between the hypothalamic-pituitary-adrenal axis and the female reproductive system: clinical implications. Ann. Intern. Med. **129:** 229–240.
33. FINKELSTEIN, M. & J.M. SHAEFER. 1979. Inborn errors of steroid biosynthesis. Physiol. Rev. **59:** 353–406.
34. MASTORAKOS, G. *et al.* 1993. Immunoreactive corticotropin-releasing hormone and its binding sites in the rat ovary. J. Clin. Invest. **92:** 961–968.
35. GHIZZONI, L. *et al.* 1997. Corticotropin-releasing hormone (CRH) inhibits steroid biosynthesis by cultured human granulosa-lutein cells in a CRH and interleukin-1 receptor-mediated fashion. Endocrinology **138:** 4806–4811.
36. MASTORAKOS, G. *et al.* 1996. Presence of immunoreactive corticotropin-releasing hormone in human endometrium. J. Clin. Endocrinol. Metab. **81:** 1046–1050.

37. GOLAND, R.S. et al. 1988. Biologically active corticotropin-releasing hormone in maternal and fetal plasma during pregnancy. Am. J. Obstet. Gynecol. **159:** 884–890.
38. ORTH, D.N. & C.D. MOUNT. 1987. Specific high-affinity binding protein for human corticotropin-releasing hormone in normal human plasma. Biochem. Biophys. Res. Commun. **143:** 411–417.
39. LAATIKAINEN, T. et al. 1987. Immunoreactive corticotropin-releasing factor and corticotropin during pregnancy, labor and puerperium. Neuropeptides **10:** 343–353.
40. MAGIAKOU, M.A. et al. 1996. The maternal hypothalamic-pituitary-adrenal axis in the third trimester of human pregnancy. Clin. Endocrinol. (Oxf.) **44:** 419–428.
41. WINTOUR, E.M. et al. 1986. Effect of long-term infusion of ovine corticotrophin-releasing factor in the immature ovine fetus. J. Endocrinol. **111:** 469–475.
42. MCLEAN, M. et al. 1995. A placental clock controlling the length of human pregnancy. Nat. Med. **1:** 460–463.
43. FLORES, D.L. & V.C. HENDRICK. 2002. Etiology and treatment of postpartum depression. Curr. Psychiatry Rep. **4:** 461–466.
44. VAMVAKOPOULOS, N.C. & G.P. CHROUSOS. 1993. Structural organization of the 5′ flanking region of the human corticotropin releasing hormone gene. DNA Seq. **4:** 197–206.
45. GREGOIRE, A.J. et al. 1996. Transdermal oestrogen for treatment of severe postnatal depression. Lancet **347:** 930–933.
46. NORWITZ, E.R., D.J. SCHUST & S.J. FISHER. 2001. Implantation and the survival of early pregnancy. N. Engl. J. Med. **345:** 1400–1408.
47. LIN, H. et al. 1993. Synthesis of T helper 2-type cytokines at the maternal-fetal interface. J. Immunol. **151:** 4562–4573.
48. PICCINNI, M.P. et al. 1998. Defective production of both leukemia inhibitory factor and type 2 T-helper cytokines by decidual T cells in unexplained recurrent abortions. Nat. Med. **4:** 1020–1024.
49. MUNN, D.H. et al. 1998. Prevention of allogeneic fetal rejection by tryptophan catabolism. Science **281:** 1191–1193.
50. TERNESS, P. et al. 2002. Inhibition of allogeneic T cell proliferation by indoleamine 2,3-dioxygenase-expressing dendritic cells: mediation of suppression by tryptophan metabolites. J. Exp. Med. **196:** 447–457.
51. SASAKI, Y. et al. 2004. Decidual and peripheral blood CD4+CD25+ regulatory T cells in early pregnancy subjects and spontaneous abortion cases. Mol. Hum. Reprod. **10:** 347–353.
52. ALUVIHARE, V.R., M. KALLIKOURDIS & A.G. BETZ. 2004. Regulatory T cells mediate maternal tolerance to the fetus. Nat. Immunol. **5:** 266–271.
53. GULERIA, I. et al. 2005. A critical role for the programmed death ligand 1 in fetomaternal tolerance. J. Exp. Med. **202:** 231–237.
54. ZHU, X.Y. et al. 2005. Blockade of CD86 signaling facilitates a Th2 bias at the maternal-fetal interface and expands peripheral CD4+CD25 +regulatory T cells to rescue abortion-prone fetuses. Biol. Reprod. **72:** 338–345.
55. POEHLMANN, T.G. et al. 2005. The possible role of the Jak/STAT pathway in lymphocytes at the fetomaternal interface. Chem. Immunol. Allergy **89:** 26–35.
56. AYATOLLAHI, M., B. GERAMIZADEH & A. SAMSAMI. 2005. Transforming growth factor beta-1 influence on fetal allografts during pregnancy. Transplant Proc.

37: 4603–4604.
57. BLOIS, S. et al. 2005. Intercellular adhesion molecule-1/LFA-1 cross talk is a proximate mediator capable of disrupting immune integration and tolerance mechanism at the feto-maternal interface in murine pregnancies. J. Immunol. **174:** 1820–1829.
58. CLARK, D.A., D. BANWATT & G. CHAOUAT. 1993. Stress-triggered abortion in mice prevented by alloimmunization. Am. J. Reprod. Immunol. **29:** 141–147.
59. WONNACOTT, K.M. & R.H. BONNEAU. 2002. The effects of stress on memory cytotoxic T lymphocyte-mediated protection against herpes simplex virus infection at mucosal sites. Brain Behav. Immun. **16:** 104–117.
60. ELENKOV, I.J. et al. 1999. Stress, corticotropin-releasing hormone, glucocorticoids, and the immune/inflammatory response: acute and chronic effects. Ann. N.Y. Acad. Sci. **876:** 1–11; discussion 11–3.
61. VITORATOS, N. et al. 2006. Association between serum tumor necrosis factor-alpha and corticotropin-releasing hormone levels in women with preterm labor. J. Obstet. Gynaecol. Res. **32:** 497–501.
62. DAHLSTROM, A. et al. 1965. Observations on adrenergic innervation of dog heart. Am. J. Physiol. **209:** 689–692.
63. CACIOPPO, J.T. et al. 1998. Autonomic, neuroendocrine, and immune responses to psychological stress: the reactivity hypothesis. Ann. N.Y. Acad. Sci. **840:** 664–673.
64. DHABHAR, F.S. 2000. Acute stress enhances while chronic stress suppresses skin immunity. The role of stress hormones and leukocyte trafficking. Ann. N.Y. Acad. Sci. **917:** 876–893.
65. SANDERS, V.M. et al. 1997. Differential expression of the beta2-adrenergic receptor by Th1 and Th2 clones: implications for cytokine production and B cell help. J. Immunol. **158:** 4200–4210.
66. ALOE, L., E. ALLEVA & M. FIORE. 2002. Stress and nerve growth factor: findings in animal models and humans. Pharmacol. Biochem. Behav. **73:** 159–166.
67. LEVI-MONTALCINI, R. et al. 1996. Nerve growth factor: from neurotrophin to neurokine. Trends Neurosci. **19:** 514–520.
68. TOMETTEN, M. et al. 2006. Nerve growth factor translates stress response and subsequent murine abortion via adhesion molecule-dependent pathways. Biol. Reprod. **74:** 674–683.
69. BLOIS, S.M. et al. 2004. Depletion of CD8+ cells abolishes the pregnancy protective effect of progesterone substitution with dydrogesterone in mice by altering the Th1/Th2 cytokine profile. J. Immunol. **172:** 5893–5899.
70. JOACHIM, R. et al. 2003. The progesterone derivative dydrogesterone abrogates murine stress-triggered abortion by inducing a Th2 biased local immune response. Steroids **68:** 931–940.
71. CREEL, S. et al. 2007. Predation risk affects reproductive physiology and demography of elk. Science **315:** 960.
72. XIAO, E., L. XIA-ZHANG & M. FERIN. 2000. Inhibitory effects of endotoxin on LH secretion in the ovariectomized monkey are prevented by naloxone but not by an interleukin-1 receptor antagonist. Neuroimmunomodulation **7:** 6–15.
73. XIAO, E., L. XIA-ZHANG & M. FERIN. 2002. Inadequate luteal function is the initial clinical cyclic defect in a 12-day stress model that includes a psychogenic component in the Rhesus monkey. J. Clin. Endocrinol. Metab. **87:** 2232–2237.

74. GARDNER, L. & A. MOFFETT. 2003. Dendritic cells in the human decidua. Biol. Reprod. **69:** 1438–1446.
75. KAMMERER, U. *et al.* 2000. Human decidua contains potent immunostimulatory CD83(+) dendritic cells. Am. J. Pathol. **157:** 159–169.
76. SPONG, C.Y. *et al.* 1999. Regulation of postimplantation mouse embryonic growth by maternal vasoactive intestinal peptide. Ann. N.Y. Acad. Sci. **897:** 101–108.
77. SHEN, G.M. *et al.* 2006. Role of vasoactive intestinal peptide and nitric oxide in the modulation of electroacupuncture on gastric motility in stressed rats. World J. Gastroenterol. **12:** 6156–6160.
78. CHORNY, A., E. GONZALEZ-REY & M. DELGADO. 2006. Regulation of dendritic cell differentiation by vasoactive intestinal peptide: therapeutic applications on autoimmunity and transplantation. Ann. N.Y. Acad. Sci. **1088:** 187–194.
79. BLOIS, S.M. *et al.* 2004. Lineage, maturity, and phenotype of uterine murine dendritic cells throughout gestation indicate a protective role in maintaining pregnancy. Biol. Reprod. **70:** 1018–1023.